MECHANISM ANALYSIS

MECHANICAL ENGINEERING

A Series of Textbooks and Reference Books

EDITORS

L. L. FAULKNER

Department of Mechanical Engineering
The Ohio State University
Columbus, Ohio

S. B. MENKES

Department of Mechanical Engineering
The City College of the
City University of New York
New York, New York

OTHER VOLUMES IN PREPARATION

MECHANISM ANALYSIS

Simplified Graphical and Analytical Techniques

Lyndon O. Barton

Project Engineer
E.I. du Pont de Nemours & Co.
Wilmington, Delaware

MARCEL DEKKER, INC.   New York and Basel

Library of Congress Cataloging in Publication Data

Barton, Lyndon O., [date]
    Mechanism analysis.

    (Mechanical engineering ; 32)
    Bibliography: p.
    Includes index.
    1. Machinery, Kinematics of.   I. Title.
II. Series.
TJ175.B29   1984   621.8'11   84-11433
ISBN 0-8247-7086-2

MARCEL DEKKER, INC.
270 Madison Avenue, New York, New York 10016

Current printing (last digit):
10 9 8 7 6 5 4 3 2 1

PRINTED IN THE UNITED STATES OF AMERICA

To my wife Olive, my children Rhonda, Loren, Carol, and
Leon, and my mother Clarice.

Preface

This book is written primarily for mechanical design engineers and mechanical design engineering students who are concerned with the design of machines, in general, and in particular with mechanism analysis, a subject which forms a principal part of the study of kinematics of mechanisms. The principal aim of this volume is to place at the disposal of the reader a practical book that will serve (1) as a handy reference for simplified approaches to problems typically encountered in the analysis and synthesis of a mechanism, and (2) as a supplementary textbook for independent study or quick review of both principles and applications in mechanism analysis.

The book presents a wide assortment of graphical and analytical techniques, as well as complete listings in FORTRAN of computer programs and programmable calculator programs for the Hewlett Packard HP-41C for analysis of basic classes of mechanisms. Special emphasis has been given to relatively simple kinematic chains such as slider-crank, four-bar, quick-return, and sliding coupler mechanisms. These mechanisms have been selected because they form the basic elements of most machines and because they are easily adaptable to the teaching of fundamental kinematic principles. Once these principles are fully understood, it is comparatively easy to apply this knowledge to the analysis of more complex mechanisms.

Several novel approaches for simplification of the analytic process are presented. These include the rectilinear and angular motion diagrams presented in Chapter 2, the link extension concept for velocity analysis by instant centers in Chapter 5, the generalized procedure for constructing the acceleration polygon in Chapter 9, the Parallelogram Method for slider-crank analysis in Chapter 12, and the Simplified Vector Method and modified version of same in Chapters 14-18.

One important feature is that the Simplified Vector Method, unlike conventional methods which rely on calculus and other forms of sophisticated mathematics, relies mainly on basic algebra and trigonometry to obtain an analytical solution. This simplified mathematical approach has made it possible to include several analytical problem solutions rarely found in kinematic textbooks. Hopefully, this approach will not only make this material accessible to a wide body of readers, but will also help to provide the quick insight often needed by designers in the analysis of a linkage.

The book is written for easy readability and comprehension, without reliance on any other source. Needed background material on topics such as Uniformly Accelerated Motion (Chapter 2), Properties of Vectors (Chapter 3), Complex Algebra (Chapter 13), and Trigonometry (Appendix B) is provided for review. Concepts are presented as concisely as possible, employing numerous illustrative examples and graphical aids, as well as step-by-step procedures for most graphical constructions. In addition, the topics are arranged in a logical sequence corresponding to that ordinarily followed in teaching a course in kinematics.

Some of the material in this book is based on several technical papers which the author has previously published (see References). Much of the material has been drawn from class notes which have been developed and used over several years of teaching kinematics of mechanisms as part of an Engineering Technology college curriculum.

The author gratefully acknowledges his indebtedness to the E. I. DuPont de Nemours and Company Engineering Department and the Delaware Technical Community College Mechanical Engineering Department for providing the engineering and teaching opportunities, respectively, that have enabled him to pursue and accumulate the knowledge and experience that form the basis of this book.

Grateful acknowledgments and appreciation are also extended to

- Penton Publishing Company, publishers of <u>Machine Design</u> magazine, for permission to reprint portions of previously published articles (including illustrations);
- American Society for Engineering Education, publishers of <u>Engineering Design Graphics</u> journal for permission to reprint portions of previously published articles (including illustrations);
- Mr. Albert A. Stewart, for the valuable assistance he has rendered in the calculator program development;
- Teachers, relatives, and friends who have been a source of inspiration and encouragement in the author's career; and
- A devoted family, for their love, understanding, and support, always.

Lyndon O. Barton

Contents

MECHANISM ANALYSIS

I
INTRODUCTORY CONCEPTS

Mechanism analysis (or kinematics of machines) is inherently a vital part in the design of a new machine or in studying the design of an existing machine. For this reason, the subject has always been of considerable importance to the mechanical engineer. Moreover, considering the tremendous advances that have been made within recent years in the design of high-speed machines, computers, complex instruments, automatic controls, and mechanical robots. It is not surprising that the study of mechanisms has continued to attract greater attention and emphasis than ever before.

Mechanism analysis may be defined as a systematic analysis of a mechanism based on principles of kinematics, or the study of motion of machine components without regard to the forces that cause the motion. To better appreciate the role of mechanism analysis in the overall design process, consider the following. Typically, the design of a new machine begins when there is a need for a mechanical device to perform a specific function. To fulfill this need, a conceptual or inventive phase of the design process is required to establish the general form of the device. Having arrived at a concept, the designer usually prepares a preliminary geometric layout of the machine or mechanism for a complete kinematic analysis. Here the designer is concerned not only that all components of the machine are properly proportioned so that the desired motions can be achieved (synthesis phase), but also with the analysis of the components themselves to determine such characteristics as displacements, velocities, and accelerations (analysis phase). At the completion of this analysis, the designer is ready to proceed to the next logical step in the design process: kinetic analysis, where individual machine members are analyzed further to determine the forces resulting from the motion.

Mechanism analysis therefore serves as a necessary prerequisite for the proper sizing of machine members, so that they can withstand the

1

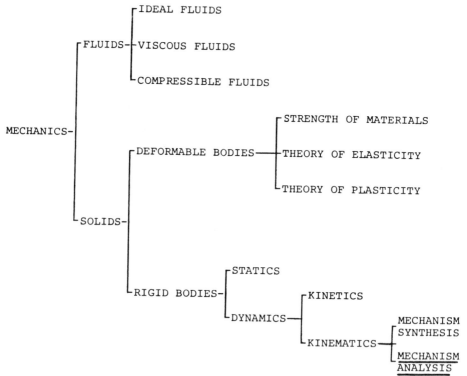

Figure I.1   Mechanism analysis and other branches of mechanics.

loads and stresses to which they will be subjected. Figure I.1 shows the
relationship of mechanism analysis to other branches of mechanics.

1
Kinematic Terminology

1.1 MECHANICAL CONCEPTS

A <u>mechanism</u> is a combination of rigid bodies so connected that the motion of one will produce a definite and predictable motion of others, according to a physical law. Alternatively, a mechanism is considered to be a kinematic chain in which one of the rigid bodies is fixed. An example of a mechanism is the slider crank shown in Figure 1.1. Instruments, watches, and governors provide other examples of mechanisms.

A <u>machine</u> is a mechanism or group of mechanisms used to perform useful work. A machine transmits forces. An example of a machine is the internal combustion engine shown in Figure 1.1.

The term "machine" should not be confused with "mechanism" even though in actuality, both may refer to the same device. The difference in terminology is related primarily to function. Whereas the function of a machine is to transmit energy, that of a mechanism is to transmit motion. Stated in other words, the term "mechanism" applies to the geometric arrangement that imparts definite motions to parts of a machine.

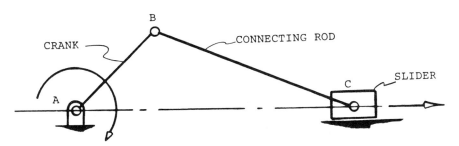

Figure 1.1   Slider-crank mechanism.

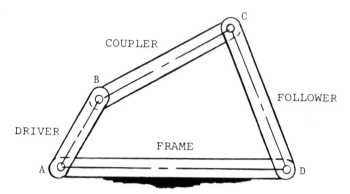

Figure 1.2   Four-bar linkage mechanism.

A <u>pair</u> is a joint between two bodies that permits relative motion.

1. A <u>lower pair</u> has surface contact, such as a hinge or pivot.
    Surface contact is a characteristic of sleeve bearings, piston
    rings, screwed joints, and ball-and-socket joints. As an
    example, joints A, B, C, and D of Figure 1.2 are lower pairs.
    These joints are called revolute or turning pairs.
2. A <u>higher pair</u> has line or point contact between the surface ele-
    ments. Point contact is usually found in ball bearings. Line
    contact is characteristic of cams, roller bearings, and most
    gears.

A <u>link</u> is a rigid body that serves to transmit force from one body to
another or to cause or control motion, such as the connecting rod or crank
arm in Figure 1.1. Alternatively, a link is defined as a rigid body having
two or more pairing elements.

A <u>kinematic chain</u> is a group of links connected by means of pairs to
transmit motion or force. There are three types of chains: locked chain,
constrained chain, and unconstrained chain.

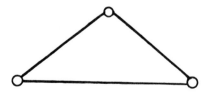

Figure 1.3   Locked chain.

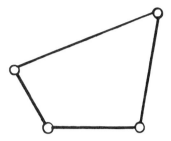

Figure 1.4   Constrained chain.

1.   A <u>locked chain</u> has no relative motion between the links. An example of this is the three-link chain shown in Figure 1.3.

2.   A <u>constrained chain</u> is one in which there is definite relative motion between the links. For example, if one link is fixed and another link is put in motion, the points on all the other links will move in definite paths and will always move in the same paths regardless of the number of lines the motion is repeated. An example is the four-bar mechanism shown in Figure 1.4.

3.   An <u>unconstrained chain</u> is one in which, with one link fixed and another link put in motion, the points on the remaining links will not follow definite paths. An example of an unconstrained chain is the five-link mechanism shown in Figure 1.5.

A <u>linkage</u> is a mechanism in which all connections are lower pairs. The four-bar and slider-crank linkages are typical examples.

A <u>simple mechanism</u> is one that consists of three or four links, whereas a <u>compound mechanism</u> consists of a combination of simple mechanisms, and is usually made up of more than four links.

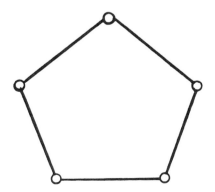

Figure 1.5   Unconstrained chain.

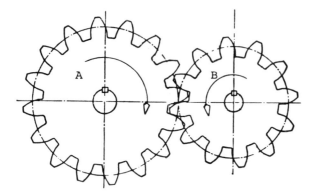

Figure 1.6   Spur gears.

A <u>structure</u> is a combination of rigid bodies capable of transmitting forces or carrying loads, but having no relative motion in them. Alternatively, a structure may be thought of as a locked kinematic chain. A <u>frame</u> is a stationary structure that supports a machine or mechanism. Normally, it is the fixed link of a mechanism (e.g., link 1 in Figure 1.2).

A <u>driver</u> is that part of a mechanism which causes motion, such as the crank in Figure 1.2. A <u>follower</u> is that part of a mechanism whose motion is affected by the motion of the driver, such as the slider in Figure 1.2.

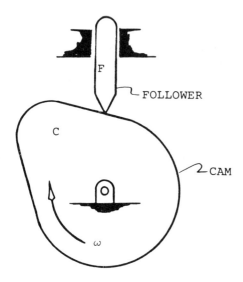

Figure 1.7   Cam-follower mechanism.

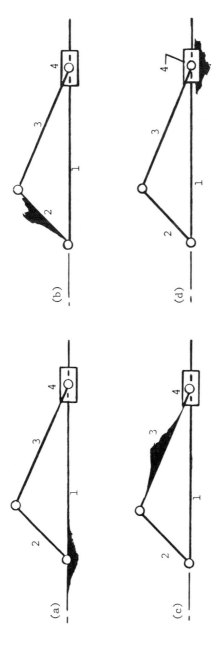

Figure 1.8 Slider-crank chain: (a) slider crank; (b) quick-return mechanism; (c) oscillating cylinder; (d) fixed piston.

Modes of Transmission

Motion can be transmitted from driver to follower by:

1. Direct contact
    a. Sliding
    b. Rolling
2. Intermediate connectors
    a. Rigid: links
    b. Flexible: belts, fluids
3. Nonmaterial: magnetic forces

A gear is a machine member, generally circular, whose active surface is provided with teeth to engage a similar member to impart rotation from one shaft to another. An example is the spur gear shown in Figure 1.6.

A cam is a rotating or sliding machine member whose function is to impart a predetermined motion to another part that rolls or slides along the surface of the member. An example is shown in Figure 1.7.

Inversion

Inversion is the process of fixing different links in a chain to create different mechanisms. Many useful mechanisms may be obtained by the inversion of various kinematic chains. An example of such inversion can be seen in the slider crank chain shown in Figure 1.8:

1. By making link 1 of the chain fixed, we obtain a steam engine mechanism (Figure 1.8a).
2. By fixing the crank, link 2, we obtain the Whitworth quick-return mechanism, used in various types of metal shapers (Figure 1.8b).
3. By fixing the connecting rod, link 3, we obtain the oscillating cylinder engine, once used as a type of marine engine (Figure 18c).
4. Finally, by fixing the slider, link 4, we obtain the mechanism shown in Figure 1.8d. This mechanism has found very little practical application. However, by rotating the figure 90° clockwise the mechanism can be recognized as part of a garden pump.

It is important to keep in mind that the inversion of a mechanism does not change the relative motion between the links, but does change their absolute motions.

1.2 MOTION CLASSIFICATION

Definitions

Motion is the act of changing position. The change of position can be
with respect to some other body which is either at rest or
moving.

Rest is a state in which the body has no motion.

Absolute motion is the change of position of a body with respect to
another body at absolute rest.

Relative motion is the change of position of a body with respect to
another body that is also moving with respect to a fixed frame
of reference.

Types of Motion

Plane Motion

In plane motion all points on a body in motion move in the same plane or
parallel planes. All points of the body or system of bodies remain at a constant distance from a reference plane throughout the motion. Typical examples are the connecting rod on a slider–crank mechanism (Figure 1.1)
and the side rod on a locomotive (Figure 1.9). There are three classes of
plane motion: (1) rotation, (2) translation, and (3) combined translation
and rotation.

1. Rotation: When one point in a body remains stationary while
    the body turns about that point, the body is said to be in rotation.
    That is, all points in the body describe circular paths about a
    stationary axis that is perpendicular to the plane of rotation.
    Crank AB is Figure 1.1 has rotary motion.
2. Translation: When a body moves without turning, that body is
    said to be in translation: or, the distances between particles

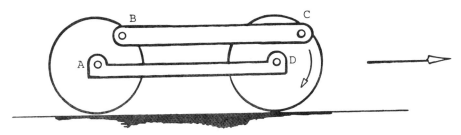

Figure 1.9   Locomotive side rod drive. Link BC undergoes curvilinear
motion; link AD undergoes rectilinear translation.

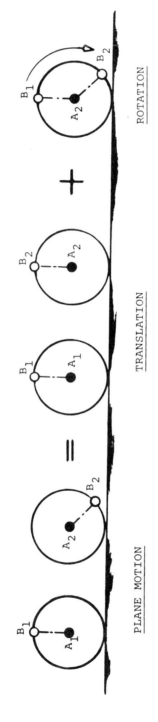

Figure 1.10   Rolling wheel.

of the body remain unchanged. There are two types of translation:

   a. Rectilinear translation, where all points in the body move along parallel straight paths, such as the slider C in Figure 1.1.

   b. Curvilinear translation, where all points in the body move along similar (or parallel) curved paths. Curvilinear translation is not to be confused with rotation, where all paths on the body are in the form of concentric circles. A good example is the locomotive side rod in Figure 1.9. Note that the paths of any two points on the rod, say B and C, have the same curvature.

3.   Combined motion: When a body undergoes simultaneous translation and rotation, that body is said to be in combined motion. That is, all points in the body change position and all lines turn as the body moves.

    A common example of a body that undergoes combined motion is the connecting rod BC in Figure 1.1, where one end, B, rotates about the crank axle A, while the other end, C, translates along a straight path as defined by the slider motion. Hence every other point on the member experiences part rotation and part translation.

    Another common example is that of the rolling wheel (Figure 1.10), where it can be seen that the resultant motion of point B is the summation of its translation and rotation motions.

Three-Dimensional Motion

There are two types of three-dimensional motion:

1.   Helical motion: When a body has rotation combined with translation along the same axis of rotation, that body is said to be in helical motion. The most common example is the turning of a nut on a screw (see Figure 1.11). The nut rotates and at the same time, translates along the axis of rotation.

2.   Spherical motion: When a body moves in three-dimensional space such that each part of that body remains at a constant distance from a fixed point, the body is said to be in spherical motion. A common example is the ball-and-socket joint, shown in Figure 1.12. A point on the ball or on the handle, which is rigidly attached to the ball, moves in space without changing its distance from the center of the sphere.

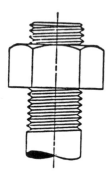

Figure 1.11   Screw and nut.

Other Types of Motion

Additional types of motion include:

1.   Continuous motion: When a point continues a move indefinitely along a given path in the same sense, such a motion is said to be continuous. An example of this is a rotating wheel, where the path of a point away from the axis returns on itself.

2.   Reciprocating motion: When a point traverses the same path and reverses its motion at the end of such path, the motion is

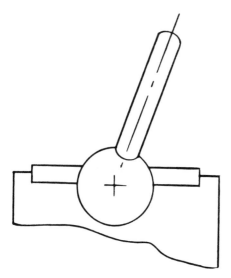

Figure 1.12   Ball joint.

said to be reciprocating. An example is the slider in a typical slider-crank mechanism (Figure 1.1). Oscillation is reciprocating circular motion, as in a pendulum.

3. Intermittent motion: When a motion of a point is interrupted by periods of rest, such motion is said to be intermittent, as in a ratchet.

1.3   MOTION CHARACTERISTICS

The path is the locus of a point as it changes from one position to another, distance is a measure of the path through which a point moves, and displacement is a measure of the net change in position of a point. There are two types of displacement:

1. Linear displacement (Δs) is the change position of a point as it moves along a straight line. Linear displacement is typically expressed in terms of inches, feet, or miles in a specified direction. Hence it is a vector quantity.

2. Angular displacement ($\Delta\theta$) is the angle between two positions of a rotating line or body. It has both magnitude and sense, either clockwise or counterclockwise. Angular displacement is typically expressed in terms of degrees, radians, or resolutions in a specified rotational sense.

Velocity is the rate of change of position of a point with respect to time, or displacement per unit time.

1. Linear velocity (v) is displacement per unit of time of a point moving along a straight line. The average linear velocity is given by the expression

$$v = \frac{\Delta s}{\Delta t}$$

where Δs is the change in linear displacement and Δt is the time interval. Linear velocity is typically expressed in terms of inches per second, feet per second, or miles per hour in a specified direction. Hence it is a vector quantity.

2. Angular velocity (ω) is the angular displacement per unit time of a line or body in rotation and has both magnitude and sense, either clockwise or counterclockwise. The average angular velocity is given by the expression

$$\omega = \frac{\Delta\theta}{\Delta t}$$

where $\Delta\theta$ is the average angular velocity and Δt is the time interval. Angular velocity is typically expressed in terms of radians per second or revolutions per minute in a specified rotational sense.

Acceleration is the rate of change of velocity with respect to time, or the rate of speedup.

1. Linear acceleration (a) is the change in linear velocity per unit of time. Average linear acceleration is given by the expression

$$a = \frac{\Delta v}{\Delta t}$$

where Δv is the change in linear velocity and Δt is the time interval. Linear acceleration is typically expressed in terms of inches per second per second or feet per second per second in a specified direction. Hence it is a vector quantity.

2. Angular acceleration (α) is the change in angular velocity per unit of time. Average angular acceleration is given by the expression

$$\alpha = \frac{\Delta\omega}{\Delta t}$$

where $\Delta\omega$ is the change in angular velocity and Δt is the time interval. Angular acceleration is typically expressed in terms of radians per second per second or revolutions per minute per minute.

Deceleration or retardation is negative acceleration or rate of slowing down.

Speed is the rate of motion in any direction, or the total distance covered in one unit of time. Speed is not to be confused with velocity, which is a vector quantity. Speed is typically expressed in magnitude terms such as feet per second, inches per second, or miles per hour, without regard to direction. Hence it is a scalar quantity.

Angular speed is the rate at which a body turns about an axis. Typical units are revolutions per minute and radians per second, without any reference to rotational sense.

Phase describes the relative positions of links in a mechanism or a machine at any instant. This is usually defined by the angle of one of the links of the mechanism, for example, the angle of crank AB in Figure 1.1.

When a mechanism moves through all its possible phases and returns to its starting position, it has completed a cycle, or a motion cycle. Virtually all mechanisms have a cyclic pattern where the cycle repeats itself over and over again. The time required for a motion to complete one motion cycle is called a period.

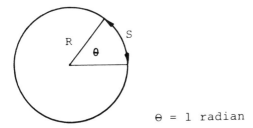

θ = 1 radian

Figure 1.13   Geometric representation of a radian.

A body that rotates in a counterclockwise direction is assumed to have a positive sense. This is because angular displacement is normally measured counterclockwise. Thus in Figure 1.1, the crank AB has a positive sense, and its angular velocity is also positive. Conversely, a body that rotates in a clockwise direction is assumed to have a negative sense, in which case the angular velocity of crank AB in Figure 1.1 is negative.

Finally, radian is the angle subtended by an arc that is equal in length to the radius of the circle (see Figure 1.13). There are 2π radians in a circle.

2

Uniformly Accelerated Motion

2.1 RECTILINEAR MOTION

Displacement, Velocity, Acceleration, and Time Relationships

The analysis of rectilinear or straight-line motion with constant or uniform acceleration is relatively common in the study of kinematics. To establish some basic relationships, let us consider a body moving with uniform acceleration where the velocity changes from an initial value to a final value after some time, while completing some distance. Let

s = distance completed, ft
v_1 = initial velocity, ft/sec
v_2 = final velocity, ft/sec
a = acceleration, ft/sec^2
t = time, sec

By definition,

Distance = average velocity × time

$$s = \frac{v_1 + v_2}{2} t \qquad (2.1)$$

After 1 sec, initial velocity is increased to $v_1 + a$.
After 2 sec, initial velocity is increased to $v_1 + 2a$.
After 3 sec, initial velocity is increased to $v_1 + 3a$.
After t sec, initial velocity is increased to $v_1 + at$.

This means that

16

$$v_2 = v_1 + at \qquad (2.2)$$

Now, substitute for v_2 in Equation (2.1) and obtain

$$s = \frac{v_1 + (v_1 + at)}{2} t$$

$$s = \frac{2v_1 + at}{2} t$$

$$s = v_1 t + \frac{1}{2} at^2 \qquad (2.3)$$

From Equation (2.2),

$$v_2 - v_1 = at \qquad (2.4)$$

and from Equation (2.1),

$$v_1 + v_2 = \frac{2}{t} s$$

or

$$v_2 + v_1 = \frac{2}{t} s \qquad (2.5)$$

Now, multiply Equations (2.4) and (2.5) and obtain

$$(v_2 - v_1)(v_2 + v_1) = at \times \frac{2}{t} s$$

which gives

$$v_2^2 - v_1^2 = 2as$$

or

$$v_2^2 = v_1^2 + 2as$$

In summary, for uniform linear acceleration:

$$s = \frac{v_1 + v_2}{2} t$$

$$s = v_1 t + \frac{1}{2} at^2$$

$$v_2 = v_1 + at$$

$$v_2 = v_1^2 + 2as$$

For uniform linear velocity ($a = 0$):

$$s = v_1 t$$

Also, if the body is accelerating, a is positive; if the body is decelerating, a is negative.

EXAMPLE 2.1

A car passes a certain point A with a velocity of 30 ft/sec and another point B 1 mile away with a velocity of 60 ft/sec. If the acceleration is uniform, determine:

      a.   The average velocity of the car
      b.   The time taken to travel from A to B
      c.   The acceleration of the car

Given

$$v_1 = 30 \text{ ft/sec}$$
$$v_2 = 60 \text{ ft/sec}$$
$$s = 5280 \text{ ft}$$

Required:

$$\bar{v} = ? \qquad t = ? \qquad a = ?$$

SOLUTION

Average velocity:

$$\bar{v} = \frac{v_1 + v_2}{2}$$

$$= \frac{30 + 60}{2}$$

$$= 45 \text{ ft/sec}$$

Time taken:

$$t = \frac{\text{distance}}{\text{average velocity}}$$

$$= \frac{5280}{45}$$

$$= 117.3 \text{ sec or } 1 \text{ min } 57.3 \text{ sec}$$

Acceleration:

$$v_2^2 = v_1^2 + 2as$$

$$60^2 = 30^2 + 2a(5280)$$

$$a = \frac{60^2 - 30^2}{2(5280)}$$

$$= 0.256 \text{ ft/sec}^2$$

EXAMPLE 2.2

In coming to a stop, a train passes one signal with a speed of 60 mph and a second signal 30 sec later. During this period the brakes are applied to give a uniform acceleration. If the signals are 2400 ft apart, find:

a.   The velocity of the train passing the second sign
b.   The magnitude of the deceleration

Given:

$$v_1 = 60 \text{ mph} = 88 \text{ ft/sec}$$
$$t = 30 \text{ sec}$$
$$s = 2400 \text{ ft}$$

Required:

$$v_2 = ? \qquad a = ?$$

SOLUTION

Final velocity v_2 is obtained from

$$s = \frac{v_1 + v_2}{2} t$$

$$2400 = \frac{88 + v_2}{2} 30$$

$$v_2 = \frac{2400(2)}{30} - 88$$

$$= 72 \text{ ft/sec}$$

Deceleration is obtained from

$$s = v_1 t + \frac{1}{2} a t^2$$

$$2400 = 88(30) + \frac{1}{2} a(30^2)$$

$$a = \frac{(2400 - 2640)(2)}{900}$$

$$= -0.53 \text{ ft/sec}^2$$

Rectilinear Motion Relationship Diagram

The importance to the study of kinematics of the rectilinear relationships just derived cannot be overemphasized. Yet for many students, and even designers, quick recall of these expressions can be difficult. Furthermore, if the appropriate references are not readily available, precious time can be lost in trying to derive the desired expressions. To aid in such situations, the simple diagram presented in Figure 2.1 can be useful. Here:

v_1 (side of the smaller square) represents the initial linear velocity
v_2 (side of the larger square) represents the final linear velocity
at (a times t) represents the difference between v_1 and v_2, where
t is time
as (a times s) represents the area of each trapezoidal section, where
s is the displacement

Considering the larger square, we can write the expression for any one of its sides as follows:

$$v_2 = v_1 + at$$

An expression for the midpoint of the side (or average linear velocity $\bar{v}$) is given by

$$\bar{v} = \frac{v_1 + (v_1 + at)}{2}$$

or

$$\bar{v} = \frac{v_1 + v_2}{2}$$

Also, considering the area of the larger square, we can write an expression for this area in terms of its constituent parts:

$$v_2^2 = v_1^2 + 2as$$

Finally, considering one of the trapezoids, we can determine its area from the relationship

$$\text{Area} = \frac{1}{2}(\text{sum of parallel sides}) \times \text{width}$$

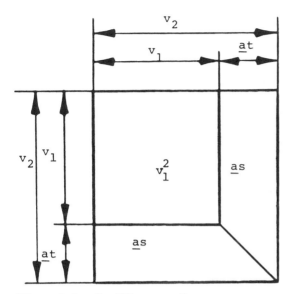

Figure 2.1   Rectilinear motion.

Therefore,

$$as = \frac{1}{2}(v_1 + v_2)at$$

or

$$s = \frac{1}{2}(v_1 + v_2)t$$

Alternatively, the area of a trapezoid may also be expressed as the sum of its rectangular and triangular portions, as follows:

$$as = v_1at + \frac{1}{2}(at)^2$$

$$s = v_1t + \frac{1}{2}at^2$$

2.2   ANGULAR MOTION

Displacement, Velocity, Acceleration, and Time Relationships

As with rectilinear motion, to establish the basic angular relationships, let us consider a rotating body a uniform angular acceleration where the angular velocity changes from an initial value to a final value after some time t, while turning through some angle. Let

θ = angle turned through, rad
ω_1 = initial angular velocity, rad/sec
ω_2 = final angular velocity, rad/sec, after t sec
α = angular acceleration, rad/sec^2
t = time, sec

By definition,

Displacement = average velocity × time

$$\theta = \frac{\omega_1 + \omega_2}{2} t \tag{2.6}$$

After 1 sec, initial velocity is increased to $\omega_1 + \alpha$.
After 2 sec, initial velocity is increased to $\omega_1 + 2\alpha$.
After 3 sec, initial velocity is increased to $\omega_1 + 3\alpha$.
After t sec, initial velocity is increased to $\omega_1 + t\alpha$.

This means that

$$\omega_2 = \omega_1 + \alpha t \tag{2.7}$$

Now substitute for ω_2 In Equation (2.6) and obtain

$$\theta = \frac{(\omega_1 + (\omega_1 + \alpha t)}{2} t$$

$$\theta = \frac{2\omega_1 + \alpha t}{2} t$$

$$\theta = \omega_1 + \frac{1}{2} \alpha t^2 \tag{2.8}$$

Again from Equation (2.7),

$$\omega_2 - \omega_1 = \alpha t \tag{2.9}$$

and from Equation (2.6),

$$\omega_1 + \omega_2 = \frac{2\theta}{t}$$

or

$$\omega_2 + \omega_1 = \frac{2\theta}{t} \tag{2.10}$$

Now, multiply Equations (2.9) and (2.10) and obtain

$$(\omega_2 - \omega_1)(\omega_2 + \omega_1) = \alpha t \times \frac{2\theta}{t}$$

which gives

$$\omega_2^2 - \omega_1^2 = 2\alpha\theta$$

or

$$\omega_2^2 = \omega_1^2 + 2\alpha\theta \qquad\qquad (2.11)$$

In summary, for uniform angular acceleration:

$$\theta = \frac{\omega_1 + \omega_2}{2} t$$

$$\omega_2 = \omega_1 + \alpha t$$

$$\theta = \omega_1 t + \frac{1}{2}\alpha t^2$$

$$\omega_2^2 = \omega_1^2 + 2\alpha\theta$$

For uniform angular velocity ($\alpha = 0$):

$$\theta = \omega_1 t$$

Also, if the body is accelerating, α is positive; if the body is decelerating, α is negative.

EXAMPLE 2.3

A motor starting from rest develops a speed of 3000 rpm in 15 sec. If the acceleration is uniform, determine:

    a.   The rate of acceleration
    b.   The number of revolutions made in coming up to speed

Given:

$$\omega_1 = 0 \ \text{(rest)}$$

$$\omega_2 = 3000 \ \text{rpm} = 3000\left(\frac{2\pi}{60}\right) = 314.2 \ \text{rad/sec}$$

$$t = 15 \ \text{sec}$$

Required:

$$\theta = ? \qquad \alpha = ?$$

SOLUTION

Acceleration:

$$\omega_2 = \omega_1 + \alpha t$$

$$314.2 = 0 + \alpha(15)$$

$$\alpha = \frac{314.2}{15}$$

$$= 20.9 \text{ rad/sec}^2$$

Number of revolutions:

$$\theta = \frac{\omega_1 + \omega_2}{2} t$$

$$= \frac{0 + 314.2}{2}(15)$$

$$= 2356.5 \text{ rad}$$

$$= \frac{2356.5}{2\pi}$$

$$= 375 \text{ rev}$$

EXAMPLE 2.4

A crank shaft rotating at 50 rpm has an angular deceleration of 1 rad/min/ sec. Calculate its angular velocity after 20 sec and the number of revolutions it makes

    a.   In 40 sec
    b.   In coming to rest

Given:

$$\omega_1 = 50 \text{ rpm} = 50\left(\frac{2\pi}{60}\right) = 5.24 \text{ rad/sec}$$

$$\alpha = -1 \text{ rad/min/sec} = \frac{-1}{60} \text{ rad/sec/sec}$$

$$t = 20 \text{ sec}$$

Required:

$$\omega_2 = ? \qquad \theta = ?$$

SOLUTION

Angular velocity after 20 sec:

$$\omega_2 = \omega_1 + \alpha t$$

$$= 5.24 + \frac{-1}{60}(40) = 5.24 - 0.33$$

$$= 4.9 \text{ rad/sec}$$

Number of revolutions in 40 sec:

$$\theta = \omega_1 t + \frac{1}{2}\alpha t^2$$

$$= 5.24(40) + \frac{1}{2}\left(\frac{-1}{60}\right)(40)^2$$

$$= 196.26 \text{ rad}$$

$$= \frac{196.26}{2\pi}$$

$$= 31.23 \text{ rev}$$

Number of revolutions in coming to rest:

Data:

$$\omega_1 = 5.24 \text{ rad/sec}$$

$$\omega_2 = 0 \quad (\text{rest})$$

$$\alpha = \frac{-1}{60} \text{ rad/sec}^2$$

Required:

$$= ? \quad (\text{angular displacement})$$

Equations:

$$\omega_2^2 = \omega_1^2 + 2\alpha\theta$$

$$\theta = 5.24^2 + 2\left(\frac{-1}{60}\right)\theta$$

$$= 5.24^2\left(\frac{60}{2}\right)$$

$$= 823.7 \text{ rad}$$

$$= \frac{823.7}{2\pi}$$

$$= 131.1 \text{ rev}$$

Angular Motion Relationship Diagram

As in the rectilinear case, a similar diagram (Figure 2.2) can be used for the angular relationships. Here,

ω_1 (side of the smaller square) represents the initial angular velocity

ω_2 (side of the larger square) represents the final angular velocity

αt (α times t) represents the difference between ω_1 and ω_2, where

α is the uniform angular acceleration, and

t is time

$\alpha\theta$ (α times θ) represents the area of each trapezoidal section, where θ is the angular displacement

Considering the larger square, we can write an expression for any one of its sides as follows:

$$\omega_2 = \omega_1 + \alpha t$$

and an expression for the midpoint of the side (or average angular velocity $\bar{\omega}$) is given by

$$\bar{\omega} = \frac{\omega_1 + (\omega_1 + \alpha t)}{2}$$

or

$$\bar{\omega} = \frac{\omega_1 + \omega_2}{2}$$

Also, considering the area of the larger square, we can write an expression for this area in terms of its constituent parts:

$$\omega_2^2 = \omega_1^2 + 2\alpha\theta$$

Finally, considering the trapezoidal area, we can write

$$\alpha\theta = \frac{1}{2}(\omega_1 + \omega_2)\alpha t$$

from which

$$\theta = \frac{1}{2}(\omega_1 + \omega_2)t$$

Alternatively, the area of one trapezoid can be expressed as the sum of its rectangular and triangular sections. Thus

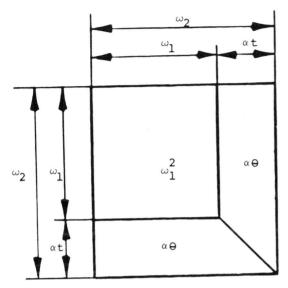

Figure 2.2   Angular motion

$$\alpha\theta = \omega_1 \alpha t + \frac{1}{2}(\alpha t)^2$$

or

$$\theta = \omega_1 t + \frac{1}{2}\alpha t^2$$

2.3   CONVERSION BETWEEN ANGULAR AND RECTILINEAR MOTION

There are many situations where it is necessary to convert from angular to rectilinear motion, and vice versa. To establish the basic relationships, consider the pulley-and-belt arrangement in Figure 2.3. If the pulley turns through an angle θ, assuming that the belt moves without slipping, the corresponding displacement of the belt is equivalent to the length of arc $r\theta$, or

$$s = r\theta \tag{2.12}$$

If the pulley is rotating at uniform velocity ω where the angular displacement $\theta = \omega t$, the total belt displacement is

$$s = r(\omega t) \tag{2.13}$$

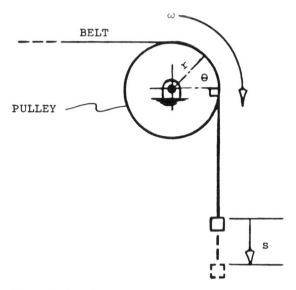

Figure 2.3   Conversion from angular to rectilinear motion.

Also, the belt moves at uniform speed; its velocity v is related to the displacement by

$$s = vt$$

Therefore, Equation (2.13) can be written as

$$vt = r\omega t$$

or

$$v = r\omega$$

This expression is probably the most useful in velocity analysis. In effect, it states that <u>if a point has uniform motion in a circular path, the linear speed at any instant is equal to the distance of that point from the center of rotation multiplied by the angular speed</u>. Also, since velocity is a directed speed, it is clear from the illustration that the only direction the velocity of the point can have at that instant is tangential to the circular path in the same sense as the angular motion.

Another way of looking at the velocity direction is to consider any object, such as a ball, being rotated about one end of a string while the other end is held fixed. Suppose that the string were to be suddenly released, the ball will then "take off" in a direction tangential to the circular path that it maintained before the release.

Similarly, it can be shown that for uniform acceleration of the pulley where the angular acceleration is α, the linear acceleration, a, of the belt is given by

$$a = r\alpha$$

This acceleration is known as the <u>tangential acceleration</u> since the linear velocity acts in a direction tangential to the path of rotation. Hence, to convert from angular motion to linear motion, we multiply the respective values by radius r.

EXAMPLE 2.5

In Figure 2.4 the pulley D is belt-driven by pulley B, which is fastened to pulley C. Starting from rest, the body A falls 60 ft in 4 sec. For each pulley, determine:

a.   The number of revolutions
b.   The angular velocity
c.   The angular acceleration

SOLUTION

Calculate linear values for the pulleys, then convert to angular values.

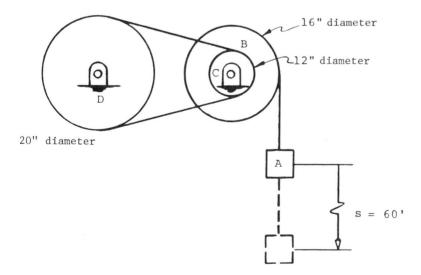

Figure 2.4   Example problem.

Given:

$$r_B = 8 \text{ in.}$$
$$r_C = 6 \text{ in.}$$
$$r_D = 10 \text{ in.}$$
$$S = 60 \text{ ft}$$
$$v_1 = 0 \text{ ft/sec}$$

Required:

θ_B = ? θ_C = ? θ_D = ?

α_B = ? α_C = ? α_D = ?

Revolutions:
   Pulley B:

$$s = r_B \theta_B$$

$$60 = \left(\frac{8}{12}\right)\theta_B$$

$$\theta_B = 60\left(\frac{12}{8}\right)$$

$$= 90 \text{ rad}$$

$$= \frac{90}{2\pi}$$

$$= 14.3 \text{ rev}$$

Pulley C:

$$\theta_C = 14.3 \text{ rev} \quad (\text{same as } \theta_B)$$

Pulley D:

$$\frac{\theta_D}{\theta_C} = \frac{\text{radius of pulley C}}{\text{radius of pulley D}}$$

$$\frac{\theta_D}{14.3} = \frac{6}{10}$$

$$\theta_D = 14.3\left(\frac{6}{10}\right)$$

$$= 8.58 \text{ rev}$$

Angular velocities:

Pulley B: The angular velocity of pulley B is obtained from the following relationship:

$$v_B = r_B \omega_B$$

where

v_B = linear velocity after 4 sec (unknown)

ω_B = angular velocity after 4 sec (required

Therefore, to find v_B, we first need to determine the linear acceleration a_B. Since the motion is not uniform or free-falling, we use the relationship

$$s = v_1 t + \frac{1}{2} at^2$$

where

s = 60 ft

$v_1 = 0$

t = 4 sec

$a = a_B$

Therefore,

$$60 = 0 + \frac{1}{2} a_B (4)^2$$

and

$$a_B = \frac{60(2)}{16}$$

$$= 7.5 \text{ ft/sec}^2$$

Linear velocity v_B is given by the relationship

$$v_2 = v_1 + at$$

where

$$v_2 = v_B$$
$$v_1 = 0$$
$$a = a_B = 7.5 \text{ ft/sec}^2$$
$$t = 4 \text{ sec}$$

Therefore,

$$v_B = 0 + 7.5(4)$$
$$= 30 \text{ ft/sec} \qquad\qquad (2.14)$$

Required angular velocity ω_B can now be determined by substituting the value found for v_B.

$$30 = \left(\frac{8}{12}\right)\omega_B$$
$$\omega_B = 30\left(\frac{12}{8}\right)$$
$$= 45 \text{ rad/sec}$$

Pulley C:

$$\omega_C = 45 \text{ rad/sec} \quad (\text{same as } \omega_B, \text{ since pulleys C and B are attached})$$

Pulley D:

$$\frac{\omega_D}{\omega_C} = \frac{\text{radius of pulley C}}{\text{radius of pulley D}}$$

$$\frac{\omega_D}{45} = \frac{6}{10}$$
$$\omega_D = 27 \text{ rad/sec}$$

Angular acceleration:
   Pulley B:

$$a_B = r_B \alpha_B$$

where

$$a_B = 7.5 \text{ ft/sec}^2 \quad (\text{found})$$

$$r_B = \frac{8}{12}$$

Therefore,

$$7.5 = \frac{8}{12}\alpha_B$$

and

$$\alpha_B = 7.5\left(\frac{12}{8}\right)$$

$$= 11.25 \, \text{rad/sec}^2$$

Pulley C:

$$\alpha_C = 11.25 \, \text{rad/sec}^2 \quad \text{(same as } \alpha_B \text{ since pulleys B and C are attached)}$$

Pulley D:

$$\frac{\alpha_D}{\alpha_C} = \frac{\text{radius of pulley C}}{\text{radius of pulley D}}$$

$$\frac{\alpha_D}{11.25} = \frac{6}{10}$$

$$\alpha_D = 11.25\left(\frac{6}{10}\right)$$

$$= 6.75 \, \text{rad/sec}^2$$

2.4 VELOCITY-TIME GRAPH SOLUTIONS

In many cases it has been found more convenient and simpler to solve some motion problems graphically, using the velocity-time diagram concept. The velocity diagram is a graph in which the velocity of a point is plotted against a time base. Figure 2.5 shows the three possible conditions that can exist:

1. Uniform velocity (Figure 2.5a)
2. Uniform acceleration (Figure 2.5b)
3. Variable velocity (Figure 2.5c)

Note that in case 1 the velocity-time curve has no slope, and therefore the point has no acceleration. This motion is normally referred to as uniform motion.

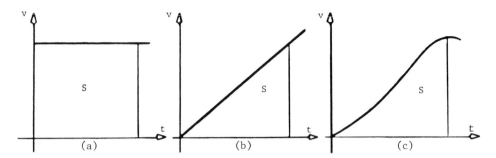

Figure 2.5   Velocity-time curves: (a) uniform velocity; (b) uniform acceleration; (c) variable velocity.

If the motion is not uniform, but the acceleration is constant, the point is said to have <u>uniformly accelerated motion</u>, as in case 2. Otherwise, the motion is variable, as in case 3, where the acceleration changes from one instant to another.

Also, note that acceleration may be depicted as positive or negative depending on whether the slope of the velocity curve is positive (upward to the right or downward to the left) or negative (upward to the left or downward to the right). In each case, the distance covered by the point is given by the area under the curve corresponding to the time period during which the motion took place.

EXAMPLE 2.6

The maximum acceleration of a Ferris wheel at a park is 1 rad/min/sec and the maximum deceleration is 2 rad/min/sec. Determine the minimum time it will take the wheel to complete 15 revolutions going from rest to rest.

SOLUTION

Let

$$T = \text{total time}$$
$$t = \text{time to accelerate}$$
$$T - t = \text{time to decelerate}$$

Given:

$$\alpha_A = 1 \text{ rad/min/sec} = \frac{1}{60} \text{ rad/sec}^2$$

$$\alpha_D = -2 \text{ rad/min/sec} = \frac{-1}{30} \text{ rad/sec}^2$$

$$\theta = 15 \text{ rev} = 15(2\pi) = 94.24 \text{ rad}$$

Required:

$$T = ?$$

Let triangle imf in Figure 2.6 represent the starting velocity v_i through maximum velocity v_m to final velocity v_f. This triangle then represents the total angular displacement θ of the wheel in time T. That is,

$$\theta = \text{area of triangle imf}$$

Now for the acceleration portion of the curve, consider triangular segment ima and find ω_m in terms of t' using the relationship

$$\omega_2 = \omega_1 + \alpha t$$

where

$$\omega_2 = \omega_m$$
$$\omega_1 = \omega_i = 0$$
$$\alpha = \alpha_A = \frac{1}{60} \, \text{rad/sec}^2$$
$$t = t' \quad \text{(acceleration time)}$$

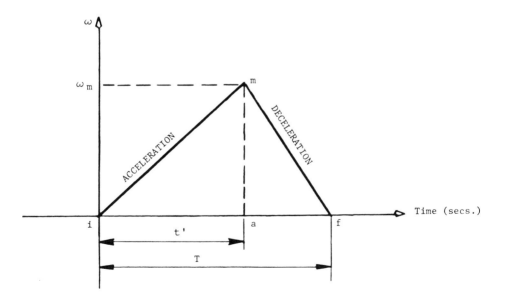

Figure 2.6   Velocity-time graph.

Hence

$$\omega_m = 0 + \left(\frac{1}{60}\right)t'$$

$$= \frac{t'}{60} \text{ rad/sec} \qquad\qquad (2.15)$$

Similarly, for the deceleration portion of the curve, consider triangular segment amf and find t' in terms of T, using the relationship

$$\omega_2 = \omega_1 + \alpha t$$

where

$$\omega_2 = \omega_f = 0 \text{ rad/sec}$$

$$\omega_1 = \omega_m = \frac{t'}{60} \text{ rad/sec}$$

$$\alpha = \alpha_D = \frac{-1}{30} \text{ rad/sec}^2$$

$$t = (T - t') \text{ sec}$$

Hence

$$0 = \frac{t'}{60} + \frac{-1}{30}(T - t')$$

$$= \frac{t'}{60} - \frac{1}{30}(T - t')$$

from which

$$T - t' = \frac{t'}{2}$$

$$T = \frac{t'}{2} + t'$$

$$= \frac{3}{2}t' \qquad\qquad (2.16)$$

$$t' = \frac{2}{3}T \qquad\qquad (2.17)$$

Using the area relationship for a triangle to define angular displacement, we note that

$$\theta = \frac{1}{2}\omega_m T$$

or

$$94.24 = \frac{1}{2}\omega_m T$$

After substituting for ω_m and T from Equations (2.15) and (2.17), we obtain

$$94.24 = \frac{1}{2}\left(\frac{1}{60}\right)\left(\frac{2}{3}T\right)T$$

from which

$$T^2 = 94.24\left(\frac{2}{1}\right)\left(\frac{60}{1}\right)\left(\frac{3}{2}\right)$$

$$= 16,963.2 \text{ sec}^2$$

$$T = 130.2 \text{ sec or 2 min 10.2 sec}$$

EXAMPLE 2.7

A car, traveling between two stoplights 4 miles apart, does the distance in 10 min. During the first minute, the car moves at a constant acceleration, and during the last 40 sec, it comes to a rest with uniform deceleration. For the remainder of the journey the car moves at a uniform speed. Find:

a.   The uniform speed
b.   The acceleration
c.   The deceleration
d.   The distances covered during uniform velocity, acceleration, and deceleration

SOLUTION

Construct a velocity-time curve (Figure 2.7) to describe the motion of the car. Given:

$$v_a = 0 \text{ (at rest)}$$

$$T = 600 \text{ sec}$$

$$t_1 = 60 \text{ sec}$$

$$t_2 = 600 - 60 - 40 = 500 \text{ sec}$$

$$S_T = 4(5280) \text{ ft (total area under curve)}$$

Let the uniform velocity be v. Then the total distance covered, S_T, may be computed from

$$S_T = \text{area of trapezium abcd}$$

$$= \frac{1}{2}(bc + ad)v$$

$$4(5280) = \frac{1}{2}(500 + 600)v$$

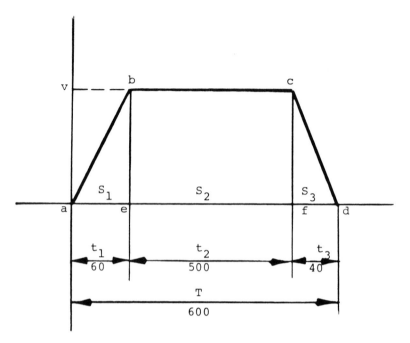

Figure 2.7   Example problem.

$$v = \frac{4(5280)(2)}{1100}$$

$$= 38.4 \text{ ft/sec}$$

Acceleration:

$$v_2 = v_1 + at$$

where

$$v_1 = v_a = 0$$
$$v_2 = 38.4 \text{ ft/sec}$$
$$t = t_1 = 60 \text{ sec}$$
$$38.4 = 0 + a(60)$$

and

$$a = \frac{38.4}{60}$$

$$= 0.64 \text{ ft/sec}^2$$

Deceleration:

$$v_2 = v_1 + at$$

where

$$v_2 = 0$$
$$v_1 = v_f = 38.4 \text{ ft/sec}$$
$$t = t_2 = 40 \text{ sec}$$
$$0 = 38.4 + a(40)$$

and

$$a = -\frac{38.4}{40}$$
$$= -0.96 \text{ ft/sec}^2$$

Note: Negative sign denotes deceleration.
Distance covered during acceleration:

$$S_1 = \text{area of triangle abe}$$
$$= \frac{1}{2}vt_1$$
$$= \frac{1}{2}(60)(38.4)$$
$$= 1152 \text{ ft}$$

Distance covered during uniform velocity:

$$S_2 = \text{area of rectangle ebcf}$$
$$= vt_2$$
$$= 38.4(500)$$
$$= 19,200 \text{ ft}$$

Distance covered during deceleration:

$$S_3 = \text{area of triangle fcd}$$
$$= \frac{1}{2}vt_3$$
$$= \frac{1}{2}(38.4)(40)$$
$$= 768 \text{ ft}$$

2.5   SUMMARY OF MOTION FORMULAS

Linear and Angular Relationships

	Linear		Angular	
	Symbol	Units	Symbol	Units
Displacement	s	ft	θ	rad
Initial velocity	v_1	ft/sec	ω_1	rad/sec
Final velocity	v_2	ft/sec	ω_2	rad/sec
Average velocity	$\bar{v}$	ft/sec	$\bar{\omega}$	rad/sec
Acceleration	a	ft/sec^2	α	rad/sec^2
Time	t	sec	t	sec

$$v_2 = v_1 + at \qquad\qquad \omega_2 = \omega_1 + \alpha t$$

$$v_2^2 = v_1^2 + 2as \qquad\qquad \omega_2^2 = \omega_1^2 + 2\alpha\theta$$

$$\bar{v} = \frac{v_1 + v_2}{2} \qquad\qquad \bar{\omega} = \frac{\omega_1 + \omega_2}{2}$$

$$s = \frac{v_1 + v_2}{2}t \qquad\qquad \theta = \frac{\omega_1 + \omega_2}{2}t$$

$$s = v_1 t + \frac{1}{2}at^2 \qquad\qquad \theta = \omega_1 t + \frac{1}{2}\alpha t^2$$

Conversion from Angular to Linear

Displacement: $s = r \times \theta$ or $(r\theta)$

Velocity: $v = r \times \omega$ or $(r\omega)$

Acceleration: $a = r \times \alpha$ or $(r\alpha)$

3
Vectors

3.1 PROPERTIES OF VECTORS

In mechanics, quantities are classified as either vectors or scalars. A vector has been defined as a quantity that has magnitude and direction. Examples of vector quantities are displacement, velocity, acceleration, and force.

A quantity that has magnitude but no direction is called a scalar. Examples of scalar quantities are time, volume, area, speed, and distance.

Graphically, a vector is represented by an arrow, as in Figure 3.1, where the length, usually drawn to scale, represents the magnitude and the arrowhead indicates the direction. The arrowhead is commonly called the head or terminus of the vector, while the opposite end is called the tail or origin. The direction is usually specified as the angle in degrees which the arrow makes usually with some known reference line. For example, in Figure 3.1a the vector $\bar{V}$ represents a velocity having a magnitude of 5 units in the direction of θ and in Figure 3.1b the vector $-\bar{V}$ represents a velocity of the same magnitude but in opposite direction. Thus, to convert a vector from positive to negative, we reverse its direction.

In this text, vector quantities are normally denoted by capital letters with bars above (e.g., $\bar{R}$, $\bar{V}$, and $\bar{A}$) to distinguish them from their scalar counterparts, denoted by capital letters without bars. Using this notation, the normal addition and subtraction signs, + and -, can be used without risk of confusion between vector and scalar operations. Lowercase letters (e.g., v and a) are also used in some instances to denote scalar quantities, particularly where vectors are not involved.

3.2 VECTOR ADDITION

There are two methods for adding two vectors.

(a)

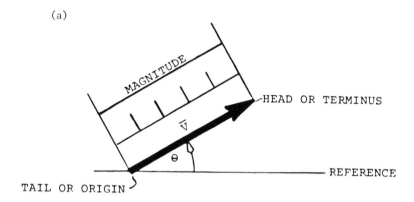

(b)

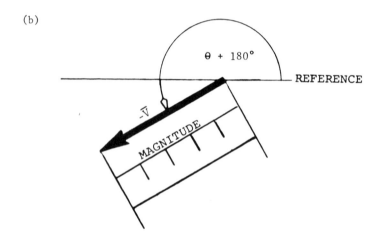

Figure 3.1   Properties of a vector: (a) vector $\overline{V}$; (b) vector $-\overline{V}$.

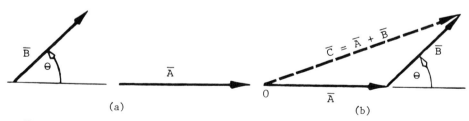

Figure 3.2   Triangular method.

1. The triangular method, where the vectors are connected head
    to tail and the resultant is determined by a third vector, which
    extends from the tail to the head of the connected vectors.

For example, consider the vectors $\bar{A}$ and $\bar{B}$ in Figure 3.2a. To
determine the sum or resultant of these vectors, start at some point O,
called the origin, and connect vector $\bar{A}$ to vector $\bar{B}$ as shown in Figure 3.2b.
Then, from the same origin or the tail of vector $\bar{A}$, draw a third vector, $\bar{C}$,
to close the triangle at the arrowhead of vector $\bar{B}$. Vector $\bar{C}$ is therefore
the required sum or resultant of vectors $\bar{A}$ and $\bar{B}$.

Note that the resultant vector always tends to oppose the general
sense of the summed vectors. We could think of it as a "counterbalancing
vector," since it appears to have a counterbalancing effect on the loop, which
in this case is a triangle.

Note also that the resultant is the same for $\bar{B} + \bar{A}$ as for $\bar{A} + \bar{B}$. This
means that vector addition is commutative.

2. The parallelogram method, where the vectors are connected
    tail to tail so that they form two adjacent sides of a parallelo-
    gram. The resultant is found by drawing a third vector extend-
    ing from the connected point to form a diagonal of the parallel-
    ogram.

For example, consider the same vectors $\bar{A}$ and $\bar{B}$ in Figure 3.2a.
To determine the sum or resultant of these vectors, connect the vectors $\bar{A}$
and $\bar{B}$ tail to tail as shown in Figure 3.3. Define this point of connection as
O (origin). Using these vectors as adjacent sides, complete the parallelogram as shown. Then from the point of origin O, draw a third vector, $\bar{C}$,
also originating at O, to form a diagonal of the parallelogram. This vector,
$\bar{C}$, is the required resultant of the summed vectors.

Both the parallelogram and triangular methods are also applicable
to vector addition involving more than two vector quantities. For example,
consider the summation and vectors $\bar{A}$, $\bar{B}$, $\bar{C}$, and $\bar{D}$ shown in Figure 3.4.
From the triangular method, we note that

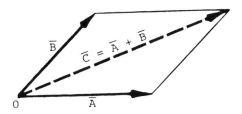

Figure 3.3   Parallelogram method.

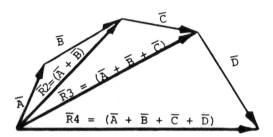

Figure 3.4   Triangular method.

$$\bar{R}_2 = \bar{A} + \bar{B}$$

$$\bar{R}_3 = (\bar{A} + \bar{B}) + \bar{C}$$

$$\bar{R}_4 = (\bar{A} + \bar{B} + \bar{C}) + \bar{D}$$

$$= \bar{A} + \bar{B} + \bar{C} + \bar{D}$$

Also, in Figure 3.5, we note, from the parallelogram method, that

$$S_2 = \bar{A} + \bar{B}$$

$$S_3 = (\bar{A} + \bar{B}) + \bar{C}$$

$$S_4 = (\bar{A} + \bar{B} + \tilde{C}) + \bar{D}$$

$$= \bar{A} + \bar{B} + \bar{C} + \bar{D}$$

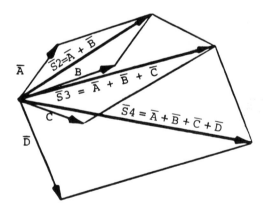

Figure 3.5   Parallelogram method.

3.3   VECTOR SUBTRACTION

Triangular Method

To subtract one vector from another we simply reverse the direction of that vector and sum both vectors normally. For example, consider vectors $\bar{A}$ and $\bar{B}$ in Figure 3.6a. To determine the resultant of $\bar{A} - \bar{B}$, we reverse the direction of vector $\bar{B}$, which in effect changes the vector from $+\bar{B}$ and $-\bar{B}$. With this change made, we now add both vectors by placing the tail of $-\bar{B}$ at the head of $\bar{A}$ and connecting the tail of $\bar{A}$ and head of $\bar{B}$ to obtain vector $\bar{A} - \bar{B}$, the resultant (see Figure 3.6b). In equation form, vector subtraction can be expressed as

$$\bar{A} - \bar{B} = \bar{A} + (-\bar{B})$$

Note that vector subtraction is just a specialized case of vector addition, where the subtracted vector is reversed in direction and treated as a positive vector.

Parallelogram Method

An alternative method of vector subtraction is to join both vectors tail to tail, then draw a third vector to connect the termini of the given vectors. This third vector, when properly directed, represents the difference of the two vectors. The direction of the third vector is obtained by directing the arrow toward the vector from which the subtraction is made.

As an example, consider again the vectors $\bar{A}$ and $\bar{B}$ in Figure 3.7a. To find $\bar{A} - \bar{B}$, we join the tail of $\bar{A}$ to that of $\bar{B}$, and the magnitude of vector $\bar{A} - \bar{B}$ is given by a line connecting the terminus of $\bar{A}$ to that of $\bar{B}$ (see Figure 3.7b). The direction is given by directing the vector $\bar{A} - \bar{B}$ from the

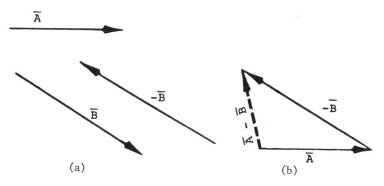

Figure 3.6   Vector subtraction: triangular method.

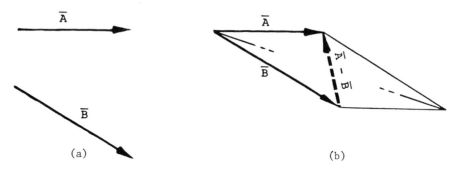

Figure 3.7   Vector subtraction: parallelogram method.

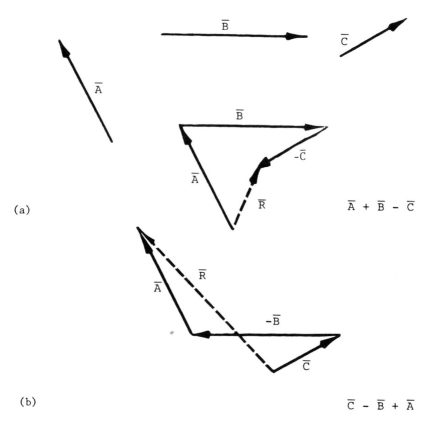

Figure 3.8   Determining resultants of vector systems.

terminus of $\bar{B}$ to the terminus of $\bar{A}$. In other words, the vector $\bar{A} - \bar{B}$ is directed from $\bar{B}$ to $\bar{A}$.

Note that this method, in effect, is a variation of the parallelogram method used in vector addition. In that case, the difference is obtained by completing the "other" diagonal that joins the termini of the given vectors.

EXAMPLE 3.1

Given vectors $\bar{A}$, $\bar{B}$, and $\bar{C}$ in Figure 3.8, where magnitudes and directions are as shown, determine the following:

(a) $\bar{A} + \bar{B} - \bar{C}$

(b) $\bar{C} - \bar{B} + \bar{A}$

SOLUTION

See construction given in Figures 3.8a and b.

3.4 THE VECTOR POLYGON

The vector polygon is the configuration that results from addition or sub-traction of more than two vectors graphically. The polygon can be considered to be a closed loop consisting of the vectors that are added or subtracted and the resultant vectors. Each vector polygon can therefore be represented by an algebraic expression in terms of the vector components and their resultant. For example, consider the vector polygon shown in Figure 3.9. Let it be required to write an algebraic expression for vector $\bar{E}$.

For convenience, it may be assumed that all vectors having one sense (clockwise or counterclockwise) with respect to the closing of the loop are positive, and vectors having the opposite sense are negative. Then it is easy to see that the sense of vector $\bar{E}$ opposes that of vectors $\bar{D}$, $\bar{B}$, and $\bar{A}$, whereas it is the same as that of vector $\bar{C}$. Therefore, from the rules of vector addition and subtraction discussed, we can immediately write the equation

$$\bar{E} - \bar{D} + \bar{C} - (-\bar{B}) - \bar{A} = 0 \qquad\qquad (3.1)$$

from which

$$\bar{E} - \bar{D} + \bar{C} + \bar{B} - \bar{A} = 0$$

or

$$\bar{E} = \bar{D} - \bar{C} - \bar{B} + \bar{A} \qquad\qquad (3.2)$$

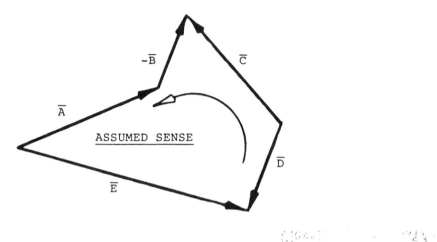

$$\bar{E} = \bar{D} - \bar{C} + (-\bar{B}) + \bar{A}$$

Figure 3.9   Vector polygon.

Similarly, the expressions for vectors $\bar{A}$, $\bar{B}$, $\bar{C}$, and $\bar{D}$ can be derived from first principles, as follows:

$$-\bar{A} - (-\bar{B}) + \bar{C} + (-\bar{D}) + \bar{E} = 0 \tag{3.3}$$

$$\bar{A} = B + \bar{C} - \bar{D} + \bar{E} \tag{3.4}$$

$$-(-\bar{B}) + \bar{C} - \bar{D} + \bar{E} - \bar{A} = 0 \tag{3.5}$$

$$\bar{B} = \bar{D} - \bar{E} + \bar{A} - \bar{C} \tag{3.6}$$

$$\bar{C} - \bar{D} + \bar{E} - \bar{A} - (-\bar{B}) = 0 \tag{3.7}$$

$$\bar{C} = \bar{D} - \bar{E} + \bar{A} - \bar{B} \tag{3.8}$$

$$-\bar{D} + \bar{C} - (-\bar{B}) - \bar{A} + \bar{E} = 0 \tag{3.9}$$

$$\bar{D} = \bar{C} + \bar{B} - \bar{A} + \bar{E} \tag{3.10}$$

Alternatively, the expressions for $\bar{A}$, $\bar{B}$, $\bar{C}$, and $\bar{D}$ can be obtained directly from Equation (3.2).

3.5   VECTOR RESOLUTION

Recalling the rule on summation of vector quantities, we saw that a number of vectors added together was equivalent to a single vector called a resultant.

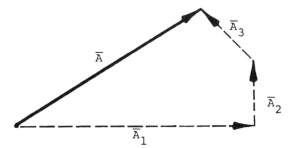

Figure 3.10   Vector components.

Or, stated another way, we can say that the resultant vector is a summation of a number of component vectors. When a vector is represented as a summation of other vectors, that vector is said to be resolved, and the vectors being summed are the <u>components</u> of the resolved vector. The vectors $\bar{A}_1$, $\bar{A}_2$, and $\bar{A}_3$ in Figure 3.10, for example, are components of resolved vector $\bar{A}$.

Although the components of a vector can be limitless, it is normally more useful to resolve a vector into just two components along specific axes. In such a case, it is useful to recall the two methods of vector addition—the triangular method and the parallelogram method—and note that the process of vector resolution is simply a reversal of the addition process.

Suppose that we wish to resolve vector $\bar{A}$ in Figure 3.11a so that its two components, $\bar{C}$ and $\bar{B}$, have the orientation shown by the dashed lines b-b and c-c. Either of the following methods can be employed.

1.   <u>Triangular method</u> (Figure 3.11b): Through the origin and terminus of vector $\bar{A}$, draw lines parallel to b-b and c-c to intersect at some point. This point, according to the triangular method, defines the terminus of vector $\bar{C}$, which extends from the tail of vector $\bar{A}$ to the tail of vector $\bar{B}$, which in turn extends to the terminus of vector $\bar{A}$.

2.   <u>Parallelogram method</u> (Figure 3.11c): An alternative resolution approach is to draw through the tail of vector $\bar{A}$ two axes, each parallel to lines b-b and c-c, then through the terminus or head of $\bar{A}$ draw a line parallel to axis c-c to intersect b-b and similarly, another line parallel to axis b-b through the terminus to intersect axis c-c. These points of intersection on axes b-b and c-c define, respectively, the termini of vector components $\bar{B}$ and $\bar{C}$.

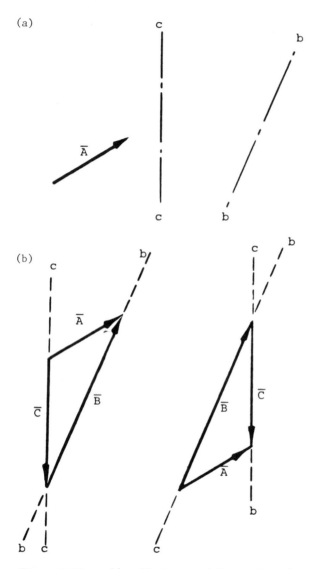

Figure 3.11a and b   Vector resolution: triangular method.

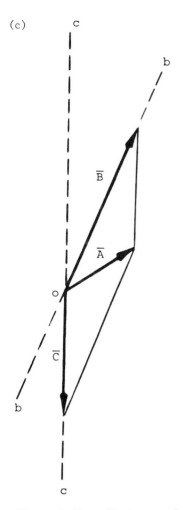

Figure 3.11c   Vector resolution: parallelogram method.

3.6   ORTHOGONAL COMPONENTS

A specialized, but commonly encountered case is the resolution of a vector
into orthogonal or rectangular components along mutually perpendicular
axes. For example, given orthogonal axes x-x and y-y and vector $\bar{V}$ in
Figure 3.12, suppose that we wish to determine the components of $\bar{V}$ along
x-x and y-y.

Using the parallelogram method discussed above, we can readily
determine these components to be $\bar{V}_x$ along axis x-x, and $\bar{V}_y$ along axis y-y,

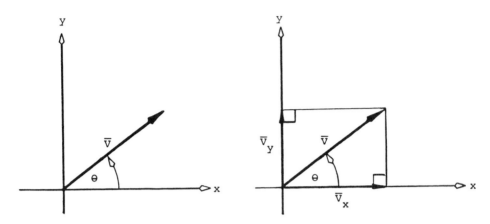

Figure 3.12   Orthogonal components of a vector.

by dropping perpendicular lines from the terminus of vector $\bar{V}$ to axes x-x and y-y, respectively. Here it is useful to note, from trigonometry, that the magnitudes of component vectors $\bar{V}_x$ and $\bar{V}_y$ are

$$V_x = V \cos \theta$$

and

$$V_y = V \sin \theta$$

where θ is the angle that vector $\bar{V}$ makes with the x axis.

3.7   TRANSLATIONAL AND ROTATIONAL COMPONENTS

Orthogonal components, when applied to link motion, may be described in terms of translational and rotational components, where

> The translational component is defined as that component of the vector which tends to cause translation of the link along its own axis.
> The rotational component is defined as that component of the vector which tends to cause rotation of the link about some center located on the link axis.

Consider the vector $\bar{V}$ depicted in Figure 3.13a and b. Observe that the translational component $(\bar{V}^t)$ in Figure 3.13b is equivalent to the x component

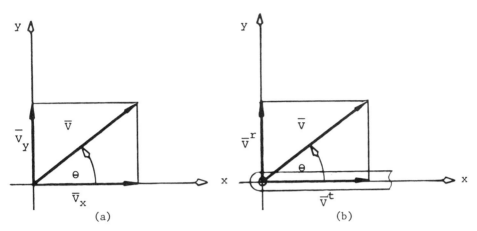

Figure 3.13   Translational and rotational components of a vector.

$(\bar{V}_x)$ in Figure 3.13a, or

$$\bar{V}_x = \bar{V}^t \quad \text{translational component}$$

and the rotational component $(\bar{V}^r)$ (Figure 3.13b) is equivalent to the y component $(\bar{V}_y)$ in Figure 3.13a, or

$$\bar{V}_y = \bar{V}^r \quad \text{rotational component}$$

3.8   EFFECTIVE COMPONENTS

The effective component of a vector may be defined as the projection of that
vector along the axis where the effect is to be measured. To determine the
effective component of any vector, we simply drop a perpendicular line
from the terminus of the vector to a line drawn through the origin of the
vector along which the effect is to be measured. The point at which these
two lines meet defines the terminus of the required effective component.
For example, if we consider again the vector $\bar{V}$ shown in Figure 3.13, $\bar{V}_x$
is the effective component of this vector along the x axis. In other words,
the effective component of vector $\bar{V}$ along axis x-x is the value $\bar{V}_x$. Similarly,
if we consider the same vector with respect to another axis, y-y, $\bar{V}_y$ is the
effective component of $\bar{V}$ along that axis.

    In general, if we consider vector $\bar{V}_p$ oriented with respect to axes
a-a and b-b as shown in Figure 3.14, then by dropping perpendiculars from
the terminus of this vector to these axes, we obtain, respectively,

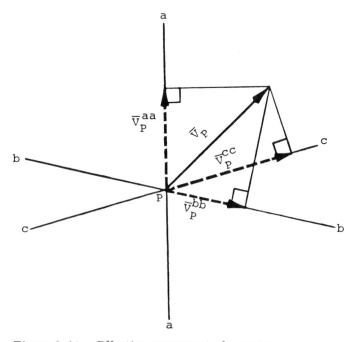

Figure 3.14   Effective component of a vector.

$\bar{V}_P^{aa}$ effective component along a–a

$\bar{V}_P^{bb}$ effective component along b–b

where the superscripts aa and bb are used to indicate the axes along which the effective components are considered.

From this example, the following observations can be made:

1.   A perpendicular drawn from the terminus of an effective component of a vector to intersect the line of action of that vector will define the magnitude of that vector. Therefore, if one effective component and direction of a vector are known, the vector can readily be determined.

2.   The line drawn from the terminus of the absolute vector must be at a right angle to the effective component, not to the absolute vector. This is because the latter construction will yield an effective component that is "greater" in magnitude than the

absolute vector, which is impossible. No component can be greater than its whole.

3. A vector can have an effective component in any conceivable direction, except that which is perpendicular to the vector itself. In this context, it might be well to note that the effective component in the direction of an absolute vector is the vector itself.

4. The orthogonal components of a vector are merely a specialized set of effective components of that vector along mutually perpendicular axes. Compare Figures 3.12 and 3.13.

5. If two effective components of a vector are known, that vector can be completely determined by constructing perpendiculars from the terminus of each effective component vector such that they intersect each other. This point of intersection defines the terminus or magnitude of the required vector. For example, in Figure 3.14, the perpendiculars drawn from the termini of $\bar{V}_P^{aa}$ and $\bar{V}_P^{bb}$ define the terminus $\bar{V}_P$.

Note that this is precisely the reverse of the procedure that is followed in finding the two effective components $\bar{V}_P^{bb}$ and $\bar{V}_P^{aa}$ when the absolute vector $\bar{V}_P$ is known.

II
GRAPHICAL TECHNIQUES

Graphical techniques offer a convenient way to solve velocity and acceleration problems relating to mechanism. Compared to analytic techniques, they are simpler, faster, and more easily understood. Accuracy, however, depends of precision of line work, measurements, and a wise choice of scales for the vectors.

In this section a variety of graphical methods are presented for analyzing mechanisms to determine the velocities and accelerations of their members. These include such methods as instant centers, effective component of velocity and acceleration, relative velocity and acceleration, graphical differentiation and integration, and some special constructions.

As one would expect, each method has advantages and disadvantages. No single method is suitable for solving all problems, even though some are more versatile than others. Generally, the most suitable method is one that yields the required information in the shortest time with the least amount of effort. The objective here is to illustrate the variety of graphical methods used to analyze a mechanism which are available to mechanical design engineers and students—hopefully, enabling them to make an appropriate selection when confronted with a practical problem.

II.A

GRAPHICAL TECHNIQUES: VELOCITY ANALYSIS

Velocity is inherently an important factor in dynamic analysis. Since force is proportional to acceleration, which is the rate of change of velocity, velocity analysis becomes a necessary prerequisite to the acceleration and force analysis of a machine member. In high-speed machines, forces generated during impact and during sudden changes of velocity can limit the operating speed of the machine. Also, as the operating speed increases, it requires greater and greater forces to make various links move through their intended cycles. Drive torques must be increased correspondingly as speeds are increased. As with impact, this can result in increased deformation and vibration within the machine.

Also, as speeds increase, lubrication and wear become more critical. For example, in the crankshaft bearing of an automobile, wear depends on the speed of the crankshaft and the pressure between the crank pins and bearings. Similarly, the cutting speeds of machine tools and the flow rates of fluids in engines and pumps are all functions of the velocities of the output members. For these reasons, methods of determining relative and absolute velocities are of great importance in making a complete analysis of the motions of parts of a machine.

4

Effective Component of Velocity Method

4.1 INTRODUCTION

The effective component method of velocity is based on two principles. The first is the use of the effective components of a vector, and the second is the rigid body principle. As discussed earlier, the effective component of a vector in any direction is the projection of that vector along a line drawn through the vector origin in the direction of interest. This direction is usually defined for the convenience of applying the rigid body principle.

4.2 THE RIGID BODY PRINCIPLE

The rigid body principle may be stated as follows: In a rigid body, the distance between two points remains constant and the velocity components along a line joining these points must be the same at both points. This principle is easily explained, in that if the velocity components were different at the two points, the link would change in length, and would therefore not remain rigid. Thus, if we know the velocity of one point of a rigid body, we can find the velocity of any other point on that body by resolving the known velocity into components along and perpendicular to the line joining the two points and making the velocity component of the unknown velocity equal to that of the known component along the joining line.

4.3 VELOCITIES OF END POINTS ON A LINK

Consider link BC in Figure 4.1, where the velocity of point B is completely known (in magnitude as well as direction). The line of action for the velocity of point C is also known. We want to find $\bar{V}_C$.

To obtain the velocity at point C, we need the effective component of $\bar{V}_C$ along BC, which has the same magnitude as the effective component of

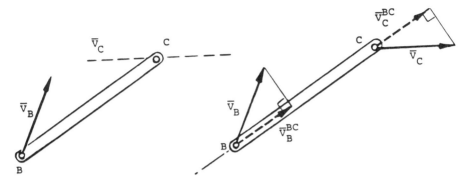

Figure 4.1   Velocities of end points on a link.

$\bar{V}_B$ along BC. We must first obtain $\bar{V}_B^{BC}$, then $\bar{V}_C^{BC}$, and finally $\bar{V}_C$.

PROCEDURE

1.   We determine the effective component of $\bar{V}_B$ along BC, that is, $\bar{V}_B^{BC}$. This is determined by dropping a perpendicular from the terminus of $\bar{V}_B$ to intersect a line joining B and C (line BC).

2.   Since the link is a rigid body and therefore all points along BC <u>must</u> experience the same velocity as $\bar{V}_B^{BC}$, we can immediately lay out the effective component of $\bar{V}_C$ along BC, that is, $\bar{V}_C^{BC}$.

$$\bar{V}_B^{BC} = \bar{V}_C^{BC}$$

3.   With the effective component $\bar{V}_C^{BC}$ completely defined, and the direction of $\bar{V}_C$ also known, we simply construct a perpendicular from the terminus of $\bar{V}_C^{BC}$ to intersect the line of action $\bar{V}_C$. This point of intersection defines the magnitude of $\bar{V}_C$.

4.4   VELOCITIES OF POINTS ON A ROTATING BODY

Now consider body D, which rotates about point O (Figure 4.2). The velocity of point B or $\bar{V}_B$ is given as shown, and the velocity of point C ($\bar{V}_C$) is required.

A close look at this problem will suggest that the procedure should be identical to that used in the preceding problem, except that now the direction of $\bar{V}_C$ is not given explicitly. However, as we noted earlier, since the body is rotating, the linear velocity of point C must be tangential to a circular path described by C. Therefore, the direction of $\bar{V}_C$ is perpendicular to the radial line AC.

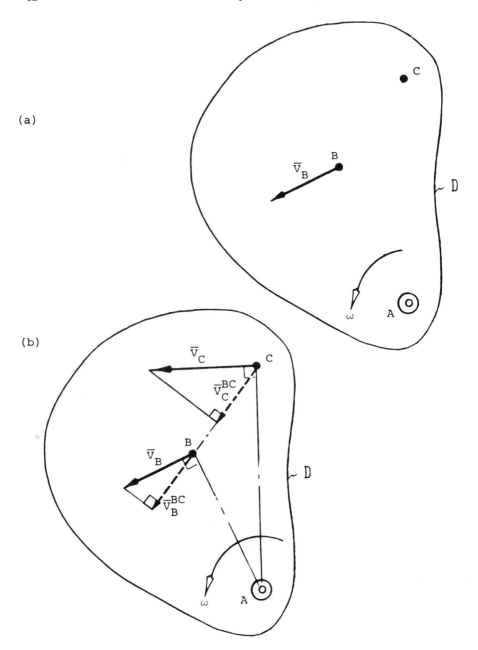

Figure 4.2   Velocities of points on a body with pure rotation.

PROCEDURE

1. Determine the effective component of $\bar{V}_B$ along BC, that is, $\bar{V}_B^{BC}$.
    Here it should be noted that the effective component is obtained
    by dropping the perpendicular from the terminus of $\bar{V}_B$ to exten-
    sion of a line joining B and C (line BC).
2. Lay out the effective component of $\bar{V}_C$ along BC, that is, $\bar{V}_C^{BC}$.

$$\bar{V}_C^{BC} = \bar{V}_B^{BC}$$

(From the rigid body principle, all points along C must have the
same velocity.)

3. Determine the required velocity, $\bar{V}_C$.

From the terminus of $\bar{V}_C^{BC}$, project a perpendicular line to intersect the line
of action of $\bar{V}_C$. This point of intersection defines the magnitude of $\bar{V}_C$.

4.5   VELOCITY OF ANY POINT ON A LINK

Sometimes it is necessary to determine the velocity of a point on a link other
than the two end points. This can easily be obtained as follows. Consider
link BC in Figure 4.3, where the velocities of points B and C are known.
Find the velocity of point D located on the link.

The velocity of point D ($\bar{V}_D$) can be obtained from the summation of
two components: (1) a translation component $\bar{V}_D^t$, which must be the same
for all points along BC (according to the rigid body principle); and (2) a rotational component $\bar{V}_D^r$, which must be proportional to $\bar{V}_B^r$ or $\bar{V}_C^r$ based on its
distance from a center of rotation (according to the rotation principle). In
summary,

$$\bar{V}_D = \bar{V}_D^t + \bar{V}_D^r$$

where

$$\bar{V}_D^t = \bar{V}_D^{BC}$$

$$\bar{V}_D^r = \frac{DO}{BO}\bar{V}_B^r$$

$$= \frac{DO}{CO}\bar{V}_C^r$$

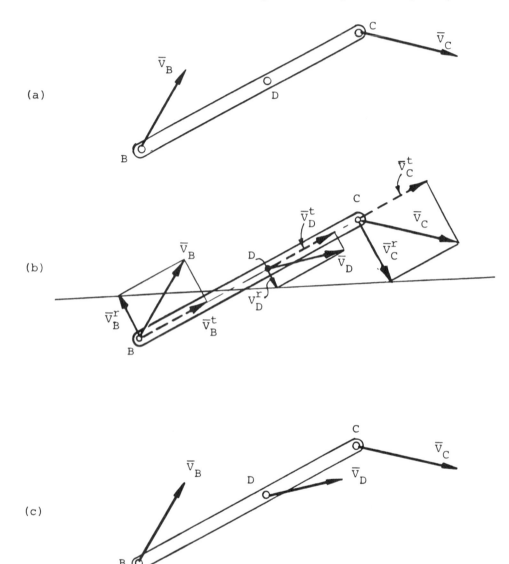

Figure 4.3   Proportionality of velocities.

PROCEDURE

1. Resolve the given velocity vectors $\bar{V}_B$ and $\bar{V}_C$ into their orthogonal (rotational and translational) components with respect to link BC. The rotational components are obtained by dropping

perpendicular lines from the termini of $\bar{V}_B$ and $\bar{V}_C$ to lines drawn normal to BC through points B and C. These components are designated $\bar{V}_B^r$ and $\bar{V}_C^r$ (see Figure 4.3b). The translational components are obtained by dropping perpendiculars from the termini of $\bar{V}_B$ and $\bar{V}_C$ to line BC or BC extended. Note that these components, designated $\bar{V}_B^t$ and $\bar{V}_C^t$, are the same as the effective components of $\bar{V}_B$ and $\bar{V}_C$. That is,

$$\bar{V}_B^t = \bar{V}_B^{BC} \quad \text{and} \quad \bar{V}_C^t = \bar{V}_C^{BC}$$

2. Lay out the translation component of $\bar{V}_D$, that is, $\bar{V}_D^t$ or $\bar{V}_D^{BC}$, along BC. From the rigid body principle, all points experience the same velocity along a straight line. Therefore,

$$\bar{V}_D^t \text{ or } \bar{V}_D^{BC} = \bar{V}_B^t \text{ or } \bar{V}_B^{BC} = \bar{V}_C^t \text{ or } \bar{V}_C^{BC}$$

3. Determine the rotational component of $\bar{V}_D$ $(\bar{V}_D^r)$. Because the point D lies on the same straight line as B and C, its rotational component must be proportional to that of point B as well as point C. Therefore, draw a straight line to connect the <u>terminus</u> of the rotational component of $\bar{V}_B$ to the <u>terminus</u> of $\bar{V}_C$. This line is the <u>line of proportionality</u> for rotational velocity components of all points on BC. Therefore, the required rotational component $\bar{V}_D^r$ is obtained by constructing a perpendicular from point D to meet the proportionality line (see Figure 4.3c).

4. Determine the velocity $\bar{V}_D$. Having determined both the rotational and translational components of $\bar{V}_D$, we can now obtain the resultant vector by graphically summing both the rotational and translational components of $\bar{V}_D$.

$$\bar{V}_D = \bar{V}_D^t + \bar{V}_D^r$$

4.6 VELOCITY ANALYSIS OF A SIMPLE MECHANISM

Crank AB of the slider-crank mechanism shown in Figure 4.4 rotates clockwise at 0.5 rad/sec. We want to determine the velocity of the slider.

In the preceding sections, the linear velocity of point B in link BC was given, and the direction of the velocity of point C was either known or determinable from the constraints of the link motion. Here, although $\bar{V}_B$ is not given directly, it is determinable since we know that its magnitude is given by

$$V_B = AB\omega_{AB}$$

and that its direction must be perpendicular to AB (in the same sense as ω_{AB}). Further, the direction of $\bar{V}_C$, although not given explicitly, is obviously horizontal, considering that the line of action of the slider of which point C is a part must be along the slot, which is horizontal. Therefore, in principle, the procedure for determining the velocity of point C is basically the same as before.

PROCEDURE

1. Lay out the velocity of point B, that is, $\bar{V}_B$ (direction and magnitude), using a convenient scale. The magnitude of this vector is given as

$$V_B = \omega_{AB} \times AB$$

$$= 0.5(5) = 2.5 \text{ in./sec}$$

2. Determine the effective component of $\bar{V}_B$ along BC. Drop a perpendicular line from the terminus of $\bar{V}_B$ to link BC.
3. Lay out the effective component of $\bar{V}_B$ along BC.

$$\bar{V}_C^{BC} = \bar{V}_B^{BC}$$

4. Determine $\bar{V}_C$.
    a. Construct a perpendicular line from the terminus of $\bar{V}_C^{BC}$ to intersect the known line of action of $\bar{V}_C$.
    b. Scale the magnitude of $\bar{V}_C$.

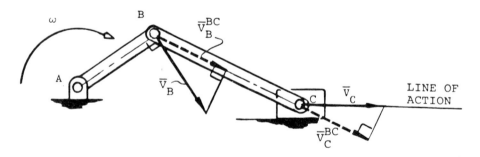

Figure 4.4   Slider-crank mechanism.

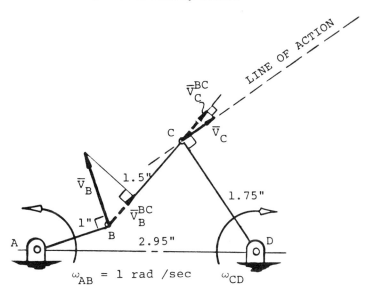

Figure 4.5   Velocity analysis of a four-bar mechanism.

$$V_C = 2.3 \text{ in./sec}$$

Hence

$$\bar{V}_C = 2.3 \text{ in./sec} \quad \text{(directed as shown)}$$

EXAMPLE 4.1

Consider the four-bar linkage in Figure 4.5, where crank AB rotates counterclockwise with an angular velocity of 1 rad/sec, as shown. Let it be required to find the angular velocity of the follower CD.

SOLUTION

To find the angular velocity of the follower, we must first determine the linear velocity of point C. The linear velocity of point C is obtained basically in the same manner as for the slider crank in Figure 4.4. There $\bar{V}_C$ was obtained using the effective component of $\bar{V}_C$ along link BC and the line of action of $\bar{V}_C$, which was known. Here the only difference to be noted is in the line of action of the velocity of point C, which must be perpendicular to follower arm CD. This is because C and rotate only about pivot D. Using this line of action of $\bar{V}_C$ and the effective component of $\bar{V}_C$ along BC, $\bar{V}_C$ is easily found as before.

PROCEDURE

1. Lay out the velocity of point B, that is, $\bar{V}_B$ (magnitude and direction), using a convenient scale. The magnitude of this vector is given as

$$V_B = \omega_{AB} \times AB$$

$$= 1(1) = 1 \text{ in./sec}$$

2. Determine the effective component of $\bar{V}_B$ along link BC ($\bar{V}_B^{BC}$). This is obtained by dropping a perpendicular line from the terminus of $\bar{V}_B$ to meet BC extended (at right angles).
3. Locate the effective component of $\bar{V}_C$ along BC. Since from the rigid body principle, the velocities of all points along BC must be the same, along BC lay out $\bar{V}_C^{BC}$ equal to $\bar{V}_B^{BC}$.
4. Determine the velocity of point C ($\bar{V}_C$). Since point C is constrained to move in a circular path, the line of action of $\bar{V}_C$ is known to be perpendicular to link CD. Therefore, $\bar{V}_C$ is found by dropping a perpendicular line from the terminus of $\bar{V}_C^{BC}$ to

    intersect the line of action of $\bar{V}_C$. This point of intersection defines the magnitude of $\bar{V}_C$.
5. Scale the magnitude of $\bar{V}_C$.

$$V_C = 0.45 \text{ in./sec}$$

6. Determine the angular velocity of CD (ω_{CD}). This is found from

$$\omega_{CD} = \frac{V_C}{CD}$$

$$= \frac{0.45}{1.75}$$

$$= 0.26 \text{ rad/sec}$$

4.7   VELOCITIES OF SLIDING CONTACT MECHANISMS

An important rule in the analysis of velocities in sliding contact states as follows: If two bodies are in sliding contact, their velocities perpendicular to the sliding path are equal. As an example, consider the Scotch yoke mechanism shown in Figure 4.6. Here slider S and yoke y are members in sliding contact, and the sliding path is T-T. Also, P is a point on slider S as well as on crank arm OP, hence no relative motion exists between these

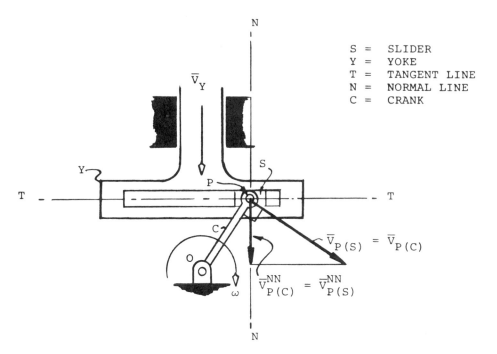

Figure 4.6   Scotch yoke mechanism.

two members at this point. This means that

$$\bar{V}_{P(C)} = \bar{V}_{P(S)}$$

Now, since Y can have only vertical motion, all points in contact with this member, including point P, must have identical velocities to that of Y in the same vertical direction. This means that $\bar{V}_Y$ must be the same as the vertical component of $\bar{V}_P$ on S or $\bar{V}_P$ on C, or

$$\bar{V}_Y = \bar{V}_{P(S)}^{NN} = \bar{V}_{P(S)}^{NN}$$

Thus, in accordance with the sliding contact rule, both bodies—slider S and yoke Y—have equal velocities in direction N-N perpendicular to sliding path T-T.

EXAMPLE 4.2

In the quick-return mechanism shown in Figure 4.7, crank OP rotates clockwise at 7 rad/sec, while slider S, to which it is attached, slides on follower F. Determine the angular velocity of the follower.

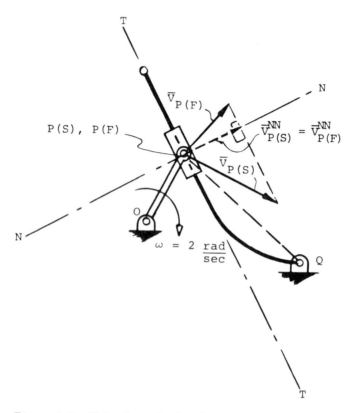

Figure 4.7   Velocity analysis of a quick-return mechanism.

SOLUTION

Like the Scotch yoke just discussed, point P on the slider is the same as
point P on the crank arm. However, note that points P on S and P on F,
although coincidentally located, do not have identical velocities. In fact, it
is for precisely this reason that sliding occurs between the members.
        Nevertheless, according to the rule on sliding, these velocities do
share a common component in the direction normal to the sliding path. Accordingly, if we were to determine the direction normal to the sliding path
and the effective component of the known velocity in this direction, we could
use this component to determine the unknown velocity.

PROCEDURE

1.   Determine and lay out the velocity of P on S ($\bar{V}_{P(S)}$). The magni-
      tude of this vector is given as

$$V_{PS} = OP\omega$$

$$= 7(1)$$

$$= 7 \text{ in.}/\text{sec}$$

2. Determine the direction of velocity of P on F ($\bar{V}_{P(F)}$). Connect P to Q with a straight line. This defines the radius arm with which point P(F) on the follower is rotating at the given instant. Therefore, through point P, draw a line perpendicularly to OP to represent the direction of $\bar{V}_{P(F)}$.

3. Determine the effective component of the velocity $\bar{V}_{P(S)}$ in the direction normal to sliding. The rule states that the velocities or velocity components of $\bar{V}_{P(F)}$ and $\bar{V}_{P(S)}$ in the direction perpendicular to sliding must be equal.

   a. Accordingly, through point P, construct the coordinate axis TT to indicate the path of sliding and another coordinate NN to indicate the path perpendicular to the path of sliding.

   b. Then drop a perpendicular line from $\bar{V}_{P(S)}$ to NN to define the effective component $\bar{V}_{P(S)}^{NN}$. Note that

$$\bar{V}_{P(S)}^{NN} = \bar{V}_{P(F)}^{NN}$$

4. Find the velocity magnitude of P on F (or $V_{P(F)}$). From the terminus of $\bar{V}_{P(S)}^{NN}$, construct a perpendicular to intersect the line of action of $\bar{V}_{P(F)}$, determined in Step 2. This point of intersection defines the magnitude of the velocity $\bar{V}_{P(F)}$.

$$V_{P(F)} = 4.4 \text{ in.}/\text{sec} \quad (\text{scaled})$$

5. Find the angular velocity of the follower, or ω_{QP}:

$$\omega_{QP} = \frac{V_{P(F)}}{QP} = \frac{4.4}{2.1}$$

$$\omega_{QP} = 2.1 \text{ rad}/\text{sec}$$

Note that the velocity of sliding is given by the vectorial difference between $\bar{V}_{P(S)}$ and $\bar{V}_{P(F)}$, or the scaled distance between the termini of these two vectors.

4.8   VELOCITY ANALYSIS OF A
COMPOUND MECHANISM

Consider the mechanism in Figure 4.8, where wheel W turns clockwise at ω rad/sec. We want to determine the velocity of pin C.

Earlier, we saw that if the direction of a velocity and its effective component were known, the magnitude of that velocity could be readily determined. Alternatively, it has been shown (see Chapter 3) that if two effective components of a vector are known, that vector can be determined completely.

In this case, since the direction of $\bar{V}_C$ is not known nor is the path of C readily determined, it is necessary to determine $\bar{V}_C$ using two effective components of this vector: (1) the effective component along link BC ($\bar{V}_C^{BC}$), and (2) the effective component along link CD ($\bar{V}_C^{CD}$). Hence we use the velocity $\bar{V}_B$ to obtain $\bar{V}_C^{BC}$ and the velocity $\bar{V}_E$ to obtain $\bar{V}_C^{CD}$ via point D.

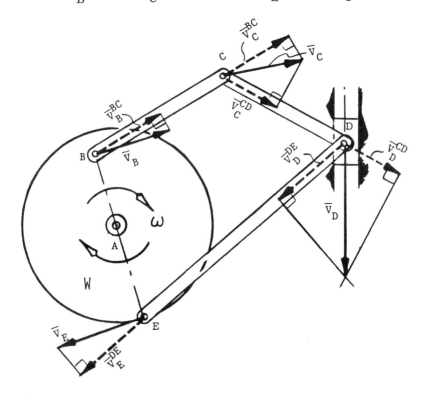

Figure 4.8   Velocities of points on a compound mechanism.

PROCEDURE

1. Starting at point B, lay out the velocity $\bar{V}_B$, whose magnitude is obtained as

$$V_B = \omega\,AB$$

2. Find the effective component $\bar{V}_B^{BC}$ along BC.
3. Locate the effective component $\bar{V}_C^{BC}$ on BC.

$$\bar{V}_C^{BC} = \bar{V}_B^{BC}$$

Since the motion of C is not known, $\bar{V}_C$ cannot be determined directly from $\bar{V}_C^{BC}$. However, by returning to the wheel and stepping off from point E in the opposite direction, we can obtain the additional information needed to define $\bar{V}_C$. Therefore:

4. Lay out the velocity $\bar{V}_E$.
5. Find the effective component $\bar{V}_E^{DE}$ along DE.
6. Locate the effective component $\bar{V}_D^{DE}$ along DE.

$$\bar{V}_D^{DE} = \bar{V}_E^{DE}$$

7. Find the velocity of slider V_D. Since the motion of slider D is known, $\bar{V}_D$ is determined directly from $\bar{V}_D^{DE}$.

8. Using $\bar{V}_D$, find the effective component of this vector along link CD, that is, $\bar{V}_D^{CD}$.

9. Locate the effective component $\bar{V}_C^{CD}$ along CD.

$$\bar{V}_C^{CD} = \bar{V}_D^{CD}$$

10. Now that we have determined two effective components for the velocity at pin C, $\bar{V}_C^{BC}$ and $\bar{V}_C^{CD}$, the absolute velocity $\bar{V}_C$ is determined by projecting perpendicular lines from the terminus of these vectors until they intersect. This point of intersection defines the magnitude of $\bar{V}_C$.

4.9 SUMMARY

The effective component method is particularly suitable for analyzing velocities of sliding members, and is very useful when the instant centers of a
mechanism are outside the limits of the drawing paper. However, a disadvantage of this method is the need for the analysis of velocities of points
from link to link. For complex mechanisms, this could result in considerable drawing time and some loss of accuracy. Another disadvantage of the
method is that it does not provide relative velocities of points on the mechanism, which are essential to the acceleration analysis.

5

Instant Center Method

5.1   INTRODUCTION

One of the most effective techniques for analyzing velocities of members or links in a mechanism is the method of instant centers. In simple terms, the instant center (or instantaneous center, as it is sometimes called) has been defined as that point about which a body may be considered to be rotating relative to another body at a given instant.

Applying this concept to a moving link of a mechanism makes it convenient to describe its motion, at any given instant, in terms of pure rotation about an instant center. In this context it may be noted that even a link which undergoes translation may be considered as rotating about an instant center located at infinity. In other words, any straight line may be considered to be an arc of a circle of infinite radius.

The ability to describe any motion in terms of pure rotation greatly simplifies the analysis of a complex mechanism by making it more convenient to determine the velocity of any point on such a mechanism. Consequently, it is clear that the key to successful application of the method of instant centers must depend on one's ability first, to locate all possible instant centers of a mechanism, and second, to use these centers effectively to determine the required velocities.

Generally, for a simple mechanism that consists of four links, this analysis presents little or no problem. However, for a complex mechanism with more than four links, experience has shown that locating all of the instant centers from first principles can be a painstaking exercise. Moreover, once the instant centers have been found and documented, the resulting diagram is often so complex that it does not allow straightforward analysis of velocities.

In this chapter we show how the velocity analysis by instant centers can be simplified by employing graphical aids such as circle diagrams and link extensions. Circle diagrams are used to help locate instant centers that cannot be found easily by inspection, while link extensions aid in the visualization of relationships between the links of a mechanism.

5.2   PURE ROTATION OF A RIGID BODY

To comprehend the concept of instant centers more fully, it is necessary first to consider some basic principles regarding pure rotation of a rigid body. We have already established the fact that if a body has rotary motion, that motion can be converted to rectilinear motion using the radius as a multiplying factor. Also, the linear velocity of any point on that body acts in a direction tangential to the path of rotation and has the same sense as the rotation. For example, consider point B on the rotating body in Figure 5.1a. If A is fixed, the linear velocity of B is given by the relationship

$$\bar{V}_B = AB\omega \quad \text{(directed perpendicular to AB)} \tag{5.1}$$

where

V_B = linear velocity of B

AB = radius of B

ω = angular velocity

Therefore, the angular velocity for the same body is given by

$$\omega = \frac{V_B}{AB} \tag{5.2}$$

Similarly,

$$V_C = AC\omega \tag{5.3}$$

and

$$\omega = \frac{V_C}{AC} \tag{5.4}$$

From Equations (5.2) and (5.4),

$$\frac{V_B}{AB} = \frac{V_C}{AC} \tag{5.5}$$

or

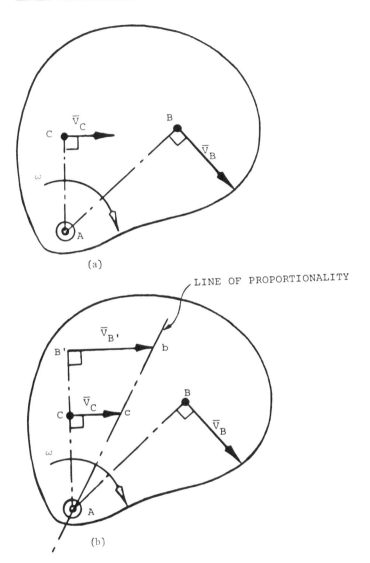

Figure 5.1   Velocities of points on a body in pure rotation.

$$\frac{V_B}{V_C} = \frac{AB}{AC} = \frac{r_B}{r_C} \tag{5.6}$$

which means that $\bar{V}_B$ is proportional to $\bar{V}_C$, as r_B is to r_C. The velocities $\bar{V}_B$ and $\bar{V}_C$ are represented vectorially in Figure 5.1b.

The proportionality of the linear velocities $\bar{V}_B$ and $\bar{V}_C$ to their respective distances from the center of rotation is easily verified graphically (see Figure 5.1b) by rotating the velocity vector $\bar{V}_B$ about center A from point B to a point B' on the radial line AC and constructing a straight line through the center A to touch the termini of $\bar{V}_C$ and $\bar{V}_{B'}$ ($\bar{V}_B$ relocated). This straight line is normally called the <u>line of proportionality</u> between triangles AB'b and ACc, where

$$\frac{B'b}{Cc} = \frac{AB'}{AC}$$

or

$$\frac{V_{B'}}{V_C} = \frac{AB'}{AC}$$

or

$$\frac{V_B}{V_C} = \frac{AB}{AC} \quad \text{(as before)}$$

EXAMPLE 5.1

Consider the 24-in.-diameter rotating disk shown in Figure 5.2, where A is the axis of rotation. B and C are points located on the radial lines AB and AC, as shown. If the linear velocity $\bar{V}_B$ is 2 ft/sec, as shown, determine the angular velocity of the disk and the linear velocity of point C. Also, determine the distance of any point E on the disk where the linear velocity of $\bar{V}_E$ is 1.3 ft/sec. Locate velocity vectors $\bar{V}_B$, $\bar{V}_C$, and $\bar{V}_E$.

SOLUTION

The angular velocity is obtained from

$$V_B = AB\omega$$

$$\omega = \frac{V_B}{AB}$$

$$= \frac{2}{12/12}$$

$$= 2 \text{ rad/sec}$$

The linear velocity of point C is obtained from

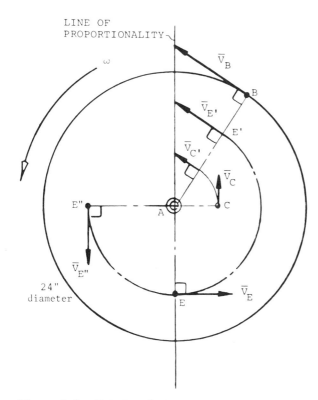

Figure 5.2   Rotating disk.

$$V_C = AC\omega$$

$$= \frac{4}{12}(2) = 0.66 \text{ ft/sec}$$

The radius of E (or AE) is obtained from

$$V_E = AE\omega$$

$$AE = \frac{V_E}{\omega}$$

$$= \frac{1.3}{2} = 0.66 \text{ ft} = 8 \text{ in.}$$

Figure 5.2 shows the location of point E and linear velocities $\bar{V}_B$, $\bar{V}_C$, and $\bar{V}_E$. Note that all other points (e.g., E' and E'') on the same path described by point E have the same linear velocity.

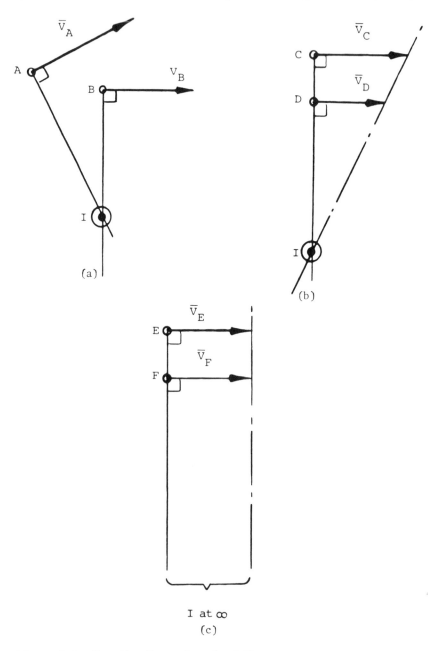

Figure 5.3   Locating the center of rotation.

In summary, then, it is useful to note the following principles:

1. The linear velocity is always tangential to the path of rotation of the point or perpendicular to the radial line joining the point to the center of rotation.
2. The direction of the velocity is always in the same general sense as that of rotation.
3. The magnitude of the velocity is always proportional to the distance of the point from the center of rotation.

Consequently, if the velocity of one point of a rotating body is known, the velocity of any other point on that body can be determined, provided that the center of rotation is known.

To illustrate these principles further, if the directions of velocities and the two points are given, the center of rotation lies at the intersection of lines drawn from the two points perpendicular to the velocity directions, as shown in Figure 5.3a. However, if these directions are the same, the location of the center of rotation also depends on the relative magnitudes of the velocities. For example:

If the velocities are unequal, the center of rotation lies at the intersection of the common perpendicular drawn from the tails of the two vectors with a line joining the termini of the same vectors. See the construction in Figure 5.3b.

If the velocities are equal, the center of rotation lies at infinity, which means that the body has no rotation and is therefore translating. See the construction in Figure 5.3c.

5.3 COMBINED MOTION OF A RIGID BODY

Although it is relatively simple to determine velocities on a body in pure rotation about a fixed axis, it is certainly not as straightforward to find the velocities on a body in combined motion. This is because a body in combined motion is simultaneously rotating and translating and therefore has no fixed axis of rotation. However, this problem can be simplified by considering the motion to conform, just for an instant, to that of pure rotation about some center of rotation, just for that instant, and thus the velocities in the body can be found using the same principles as those applicable to pure rotation. The center of rotation in this case is aptly called the instant center of rotation, since its position changes continuously from one instant to another.

The validity of this approach can be demonstrated by considering link AB in Figure 5.4a, which moves with combined motion. Imagine that the link moves from position 1 to position 2 in a very small time interval.

Then, using the construction shown in Figure 5.4b, the same motion can be seen to conform to a circular path where point I is the center. This point is the instant center of link AB. Further, if we connect points A and B (or points A' and B') to the center I by radial lines as shown in Figure 5.5, we see that the velocity directions of these points are indeed perpendicular to the four radial lines drawn, which is consistent with the principles of pure rotation.

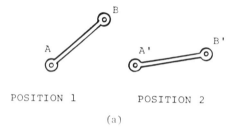

POSITION 1   POSITION 2

(a)

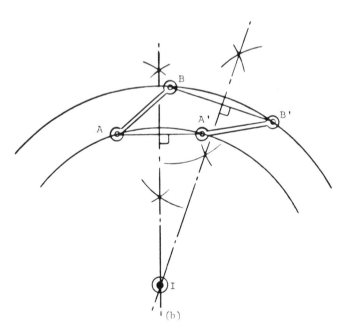

Figure 5.4   Instant center of link in plane motion: (a) link in plane motion; (b) construction.

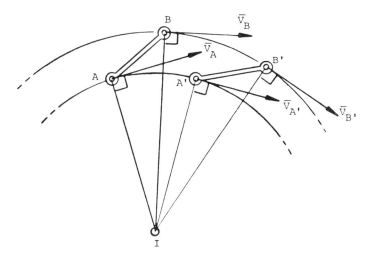

Figure 5.5   Velocity of a link in plane motion.

Thus, if we know the velocity directions of two points on a floating link, the instant center of rotation must lie at the intersection of lines drawn from those points perpendicular to the velocity directions. Therefore, using the instant center, the combined motion of a link or any rigid body can be conveniently reduced to pure rotation, thereby simplifying the velocity analysis.

5.4   VELOCITY OF A BODY WITH ROLLING CONTACT

A common example of combined motion occurs with a rolling wheel. Consider the wheel rolling to the right in Figure 5.6. Here, as the wheel moves from one position to another, there is both rotational and translational motion.

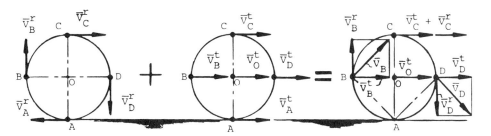

Figure 5.6   Velocities of points on a rolling wheel.

To visualize this combined motion, it is convenient to think of each motion as though it occurred independently, then superpose the two effects to obtain the final result. For example, the combined motion can be expressed graphically as shown in Figure 5.6, where $\bar{V}^r$ and $\bar{V}^t$ are the rotational and translational velocities for the respective points on the rim. From the summation of the rotational and translational components, it is noted that:

1. Point A is a point of zero velocity.
2. The velocity of each point on the rim is perpendicular to a line joining that point to point A.
3. The magnitude of the velocity of each point is proportional to its distance from point A.

Thus point A is considered the instant center of rotation of the rolling wheel.

In summary, the instantaneous center of a wheel rolling without slipping lies at the point of contact. All points of the wheel have velocities perpendicular to their radii from the instant center and these velocities are proportional in magnitude to their respective distances from that center.

5.5   TYPES OF INSTANT CENTERS

Instant centers are of three types:

1. Fixed (type 1), that is, a stationary point in one body about which another body actually turns. This is normally a fixed axis of rotation on a mechanism.
2. Permanent (type 2), that is, a point common to two bodies having the same velocity in each body, such as a hinged joint connecting two moving links of mechanism. The term "permanent" implies that the relative position between the connected links is always the same, regardless of the change in position of the mechanism.
3. Imaginary (type 3), that is, a point within or outside the mechanism which can be visualized as having the same characteristics as either a fixed center (type 1) or a permanent center (type 2) at any given instant. When this center behaves like a fixed center about which the body tends to turn, it is considered an instant axis of rotation.

The slider-crank mechanism shown in Figure 5.7 depicts these three types of instant centers. Usually, fixed and permanent instant centers can be readily identified by inspection. Typically, imaginary instant centers must be located by more detailed analysis. Generally, the circle diagram method (described in Section 5.6) is used to determine the imaginary centers.

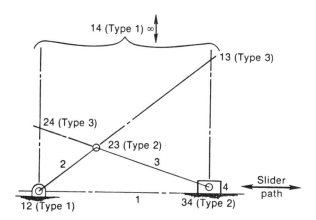

Figure 5.7   Instant centers for a slider-crank mechanism. The location of type 1 and type 2 instant centers can be determined by inspection, whereas the location of type 3 instant centers requires additional analysis.

Note that the instant center 14 (read "one-four") for the path of slider 4 on frame 1 is indeterminate because it lies at infinity. This is because the slider path is a straight line, and therefore the slider can be considered as a body that actually turns about a point located at infinity. An instant center that lies at infinity can be located along an infinite number of lines perpendicular to a straight path.

5.6   LOCATING INSTANT CENTERS

Obvious Instant Centers

Obvious instant centers are those that can be readily located (by inspection) on the mechanism. These may be of either fixed or permanent type. There are four types of obvious instant centers:

1.   Instant center for pin-connected links (see Figure 5.8).
2.   Instant center for sliding body (see Figure 5.9).
3.   Instant center for rolling body (see Figure 5.10).
4.   Instant centers for direct contact mechanisms.
     a.   For sliding contact between 2 and 4, instant center 24 lies at the intersection of the common normal through the contact point and the line of centers (see Figure 5.11a).
     b.   For rolling contact between bodies 2 and 4, both instant center 24 and the contact point are coincident and lie on the line of centers (see Figure 5.11b).

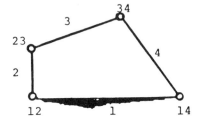

Figure 5.8   Instant center for pin-connected links.

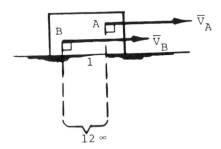

Figure 5.9   Instant center for a sliding body.

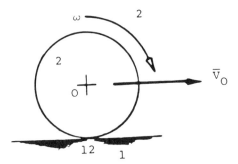

Figure 5.10   Instant center for a rolling body.

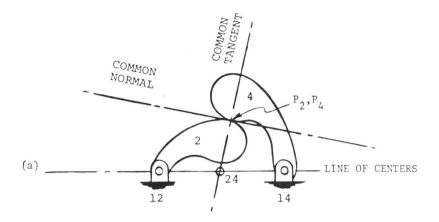

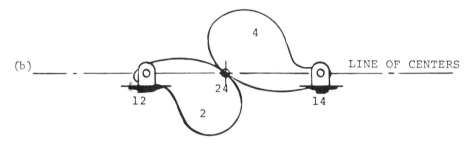

Figure 5.11   (a) Sliding contact; (b) rolling contact.

Circle Diagram Method

The circle diagram method is based on Kennedy's theorem, which states that any three bodies having plane motion relative to one another have three instant centers, and they lie on a straight line.

PROOF

Consider any three bodies 1, 2, and 3 having plane relative motion as shown in Figure 5.12. Assuming that bodies 2 and 3 are pinned to body 1, and therefore that instant centers 12 and 13 are known, the problem is to show that the third instant center of 2 and 3 must be on the straight line connecting 12 and 13. First, suppose that the instant center of 2 and 3 were at P'. Then, as a point in body 2, P' must move at right angles to line 12-P' ($\bar{V}_{23}^{(2)}$), and as a point in body 3, P' must move at right angles to line 13-P' ($\bar{V}_{23}^{(3)}$). This means that point P' moves in two different directions at the same time,

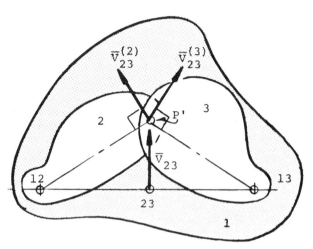

Figure 5.12   Three bodies in relative motion.

which is impossible. Therefore, in order for the instant center of 2 and 3 to have the same velocity direction in both 2 and 3, it must lie on the straight line joining 12 and 13. This proof also applies if all three bodies are moving.

Consider the four-bar linkage shown in Figure 5.13. Let it be required to find all instant centers of this mechanism.

PROCEDURE

1.   Locate all fixed and permanent instant centers on the mechanism in Figure 5.13. These centers are usually found by inspection. For example, instant center 12 is located where link 1 joins link 2, and instant center 23 is located where link 2 joins link 3.

Note that the order of the digits used to designate the instant centers is not important. That is, either 23 or 32 may be used to designate the same instant center.

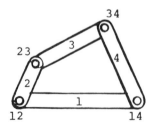

Figure 5.13   Mechanism.

Point represents
link number.

Figure 5.14   Step 2.

2.   Lay out points 1, 2, 3, and 4, approximately equally spaced
      and in sequence on a circle (see Figure 5.14), each point repre-
      senting a link on the mechanism (Figure 5.13). Any straight
      line drawn to connect any two of these points represents an in-
      stant center on the mechanism.
3.   Using solid lines, indicate on the diagram (Figure 5.15) the
      instant centers that have been located so far on the mechanism
      (i.e., the fixed and permanent centers) by connecting the points
      that represent those centers. Lines 12, 23, 34, and 14 should
      therefore be drawn in as solid.
4.   Connect all other points in the diagram (Figure 5.15) using
      dashed lines to indicate the instant centers that remain to be
      found. For example, lines 13 and 24 indicate those outstanding
      centers.
5.   To locate these centers, examine the diagram in step 4 (Fig-
      ure 5.15) to find one dashed line that, if it were solid, will com-
      plete two solid triangles. For example, the dashed line 13 com-
      pletes triangles 123 and 341 (Figure 5.15). Using these triangles,
      we can now locate instant center 13, noting that the sides of tri-
      angle 123 represent three instant centers—12, 23, and 13—
      which by Kennedy's theorem, must lie on a straight line within

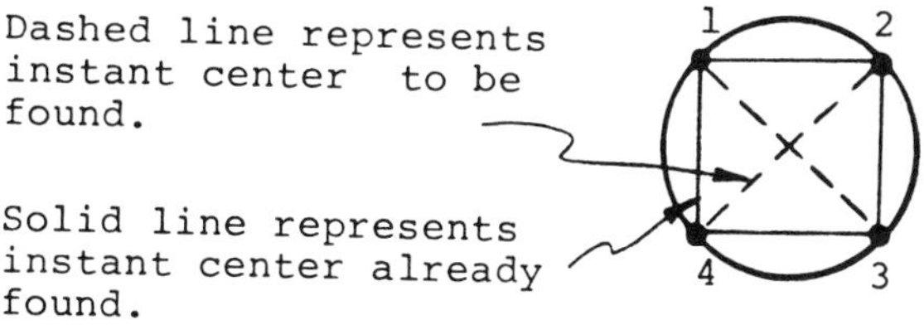

Dashed line represents
instant center   to be
found.

Solid line represents
instant center already
found.

Figure 5.15   Steps 3 and 4.

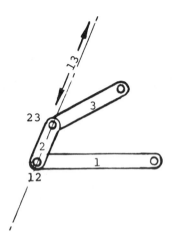

Figure 5.16   Step 5.

or outside the mechanism. Hence the instant center 13 lies somewhere on a line joining points 12 and 23, or its extension in the mechanism (see Figure 5.16).

Similarly, the sides of triangle 341 represent three instant centers—34, 14, and 13—which again by Kennedy's theorem must lie on a straight line within or outside the mechanism. Hence instant center 13 must lie on a line joining points

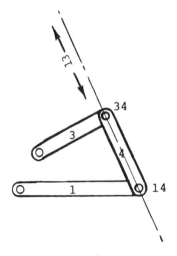

Figure 5.17   Step 5.

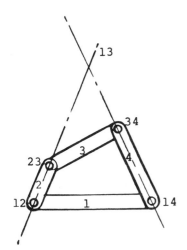

Figure 5.18   Step 6.

34 and 14, or its extension in the mechanism (see Figure 5.17).
Since by both triangles, instant center 13 is given to be somewhere along lines 12-23 and 34-14 (Figures 5.16 and 5.17), is
must be located at the intersection of these lines.

6.   Accordingly, on the mechanism (Figure 5.18) draw two straight
lines, 12-23-13 and 14-34-13, and at their intersection locate
instant center 13. Then, on the diagram, change line 13 to a
solid line before proceeding to find the other instant center 24
(Figure 5.19).

It is important that after an instant center has been located on the mechanism, it be immediately drawn in as a solid line on the diagram. Otherwise,
when locating the remaining centers, it may not be possible to find additional pairs of triangles which are solid except for a common dashed line.

Common line that
completes two triangles

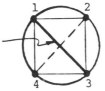

Fig. 19   Step 6.

5.7   VELOCITY PROPERTIES OF THE INSTANT AXIS

A link is considered to be rotating when all points in that link remain at
fixed distances from an axis of rotation. In a machine, each member is
rotating about a fixed axis or about a moving axis whose location varies
from one instant to another. For purposes of analysis, this moving axis
may be thought of as a stationary axis having properties similar to those
of a fixed axis, at any given instant. Also, it should be noted that:

1. There is one instant axis of velocity for each floating link of a
   machine.
2. There is no single instant axis of velocity for all links of a
   machine.
3. Instant centers are also axes of rotation when they have no absolute motion. That is, their velocity is zero with respect to the
   frame. Such instant centers can be readily identified because
   their numerical designation includes the number representing
   the frame. For instance, in the slider-crank example, link 1
   represents the frame and instant centers 12 and 13 are axes of
   rotation for links 2 and 3, respectively. Instant 14 is also an
   axis of rotation but is indeterminate and lies at infinity. Instant
   centers 24, 23, and 34 are not axes of rotation and may or may
   not have absolute motion with respect to the frame. Note that 12
   is a fixed axis of rotation, whereas 13 is a moving axis of
   rotation.
4. The instant center of velocity is not an instant center of acceleration, although it moves as the link moves and may have an
   actual zero acceleration. Neither does it necessarily have zero
   acceleration as does the fixed center.

5.8   VELOCITY ANALYSIS BY INSTANT CENTERS

The determination of velocities in mechanisms by instant centers is based
on three principles:

1. The velocity magnitude at a point in a rotating body is directly
   proportional to the radius of rotation of that point.
2. The velocity direction at a point in a rotating body is perpendicular to its radius of rotation.
3. An instant center is a point common to two bodies and has the
   same linear velocity (both magnitude and direction) in both bodies.

Consequently, if the linear velocity of any point in a body relative to an
instant center is known, the velocity of any other point in that body, relative

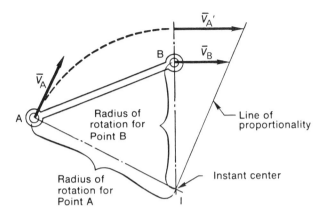

Figure 5.20   Velocity is proportional to radius of rotation. The velocity of any point in a link is proportion to its radius of rotation about an instant center. As a result, if the velocity of one point (such as B) is known, the velocity of any other point (such as A) in that link can be determined graphically by proportions.

to the same instant center, can be determined graphically by proportions, as illustrated by the radius of rotation method in Figure 5.20.

Note that instant centers must be located first before velocities can be determined. Also, in mechanism analysis, velocities are usually determined about fixed instant centers (either normally fixed or momentarily fixed) having no velocity with respect to the frame.

To illustrate how these principles are applied, let us now consider two important concepts as they apply to a four-bar linkage (Figure 5.21): (1) fixed axis (or center), and (2) link extensions.

The term <u>fixed axis</u> is used to refer to an instant center that is normally fixed or one that is momentarily fixed (an instant axis). To identify

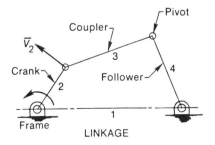

Figure 5.21   Four-bar linkage.

the fixed axis of a mechanism, we look for those centers whose numerical designations include the fixed link of the mechanism, that is, the link whose velocity is zero. Since in this example the fixed link is 1, it immediately becomes evident that the fixed axes are (12), (13), and (14).

Note also that a fixed axis typically has a numerical designation which includes both the frame and the link that rotates about the axis. For example, the fixed axes (12), (13), and (14) have, respectively, links 2, 3, and 4 rotating about them. The other centers—(24), (34), and (23)—can be considered as joints on the extended mechanism. These centers may or may not have a velocity.

Link extensions are imaginary bodies defined by three instant centers, one of which typically is a fixed axis, and by a common link that rotates

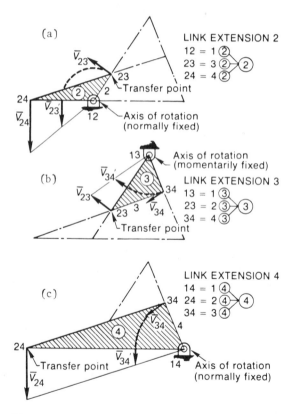

Figure 5.22   Link extensions for a four-bar linkage. Link extensions 2(a) and 4(c) have fixed axes of rotation. Link extension 3(b) has a moving axis rotation, denoted by the imaginary instant center 13, which for analytical purposes is assumed to be momentarily fixed.

about this fixed axis. To identify the link extension for a given link, we
simply locate the fixed axis (or center) about which that link rotates and the
other centers whose numerical designations include that link. These centers,
together, comprise the link extension for the given link.

The four-bar mechanism has three rotating members: links 2, 3,
and 4. Link extensions for these rotating links are defined by the following
sets of centers:

(12) (23) (24)   where link 2 is the common link and center (12) is the
                 fixed axis of rotation. This link extension is depicted
                 as triangular plate ② in Figure 5.22a.
(13) (23) (34)   where link 3 is the common link and center (13) is the
                 fixed axis of rotation. This link extension is depicted
                 as triangular plate ③ in Figure 5.22b.
(14) (24) (34)   where link 4 is the common link and center (14) is the
                 fixed axis of rotation. This link extension is depicted
                 as triangular plate ④ in Figure 5.22c.

For purposes of analysis, link extensions essentially reduce complex mechanisms to several simpler mechanisms, thereby reducing the computation
of velocities to simple graphics. Also, link extensions help to illustrate
such concepts as axes of rotation, common links, and transfer points—all
of which are essential elements of velocity analysis.

5.9   VELOCITY ANALYSIS OF A SIMPLE MECHANISM

Consider again the four-bar linkage shown in Figure 5.21. For the position
shown, the velocity $\bar{V}_{23}$ is given and the velocity $\bar{V}_{34}$ is required.

PROCEDURE

First, we determine all instant centers for the mechanism. The total number
(N) of instant centers of a mechanism is given by the equation

$$N = \frac{n(n-1)}{2}$$

where n is the number of links in the mechanism. To locate these centers,
the circle diagram method has been found to be most useful, particularly
for the more complex mechanisms. Instant centers for the four-bar linkage
are shown in Figure 5.23.

Next, we examine the figure to identify the fixed centers and the
link extensions that tend to have rotation about these centers. Again, these
link extensions are shown in Figure 5.22.

(a)

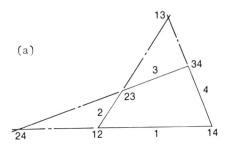

(b)

$N = \dfrac{n(n-1)}{2}$

$\quad = \dfrac{4(3)}{2}$ = 6 instant centers

Point represents
link 1

Line represents
instant center 14

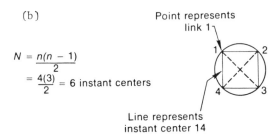

Figure 5.23   Instant centers for a four-bar linkage. Instant centers 24 and 13 are both imaginary. However, center 13 is also an axis of rotation for floating link 3. (a) Instant center diagram; (b) circle diagram.

Finally, having defined the link extensions and their respective axes of rotation, we are ready to apply principles 1 and 2, using the rotation of radius method (shown in Figure 5.20) to determine the required velocities of points in the linkage. In the present example, the velocity of link 4 ($\bar{V}_4$) can be determined by proportions using the link extension (13) (23) (34) and the known velocity $\bar{V}_{23}$. For instance, velocity $\bar{V}_{34}$ at 34 is simply proportional to the radius of rotation of 34 about 13. Therefore,

$$\bar{V}_{34} = \bar{V}_{23}\,\frac{13\text{-}34}{13\text{-}23}$$

where 13-34 and 13-23 denote the radii of rotation of 34 and 23, respectively. Note that $\bar{V}_{34}$ is a velocity common to both links 3 and 4.

This observation leads to the concept of a transfer point. A <u>transfer point</u> is an instant center that has the same velocity in two different links or link extensions. In the four-bar linkage, 34 is a transfer point for links 3 and 4 and for link extensions 3 and 4. Also, 23 is a transfer point for links 2 and 3, and 24 is a transfer point for links 2 and 4.

The concept of a transfer point provides a powerful tool for velocity analysis whereby the velocity of any point in a link can be determined without knowing the velocity of another point in the same link. For example, suppose that only $\bar{V}_{23}$ is known; then the velocity at any point in link 4, or its extension, can be determined without knowing the velocity at 34. This is accomplished by first finding the velocity at 24, using link extension 2, which contains the given velocity $\bar{V}_{23}$, then transferring the velocity found at 24 ($\bar{V}_{24}$) to link extension 4 to find the velocity of any other point on that extension. This means that the velocity at point 34, which is a point on link extension 4, can be determined as follows:

1. The velocity at 24, using link extension 2, is

$$\bar{V}_{24} = \bar{V}_{23} \frac{12\text{-}24}{12\text{-}23}$$

2. The velocity at 34, using link extension 4, is

$$\bar{V}_{34} = \bar{V}_{24} \frac{14\text{-}34}{14\text{-}24}$$

5.10   VELOCITY ANALYSIS OF A COMPOUND MECHANISM

Consider a steam locomotive driven by a compound mechanism having six links and 15 instant centers (Figure 5.24a). As indicated by the circle diagram, nine of the instant centers were identified by inspection, while the remaining six imaginary instant centers had to be located by detailed analysis. Given a velocity $\bar{V}_A$ at instant center 36 in link 6, find the velocity at point B in link 2, using extended links to simplify the analysis (Figure 5.24b).

First, the velocity at instant center 26 must be determined. Instant center 26 is the transfer point between links 2 and 6. Therefore, 26 has the same velocity in both link extensions 2 and 6.

For link extension 6 the axis of rotation is 16 and the common link is 6. By inspection it can be determined that link extension 6 consists of instant centers 16, 26, and 36 (Figure 5.25a). Thus the velocity at the transfer point 26 is determined by proportions as

$$\bar{V}_{26} = \bar{V}_{36} \frac{16\text{-}26}{16\text{-}36}$$

For link extension 2 the axis of rotation is 12, the common link is 2, and the constituent instant centers are 12, 23, and 26 (Figure 5.25b). Note that transfer point 26 is common to both link extensions. Given velocity $\bar{V}_{26}$, the velocity 23 is

$$\bar{V}_{23} = \bar{V}_{26} \frac{12\text{-}23}{12\text{-}26}$$

Finally, the velocity at point B is determined graphically by proportions as

$$\bar{V}_B = \bar{V}_{23} \frac{12\text{-}B}{12\text{-}23}$$

It should be noted that $\bar{V}_B$ could have been determined more directly by recognizing that the velocity at 23 is the same as that of 36. However, transfer points were used deliberately in the analysis, for instructive purposes.

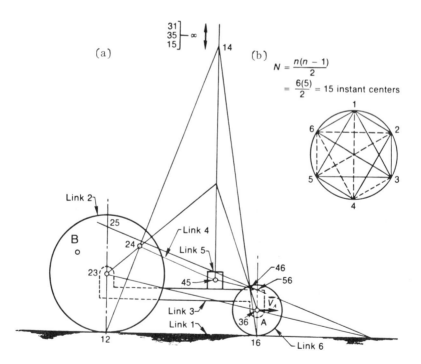

Figure 5.24   Determining velocities by link extensions: (a) mechanism; (b) circle diagram.

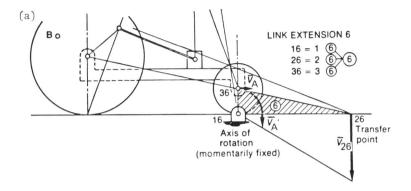

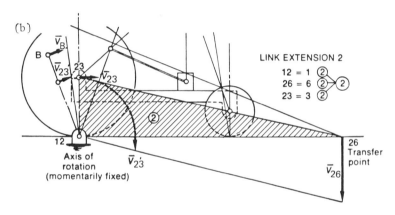

Figure 5.25   Velocity analysis by link extension. Given a velocity in link 6, velocities in link 2 can be determined once the transfer point (instant center 26) is identified and the transfer point velocity ($\bar{V}_{26}$) is determined. (a) Link extension 6; (b) link extension 2.

5.11   SUMMARY

To find the linear velocity of a point in a link, we first identify the link (or its extension) in which the point occurs and its axis of rotation. If the velocity of any point in this link (or its extension) is known, the velocity of that point can readily be determined from the principles of rotation. However, if there is no known velocity in this link (or its extension), we (1) select a point (i.e., an instant center) common to this link and a second link (or its extension), where the velocity of such a point can be determined from rotation principles; (2) determine this velocity; and (3) transfer the velocity found to the first link (or its extension) to find the required velocity, again using the principles of rotation.

6
Relative Velocity Method

6.1 INTRODUCTION

The relative velocity method is probably the most common among the graphical methods used in velocity analysis. Compared to the other graphical methods, it readily provides solutions not only for absolute velocities, but also for relative velocities of points in a mechanism without requiring the location of instant centers. This singular feature makes it most desirable to use when determining the relative velocities needed for acceleration analysis. The relative velocity method is based on two important concepts: (1) relative motion, and (2) the rigid body principle, considered earlier.

6.2 RELATIVE MOTION CONCEPT

Relative motion has been defined as the motion of a body with respect to another body that is itself moving. If the motion of the body is with respect to a stationary frame of reference or fixed point such as the earth, the motion is defined as absolute motion.

If we think of two bodies, A and B, having independent or absolute motion, the velocity of A relative to B is the velocity that A appears to have to an observer traveling on B. To illustrate the concept of relative motion, let us consider three ways in which we can observe a car A traveling at 50 mph while seated in a second car, B.

1. Our car (B) is parked on the shoulder of the road. Then, in the adjacent lane, car A comes speeding past at 50 mph and immediately we have the experience of seeing the vehicle moving at 50 mph.
2. Our car (B) is moving at 50 mph in one lane, and in another lane, car A is moving in the opposite direction with a speed of

50 mph. As it speeds past our car, it appears to be moving at
a speed of 100 mph.

3. Our car (B) is now moving at 40 mph, and along comes car A,
    still in another lane and moving at 50 mph, but in the same
    direction as our car. We now have the experience of seeing car
    A moving at 10 mph.

Note that the velocity of car A, although the same in each case, does in fact
vary, from our point of view, and that we realize the true velocity only when
we make our observation from a stationary position, as in case 1. This
proves that the velocity which a moving body appears to have is always
dependent on the observation point or frame of reference.

Consider two ships, A and B, whose velocities, $\bar{V}_A$ and $\bar{V}_B$, in still
water are in the direction shown in Figure 6.1a. It is desired to find the
velocity that B appears to have to an observer on A.

Since the required velocity is to be relative to A, we must consider
a method of bringing A to rest. To do this, we imagine the water to be a
stream moving with a velocity equal and opposite to that of A (i.e., with a
velocity in $-\bar{V}_A$). This means that as fast as A moves forward, the stream
moves backward with the same velocity, and consequently A makes no
progress as far as the earth is concerned. The same effect would be produced if one were to run forward on an endless belt moving with a velocity
equal and opposite to that of the runner. As far as the surroundings are
concerned, the runner would be stationary.

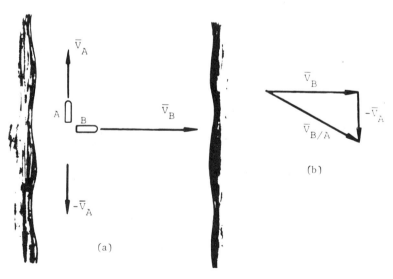

Figure 6.1   Vectorial representation of relative velocity.

Now, consider what is happening to B. Not only is it moving with its original velocity $\bar{V}_B$, but it has added to it the velocity of the stream, which is $-\bar{V}_A$. Therefore,

Velocity of B relative to A = velocity of B + velocity of stream
= velocity of B + (-velocity of A)
= velocity of B - velocity of A
= difference of velocities

Hence we can conclude that: <u>The velocity of B relative to A is the vectorial difference between the velocities of A and B.</u> Symbolically,

$$\bar{V}_{B/A} = \bar{V}_B - \bar{V}_A \qquad (6.1)$$

Note the order of subscripts on both sides of the equation. An alternative form of Equation (6.1) is

$$\bar{V}_B = \bar{V}_A + \bar{V}_{B/A}$$

It should be noted that whereas the relative velocity is the vectorial difference between two velocities, the resultant of two velocities is the vectorial sum of those velocities. Figure 6.1b illustrates the vector polygon that represents Equation (6.1).

6.3   THE VELOCITY POLYGON

The velocity polygon is an alternative form of the vector polygon presented earlier. To illustrate the procedure, let us consider ships A and B discussed in Section 6.2.

PROCEDURE (refer to Figure 6.2)

1. Define a point o, called the pole, as the origin for the construction. This is a point of zero velocity and it represents all the fixed points on the mechanism. All absolute velocities originate from this point. By "absolute velocity" we mean the real and true velocity of a body, as observed from a stationary frame of reference such as the earth.
2. Since the velocities $\bar{V}_A$ and $\bar{V}_B$ are absolute velocities, draw $\bar{V}_A$ and $\bar{V}_B$ from point o and define their respective termini, a and b.
3. From point a, the terminus of $\bar{V}_A$, draw a third vector to terminate at point b, the terminus of $\bar{V}_B$, thereby closing the polygon. This vector defines the relative velocity $\bar{V}_{B/A}$.

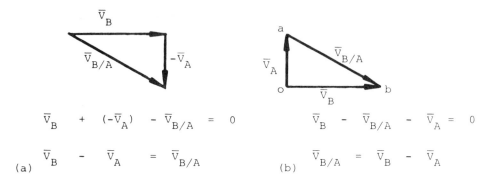

$$\overline{V}_B + (-\overline{V}_A) - \overline{V}_{B/A} = 0 \qquad\qquad \overline{V}_B - \overline{V}_{B/A} - \overline{V}_A = 0$$

$$\overline{V}_B - \overline{V}_A = \overline{V}_{B/A} \qquad\qquad\qquad \overline{V}_{B/A} = \overline{V}_B - \overline{V}_A$$

(a)   (b)

Figure 6.2   Vectorial representation of relative velocity: (a) vector polygon; (b) velocity polygon.

Note the agreement in the results between the vector polygon and the velocity polygon given in Figure 6.2a and b, particularly the fact that $\overline{V}_{B/A}$ is the same in both cases. This agreement is also validated by the vector equations shown.

6.4   VELOCITY POLYGON CONVENTION

The convention just employed in developing the velocity polygon (Figure 6.2b) can be summarized as follows:

> $\overline{oa}$ represents the velocity of A or $\overline{V}_A$.
> $\overline{ob}$ represents the velocity of B or $\overline{V}_B$.
> $\overline{ab}$ represents the velocity of B relative to A (or $\overline{V}_{B/A}$).

Note that the letter to which the arrow points indicates the velocity under consideration. In the vectors $\overline{oa}$ and $\overline{ob}$ the arrows point toward a and b, respectively. Hence the vectors represent the velocities of A and B, respectively. Similarly, in the vector $\overline{ab}$, the arrow points toward b and away from a; hence the vector represents the velocity of B relative to A. If the arrow were reversed, pointing toward a and away from b, the vector would represent the velocity of A relative to B.

6.5   VELOCITY POLYGON: LINKAGE APPLICATION

Application of the velocity polygon to linkage analysis is based on the following two important principles covered earlier.

1. If two points lie on the same rigid body, their relative velocity
    is the vectorial difference between their absolute velocities.
2. If two points lie on the same rigid body, their relative velocity
    is perpendicular to the line connecting the two points.

6.6   RELATIVE VELOCITY OF TWO POINTS
ON A RIGID BODY

Consider floating link AB in Figure 6.3. Imagine that we were seated at
end A looking toward the other end, B. Since the distance between A and B
does not change (i.e., it is a rigid body) and the link has motion, B would
appear to move about us in a circular path. This is the only motion that B
could have.

Therefore, the linear velocity of B relative to A ($\bar{V}_{B/A}$) must act
perpendicular to link AB, and proportional to the distance from A to B, or

$$\bar{V}_{B/A} = AB \times \omega \quad \text{(in one direction)}$$

Similarly, if we were seated at end B looking toward A, A would also appear
to have the same circular motion about us. Therefore, the linear velocity
of A relative to B ($V_{A/B}$) must also act perpendicular to link AB, and proportional to the distance from A to B, or

$$\bar{V}_{A/B} = AB \times \omega \quad \text{(in the opposite direction)}$$

As a result, it is easy to see that the linear velocity of B relative to A is
equal and opposite to the linear velocity of A relative to B, or

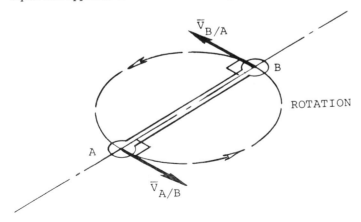

Figure 6.3   Relative velocity of points on a link.

$$\bar{V}_{B/A} = -\bar{V}_{A/B}$$

In summary, if two points, A and B, lie on the same rigid body, and the velocity of B relative to A is required, we must assume that A is fixed and that B rotates about it. Similarly, if the velocity of A relative to B is required, we must assume that B is fixed and that A rotates about it.

6.7   VELOCITIES OF END POINTS ON A FLOATING LINK

If the velocity of one end of a floating link is completely known (in direction and magnitude) and that of the other end is known only in direction, the magnitude of the latter velocity can be determined from a velocity polygon. For example, consider link AB of Figure (6.4a). Given the velocity of point A (both magnitude and direction) and the velocity of point B (direction only), determine the complete velocity, $\bar{V}_B$.

Since point A has motion, the velocity of point B is obtained from the vector equation

$$\bar{V}_B = \bar{V}_A + \bar{V}_{B/A}$$

where the vector $\bar{V}_{B/A}$ is the vectorial difference between the completely known velocity $\bar{V}_A$ and the partially known velocity $\bar{V}_B$. Also, since points A and B are on the same link, the velocity of B relative to A ($\bar{V}_{B/A}$) must act perpendicularly to the line joining A to B. Hence, by laying out the velocity of A, which is known, and the velocity directions of B and $\bar{V}_{B/A}$, the magnitudes of both $\bar{V}_B$ and $\bar{V}_{B/A}$ can be determined.

PROCEDURE

1.   Define a point o as a pole for the velocity polygon.
2.   From point o, lay out the vector $\bar{V}_A$, terminating at a point a.
3.   Through point o, draw a line "b"-"b" parallel to the given direction of vector $\bar{V}_B$. The magnitude and orientation of this vector are defined by a point b on this line, which is as yet unknown.
4.   To define $\bar{V}_B$, we must relate the velocity of point A, which is completely known, to that of B, which is only partially known. This means that we must seek to determine the velocity of B relative to A (or $\bar{V}_{B/A}$). Referring to the link, we know that the velocity vector $\bar{V}_{B/A}$ must be perpendicular to link AB; and from the velocity polygon convention, we also know that this velocity must be directed from point a to point b on the polygon. Therefore, draw a vector in the direction perpendicular to link AB, originating from point a and terminating at a point b in

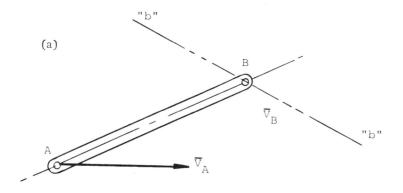

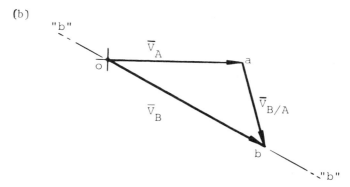

Figure 6.4   Velocities of end points on a link: (a) link; (b) velocity polygon.

line "b"-"b". This line, $\overline{ab}$, represents the vector $\bar{V}_{B/A}$, and the line ob represents the required velocity vector, $\bar{V}_B$.

A quick check of the completed velocity polygon (Figure 6.4b) reveals the balanced vector equation

$$\bar{V}_B = \bar{V}_A + V_{B/A}$$

Note that it would have made no difference in the results obtained for $\bar{V}_B$ if we had considered the velocity of A relative to B ($\bar{V}_{A/B}$) instead of the velocity of B relative to A ($\bar{V}_{B/A}$). However, as far as the velocity polygon is concerned, since $\bar{V}_{A/B}$ is directed

opposite to $\bar{V}_{B/A}$, the direction of vector $\bar{V}_{B/A}$ would have to
be reversed, in order to be consistent with the vector equation

$$\bar{V}_B = \bar{V}_A - \bar{V}_{A/B}$$

6.8   VELOCITY OF ANY POINT ON A LINK

If the velocities of two points on a link are completely known, the velocity
of any other point on that link can easily be determined from the velocity
triangle using the method of proportionality. Let us consider again link AB
in Figure 6.5a, where the velocities $\bar{V}_A$ and $\bar{V}_B$ are both known from the
velocity triangle oab (Figure 6.5b). It is required to find the velocity of a
point C in the link.

The velocity of point C is given by the vector equation

$$\bar{V}_C = \bar{V}_A + \bar{V}_{C/A}$$

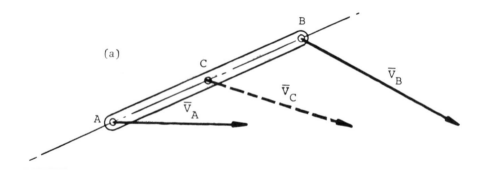

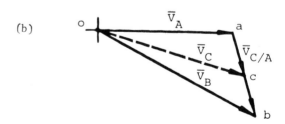

Figure 6.5   Velocity of any point on a link: (a) link; (b) velocity polygon.

where $\bar{V}_{C/A}$, the velocity of C relative to A, must be (1) perpendicular to AC and (2) proportional to AC. For example, if point C were located at the midpoint of the link between A and B, vector $\bar{V}_{C/A}$ would be defined by a distance $\overline{ac}$ on the velocity polygon, where point c is the midpoint between a and b. Hence, for any point C on link AB, the velocity $\bar{V}_{C/A}$ is obtained by selecting a point c between a and b on the velocity polygon such that

$$\frac{ac}{ab} = \frac{AC}{AB}$$

The distance $\overline{ac}$ will then satisfy both conditions of direction and magnitude of vector $\bar{V}_{C/A}$. Therefore, a line drawn from o to c, that is, $\overline{oc}$, will represent $\bar{V}_C$, the vector sum of $\bar{V}_A$ and $\bar{V}_{C/A}$.

PROCEDURE

1. Determine a point c on line $\overline{ab}$ of the triangle such that

$$\frac{ac}{ab} \text{ (on the polygon)} = \frac{AC}{AB} \text{ (on the link)}$$

2. Join $\overline{oc}$.
3. Now since all vectors originating from the pole o represent absolute velocities, line $\overline{oc}$ represents magnitude and direction of velocity of C ($\bar{V}_C$). A quick check of the velocity polygon (Figure 6.5b) will verify the vector equation $\bar{V}_C = \bar{V}_A + \bar{V}_{C/A}$.

6.9   VELOCITY OF ANY POINT ON
        AN EXPANDED LINK

In the preceding section, the point of unknown velocity considered was located on the centerline connecting two points of known velocity. Let us now consider the expanded link ABD in Figure 6.6a, where the velocities of points A and B are both given as shown, and let it be required to find the velocity of point D, located away from centerline AB.

The velocity of D ($\bar{V}_D$) is given by the vector equation

$$\bar{V}_D = \bar{V}_A + \bar{V}_{D/A}$$

or

$$\bar{V}_D = \bar{V}_B + \bar{V}_{D/B}$$

where $\bar{V}_{D/A}$ and $\bar{V}_{D/B}$ can be obtained from a velocity polygon (Figure 6.6b).

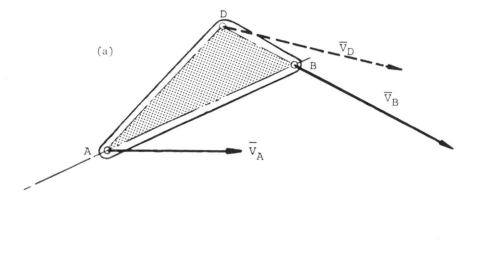

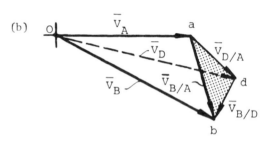

Figure 6.6   (a) Link; (b) velocity polygon.

PROCEDURE

1. Develop the velocity polygon, assuming a simple link AB. (Note
    that this polygon is the same as that of the preceding example.)
2. Through point a on the polygon, draw a line perpendicular to
    side AD on the link to indicate the direction of $\bar{V}_{D/A}$. Since A
    is rigidly connected to D, the only direction that the velocity of
    D relative to A could have is one perpendicular to a line joining
    A to D.
3. Similarly, to indicate the direction of $\bar{V}_{D/B}$, draw a line through
    point b perpendicular to DB.
4. Define the point where the two perpendiculars in steps 2 and 3
    intersect as point d. This point defines the terminus of vector $\bar{V}_D$.
5. Draw a line to connect point d to the polar origin o. This line
    defines the magnitude of the required vector $\bar{V}_D$.

The Velocity Image: Alternative Approach

In the example of Figure 6.6, we note that the relative velocity $V_{D/A}$ is given by the vector extending from point a to point d, and the relative velocity $V_{D/B}$ is given by the vector extending from b to d. We also note that triangle abd is similar to triangle ABD, since it was produced by lines drawn perpendicular to corresponding sides of the link. Because of this similarity (or proportionality), triangle abd is often referred to as the velocity image of link ABD.

The velocity image is a useful concept in velocity analysis. If the velocities of any two points on a link are known in the velocity polygon, the velocity of a third point on that link can readily be determined by constructing the velocity image, making sure that the lines which define the image are perpendicular to the corresponding lines which form the link. Also, the letters used to designate both the link and the image must run in the same cyclic order.

6.10   VELOCITY ANALYSIS OF A SIMPLE MECHANISM

Consider the four-link mechanism shown in Figure 6.7a, in which the angular velocity of link 2 or ω_2 is 10 rad/sec. Determine the angular velocities of links 4 and 3.

Since point C is on link 4, the angular velocity of link 4 is the same as that of point C and is given by

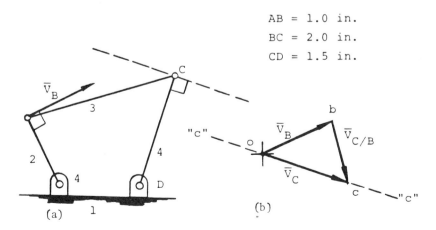

Figure 6.7   Velocity analysis of a four-bar mechanism: (a) four-bar mechanism; (b) velocity polygon.

$$\omega_4 = \omega_C = \frac{V_C}{CD}$$

where V_C is the magnitude of the linear velocity of point C. Hence we must first determine $\bar{V}_C$, which is given by the vector equation

$$\bar{V}_C = \bar{V}_B + \bar{V}_{C/B}$$

Similarly, the angular velocity of link 4 is found from

$$\omega_3 = \omega_{CB} = \frac{V_{C/B}}{CB}$$

PROCEDURE (see Figure 6.7b)

1. Define a point o as the pole or origin for a velocity polygon.
2. Starting from point o and using a convenient scale, lay out vector $\overline{ob}$ perpendicular to link 2 to represent velocity $\bar{V}_B$, whose magnitude is computed as

$$V_B = \omega_2 \times AB = 10 \text{ in./sec}$$

3. Through point o, draw a line "c"-"c", perpendicular to link 4, to indicate the direction of velocity $\bar{V}_C$. Note that $\bar{V}_C$ must be perpendicular to CD since link 4 could have circular motion only about the fixed axis D. However, the orientation and magnitude are as yet unknown.
4. To define the magnitude of $\bar{V}_C$, we must find the velocity $\bar{V}_{C/B}$, which relates point b (known) to point c (unknown) on the polygon. By definition, this velocity must be perpendicular to link BC, and by convention, must be directed to point c from point b on the polygon. Therefore, draw a line perpendicular to BC extending from point b and terminating at line "c"-"c". This point of termination defines point c or the magnitude of $\bar{V}_C$; and it also defines the magnitude of $\bar{V}_{C/B}$.
5. Scale velocity vectors $\bar{V}_C$ and $\bar{V}_{C/B}$ to obtain the magnitudes.

$$V_C = 12.0 \text{ in./sec}$$

$$V_{C/B} = 8.5 \text{ in./sec}$$

6. Finally, as required angular velocities of links 4 and 3 are given as

$$\omega_4 = \frac{V_C}{CD} = \frac{12.0}{1.5} = 8.0 \text{ rad/sec}$$

and

$$\omega_3 = \frac{V_{C/B}}{CB} = \frac{8.5}{2.0} = 4.25 \text{ rad/sec}$$

6.11   VELOCITIES OF SLIDING CONTACT MECHANISMS

When one body slides on another, the difference in their absolute velocities (or their relative velocity) is defined as the <u>velocity of sliding</u>. The velocity of sliding is always directed along the common tangent drawn through the contact point.

To illustrate this concept, consider the cam-and-follower mechanism shown in Figure 6.8a, where P is the point of contact between the two bodies. The angular velocity of the cam ω_C is known and $\bar{V}_{P(C)/P(F)}$, the velocity of sliding, is required.

By definition, the velocity of sliding is the velocity of P on C relative to the velocity P on F, or

$$\bar{V}_{P(C)/P(F)} = \bar{V}_{P(C)} - \bar{V}_{P(F)}$$

The velocity of sliding must, by definition, be along the common tangent T-T. Therefore, $\bar{V}_{P(C)/P(F)}$ can be determined as follows.

PROCEDURE

1.   Define polar origin o to start the velocity polygon (Figure 6.8b).
2.   From origin o, lay out known velocity $\bar{V}_{P(C)}$ (magnitude and direction) perpendicular to OP.

$$\bar{V}_{P(C)} = OP \times \omega \quad \text{(directed perpendicular to OP)}$$

3.   Define the terminus of $\bar{V}_{P(C)}$ as p(C).
4.   Through the origin draw line "p(F)-"p(F)" perpendicular to link OP to indicate the direction of $\bar{V}_{P(F)}$.
5.   To define $\bar{V}_{P(C)/P(F)}$, draw a line starting from p(C) extending toward line "p(F)"-"p(F)" in the known direction of sliding (i.e., parallel to tangent line T-T). The point at which the two lines intersect defines the terminus of $\bar{V}_{P(F)}$, and line p(F)-p(C) defines the magnitude of $\bar{V}_{P(C)/P(F)}$. Vector $\bar{V}_{P(C)/P(F)}$ is pointed

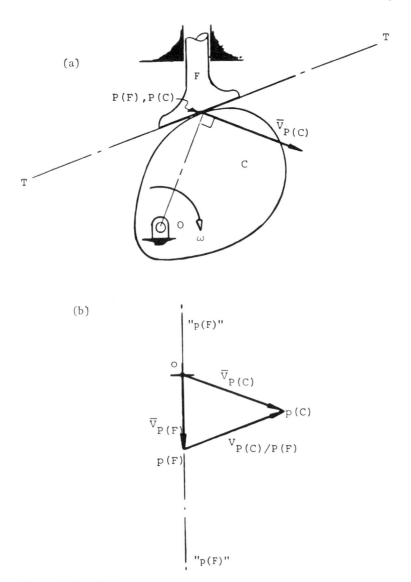

Figure 6.8   Velocities of points on a cam mechanism: (a) mechanism;
(b) velocity polygon.

toward p(C), in accordance with convention. A quick check of the completed polygon reveals the vector equation

$$\bar{V}_{P(C)/P(F)} = \bar{V}_{P(C)} - \bar{V}_{P(F)}$$

EXAMPLE 6.1

Consider the quick-return mechanism shown in Figure 6.9a. Crank OP rotates at 2 rad/sec and slider S slides on follower F. Determine the angular velocity of the follower and the velocity of sliding.

SOLUTION

To find the angular velocity of the follower, we must first find the linear

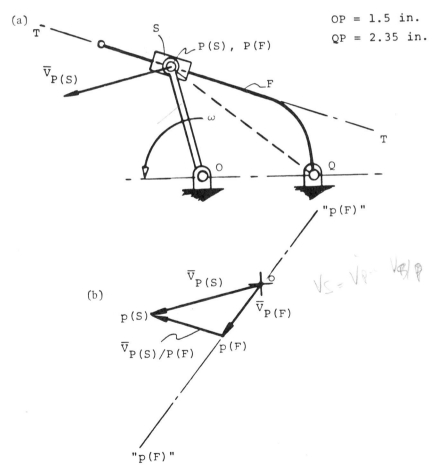

Figure 6.9   Velocity analysis of a quick-return mechanism: (a) mechanism; (b) velocity polygon.

velocity of point P on F ($\bar{V}_{P(F)}$). This unknown velocity is related to the known velocity, $\bar{V}_{P(S)}$, as follows:

$$\bar{V}_{P(F)} = \bar{V}_{P(S)} + \bar{V}_{P(F)/P(S)}$$

where $\bar{V}_{P(F)/P(S)}$ is the velocity of sliding (or the velocity of slip), which is directed along the instantaneous path of the slider.

After determining the velocity of point P on F, we obtain the required angular velocity from the relationship

$$\omega_F = \frac{V_{P(F)}}{QP}$$

PROCEDURE

1. Define polar origin o to start the velocity polygon (Figure 6.9b).
2. Lay out the velocity of point P on S ($\bar{V}_{P(S)}$) perpendicular to OP. The magnitude of this velocity is obtained from

$$V_{P(S)} = \omega(OP)$$

$$= 2(1.5) = 3 \text{ in./sec}$$

3. Define the terminus of the vector $\bar{V}_{P(S)}$ as p(S).
4. Through point o draw a line "p(F)"-"p(F)" to indicate the known direction of $\bar{V}_{P(F)}$.
5. Lay out the relative velocity (or velocity of slip) starting from p(S) and moving in the known direction of slip toward line "p(F)"-"p(F)". This relative velocity vector intersects line "p(F)"-"p(F)" and defines the magnitude of $\bar{V}_{P(F)}$ as well as that of the slip velocity $\bar{V}_{P(S)/P(F)}$ at that point. Label the point of intersection p(F). Note that the orientation of the velocity of the sliding vector depends on whether we consider the velocity of the slider relative to the follower, where vector $\bar{V}_{P(S)/P(F)}$ is is pointed toward p(S), or the velocity of the follower relative to the slider, where vector $\bar{V}_{P(F)/P(S)}$ would be pointed in the opposite direction, toward p(F).
6. Scale vectors $\bar{V}_{P(F)}$ and $\bar{V}_{P(S)/P(F)}$ to obtain their magnitudes as follows:

$$V_{P(F)} = 1.7 \text{ in./sec}$$

and

$$V_{P(S)/P(F)} = 2.0 \text{ in./sec}$$

7. Determine the angular velocity of F from the relationship

$$V_{P(F)} = \omega_{QP} \times QP$$

Therefore,

$$\omega_{QP} = \frac{V_{P(F)}}{QP}$$

$$= \frac{1.7}{2.35} = 0.72 \text{ rad/sec}$$

6.12   VELOCITIES OF A BODY WITH ROLLING CONTACT

Consider wheel W in Figure 6.10a, which rolls, without slipping, on a track. As the wheel moves to the right, the center A is considered the moving frame of reference and the velocity of every other point on the wheel is relative to the velocity of this point. Therefore, we apply the relative motion concept to determine the velocities of points B and C as follows.

PROCEDURE

1. Define polar origin o (Figure 6.10b).
2. Locate the velocity vector $\bar{V}_A$. The direction of this vector is perpendicular to a line joining A to the instant center I (or AI), and the magnitude is given by

$$V_A = AI \times \omega$$

3. Define the terminus of $\bar{V}_A$ as a.
4. Lay out the direction of the vector $\bar{V}_B$ perpendicular to a line joining B to I (or BI). Therefore, draw a line "b"-"b", of undefined length, through pole o.
5. Determine the vector $\bar{V}_B$.
    a. Determine the vector $\bar{V}_{B/A}$. By definition, $\bar{V}_{B/A}$ assumes that point B rotates about point A and is perpendicular to a line connecting B to A (i.e., AB) on the wheel. Also, this velocity is represented by a vector headed from a to b on the polygon.
    b. Therefore, from a draw a line in a direction perpendicular to AB to intersect line "b"-"b". This point of intersection defines the termini of vectors $\bar{V}_{B/A}$ and $\bar{V}_B$.
6. Label the point of intersection b. A quick check of the polygon should immediately reveal the vector equation:

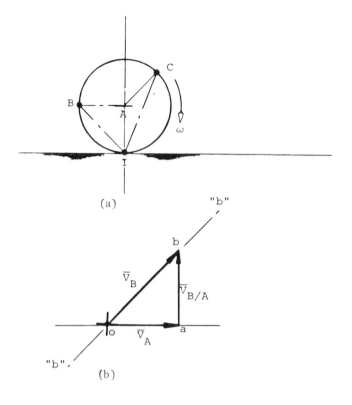

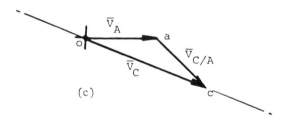

Figure 6.10   Velocities of a body with rolling contact.

$$\bar{V}_B = \bar{V}_A + \bar{V}_{B/A}$$

Following the procedure described above, the velocities of point C ($\bar{V}_C$) can be determined as shown in Figure 6.10c.

6.13   VELOCITY ANALYSIS OF A COMPOUND MECHANISM

Consider the toggle mechanism in Figure 6.11a. Wheel W rotates at 10 rpm. Find the velocities of slider S and point P.

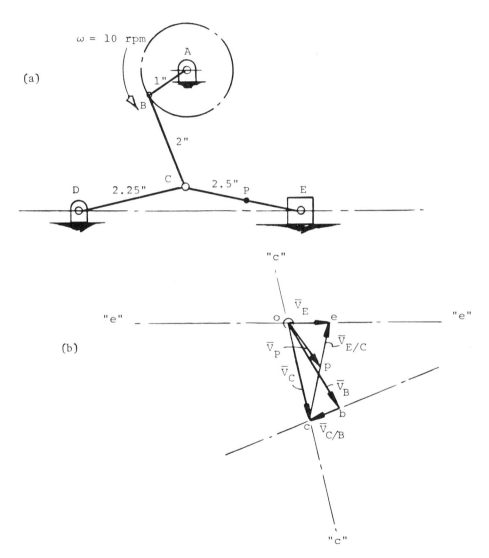

Figure 6.11   Velocities of points on a toggle mechanism: (a) mechanism; (b) velocity polygon.

We first note the following:

1. The velocity of the slider is the same as that of point E, that is, $\bar{V}_E$.
2. Linkage may be analyzed in two parts:
    a. The four-bar section ABCD, to determine $\bar{V}_C$.
    b. The slider-crank section DCE, using the value found for $\bar{V}_C$ in part (a) to find $\bar{V}_E$.
3. $\bar{V}_E$ is found using the velocity image of link CE.

PROCEDURE

1. Define polar origin o to start the velocity polygon (Figure 6.11b).
2. Lay out, using a convenient scale, velocity vector $\bar{V}_B$. The magnitude of this vector is given as

$$V_B = AB \times \omega_{AB}$$

Therefore,

$$\omega_{AB} = 10\left(\frac{2\pi}{60}\right) = 1.05 \text{ rad/sec}$$

3. Define the terminus of vector $\bar{V}_B$ as point b.
4. Through the origin draw a line "c"-"c" perpendicular to CD to indicate the orientation of velocity vector $\bar{V}_C$.
5. Lay out relative velocity $\bar{V}_{C/B}$. Starting from point b, draw a line heading toward line "c"-"c" in a direction perpendicular to link BC. This line should meet "c"-"c" at point c, which defines the termini of both $\bar{V}_C$ and $\bar{V}_{B/C}$.
6. Scale the magnitude of $\bar{V}_C$, that is, the line $\overline{oc}$.

$$V_C = 1.06 \text{ in./sec}$$

7. Again, through the origin, draw the line "e"-"e" to indicate the orientation of the absolute velocity vector $\bar{V}_E$ (horizontal).
8. Lay out relative velocity vector $\bar{V}_{E/C}$. Starting from point c, draw a line heading toward line "e"-"e" in a direction perpendicular to link CE. This line should meet line "e"-"e" at point c to define the magnitudes of both $\bar{V}_E$ and $\bar{V}_{E/C}$.
9. Scale the magnitude of $\bar{V}_E$, that is, the line $\overline{oe}$.
10. To find the velocity of point P ($\bar{V}_P$), locate a point p in line $\overline{ec}$ on the polygon such that the ratio of $\overline{cp}$ to $\overline{pe}$ is the same as ratio of CP to PE on the link, or

$$\frac{cp}{pe} = \frac{CP}{PE}$$

11. Connect point p with a vector originating from point o. This vector, $\bar{op}$, then defines the velocity $\bar{V}_P$.

12. Scale the magnitude of $\bar{V}_P$.

$$V_P = 0.58 \text{ in./sec}$$

II.B
GRAPHICAL TECHNIQUES: ACCELERATION ANALYSIS

Acceleration is a very important property of motion to a machine designer. By Newton's law, $F = ma$, the force imposed on machine members is directly proportional to the acceleration. Although fundamentally kinematics is not concerned with forces, the study of acceleration is vitally important to the dynamic analysis, because of its influence on stresses bearing loads, vibration, and noise.

In high-speed machines, the accelerations and resulting inertial forces can be very large compared to static forces which do useful work. For example, in modern automobile and aircraft engines, the stresses imposed on the connecting rods as a result of these accelerations are considerably greater than those produced on the piston by gas pressure. Also, a small imbalance of the rotating members can produce forces that are many times larger than the weights of the members. Therefore, a complete acceleration analysis is a prerequisite to stress analysis and proper design of machine members.

7

Linear Acceleration Along Curved Paths

7.1 INTRODUCTION

A body moving along a curved path is always subjected to an acceleration since its velocity changes from one position to the next. If the body moves with a constant angular speed, the velocity changes is one direction only, and the resultant acceleration is a normal acceleration. If the body moves with a variable angular speed, the velocity change is one of direction and magnitude, and the resultant acceleration is a summation of the normal acceleration (due to the direction change) and a tangential acceleration (due to the magnitude change).

7.2 NORMAL ACCELERATION

Consider a point A on a body which moves in a circular path of radius R with constant angular speed through an angle $\Delta\theta$ (Figure 7.1a). The initial velocity at point A_i is $\bar{V}_A^i$ and the final velocity at point A_f is $\bar{V}_A^f$.

The velocity polygon representing this motion is given in Figure 7.1b. Here the resultant change in velocity of point A ($\Delta\bar{V}_A = \bar{V}_A^f - \bar{V}_A^i$) is in direction only, since $\bar{V}_A^i$ and $\bar{V}_A^f$ are equal in magnitude. Also, for a small angular displacement $\Delta\theta$, $\Delta\bar{V}_A$ is oriented normal or perpendicular to the instantaneous linear velocity.

The normal acceleration magnitude of point A or A_A^N is obtained from

$$\bar{A}_A^N = \frac{\bar{V}_A^f - \bar{V}_A^i}{\Delta T} = \frac{\Delta\bar{V}_A}{\Delta T}$$

where

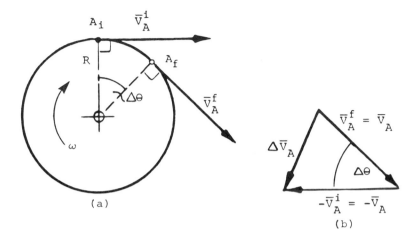

Figure 7.1   Disk with uniform rotation.

$$A_A^N = \frac{\Delta V_A}{\Delta T}$$

$$\Delta V_A = V_A \, \Delta\theta \quad \text{(for small } \Delta\theta\text{)}$$

$$A_A^N = \frac{V_A \, \Delta\theta}{\Delta T}$$

But

$$\frac{\Delta\theta}{\Delta T} = \omega$$

$$A_A^N = V_A \omega$$

$$= R\omega \times \omega = R\omega^2$$

Also,

$$\omega = \frac{V}{R}$$

$$A_A^N = V_A \frac{V_A}{R} = \frac{V_A^2}{R} \qquad (7.1)$$

This relationship states that the magnitude of the acceleration of any point on a body rotating at a constant speed is equal to the square of the angular velocity of the body multiplied by the distance of the point from the center or rotation, or the square of the linear velocity divided by the distance of the point from the axis of rotation. This acceleration called the normal acceleration, is always directed radially toward the center of rotation perpendicular or normal to the instantaneous linear velocity.

A familiar example of radial or normal acceleration is that of a car traveling around a circular track at a constant speed. Although the speedometer reading remains the same, there is acceleration because the direction of velocity is continually changing. Whenever there is a change of direction of velocity, there is acceleration. Furthermore, it is the normal acceleration that keeps the car on its circular path. If it ceases, the car will at once move in the direction tangent to the circular track.

It should be noted from Equation 7.1 that if the point has rectilinear motion, the normal acceleration is always zero. That is,

$$A_A^N = \frac{V_A^2}{R_\infty} = \frac{V_A^2}{\infty} = 0$$

7.3 TANGENTIAL ACCELERATION

Consider a point A on a body that moves in a circular path of radius R with increasing angular speed ω through an angle $\Delta\theta$ (Figure 7.2). As before, $\bar{V}_A^i$ is the initial velocity at point A_i and $\bar{V}_A^f$ is the final velocity at point A_f.

Here, for convenience, we consider the velocity change due to magnitude only. This gives the vector equation

$$\Delta\bar{V}_A = \bar{V}_A^f - \bar{V}_A^i \quad \text{(in tangential direction only)}$$

Therefore, the magnitude of the tangential acceleration of point A (A_A^T) obtained from

$$A_A^T = \frac{\Delta V_A}{\Delta T}$$

$$= \frac{R\omega^f - R\omega^i}{\Delta T}$$

$$= \frac{R\,\Delta\omega}{\Delta T}$$

But

$$\frac{\Delta\omega}{\Delta T} = \alpha$$

Therefore,

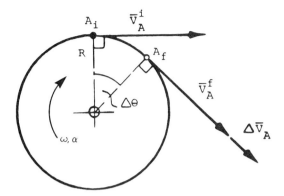

Figure 7.2   Disk with nonuniform rotation.

$$A_A^T = R\alpha \tag{7.2}$$

This relationship states that the magnitude of the tangential acceleration of any point on a rotating body whose angular velocity is changing is equal to the angular acceleration multiplied by the distance of the point from the center of rotation. (Note: This is the same relationship as that derived by an alternative method in Section 2.3.)

7.4   ABSOLUTE ACCELERATION

The absolute or resultant acceleration of a point A (or $\bar{A}_A$) (Figure 7.3) is the vectorial sum of the normal or radial acceleration ($\bar{A}_A^N$) and the tangential acceleration ($\bar{A}_A^T$). Symbolically,

$$\bar{A}_A = \bar{A}_A^N + \bar{A}_A^T$$

where the magnitude is given by

$$A_A = |\bar{A}_A| = \sqrt{(A_A^N)^2 + (A_A^T)^2}$$

$$= \sqrt{(R\omega^2)^2 + (R\alpha)^2}$$

The direction of this acceleration or the angle ϕ which the vector makes with the radius R is obtained from

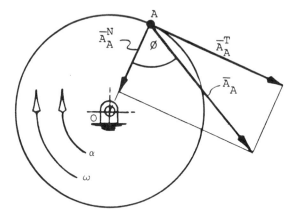

Figure 7.3   Absolute acceleration of a point with nonuniform rotation.

$$\phi = \tan^{-1} \frac{A_A^T}{A_A^N}$$

EXAMPLE 7.1

A, B, and C are points on a rigid body that rotates about center O as shown in Figure 7.4. Given that the angular velocity ω is 2 rad/sec and the angular acceleration α is 3 rad/sec^2, calculate and check graphically the magnitudes and phase angles of the absolute acceleration of A, B, and C.

SOLUTION

The normal acceleration magnitude A^N is given by

$$A^N = r\omega^2$$

Therefore,

$$A_A^N = 4(2)^2 = 16 \text{ in./sec}^2$$

$$A_B^N = 1(2)^2 = 4 \text{ in./sec}^2$$

$$A_C^N = 5(2)^2 = 20 \text{ in./sec}^2$$

The tangential acceleration magnitude A^T is given by

$$A^T = r\alpha$$

Therefore,

$$A^T_A = 4(3) = 12 \text{ in./sec}^2$$

$$A^T_B = 1(3) = 3 \text{ in./sec}^2$$

$$A^T_C = 5(3) = 15 \text{ in./sec}^2$$

The absolute acceleration magnitude A is given by

$$A = \sqrt{(A^N)^2 + (A^T)^2}$$

Therefore,

$$A_A = \sqrt{(16)^2 + (12)^2} = 20 \text{ in./sec}$$

$$A_B = \sqrt{(4)^2 + (3)^2} = 5 \text{ in./sec}$$

$$A_C = \sqrt{(20)^2 + (15)^2} = 25 \text{ in./sec}$$

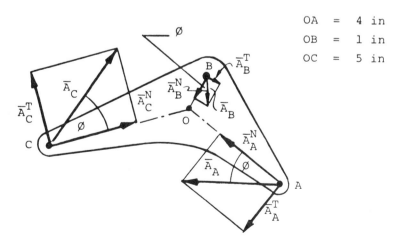

OA	=	4 in
OB	=	1 in
OC	=	5 in

Figure 7.4   Example problem.

The phase angle ϕ is given by

$$\phi = \tan^{-1} \frac{A^T}{A^N}$$

$$\phi_A = \tan^{-1} \frac{12}{16} = 36.8°$$

$$\phi_B = \tan^{-1} \frac{3}{4} = 36.8°$$

$$\phi_C = \tan^{-1} \frac{15}{20} = 36.8°$$

Figure 7.4 also shows the required graphical solution.

Note that this problem could have been solved using the fact that both normal and tangential components are proportional to their distances from the center of rotation. That is, having found the values of components for point A, we could have determined the values of the corresponding components for points A and B by simple ratio.

Also, note that the phase angle ϕ is the same for all points on the link. This is because this angle depends only on the angular velocity and angular acceleration of the link, as can be seen from the following derivation:

$$\phi = \tan^{-1} \frac{A^T}{A^N}$$

$$= \tan^{-1} \frac{r\alpha}{r\omega^2}$$

$$= \tan^{-1} \frac{\alpha}{\omega^2}$$

EXAMPLE 7.2

In Figure 7.5, pulleys P_2 (2 in. in diameter) and P_3 (6 in. in diameter) are driven by pulley P_1 (4 in. in diameter) via a continuous belt. At this instant, P_1 has an angular velocity of 1 rad/sec counterclockwise, and an angular acceleration of 5 rad/sec^2. Determine the acceleration of points A, B, C, and D on the belt.

SOLUTION

For point A on pulley P_1:

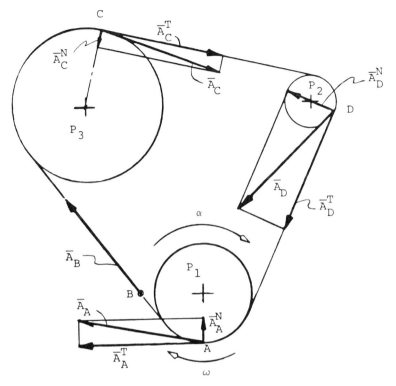

Figure 7.5   Absolute acceleration of points on a belt.

Normal acceleration:

$$\bar{A}_A^N = r_1\omega^2 \quad \text{(directed toward center of pulley)}$$

$$= 2(1)^2 = 2 \text{ in./sec}^2 \quad \text{(directed toward center of pulley)}$$

Tangential acceleration:

$$\bar{A}_A^T = r_1\alpha \quad \text{(directed along the belt)}$$

$$= 2(5) = 10 \text{ in./sec}^2 \quad \text{(directed along the belt)}$$

Absolute acceleration:

$$\bar{A}_A = \bar{A}_A^N + \bar{A}_A^T \quad \text{(vectorial sum)}$$

$$= 10.2 \text{ in./sec}^2 \quad \text{(directed as shown)}$$

For point B on the belt:

$$\bar{A}_B = \bar{A}_A^T \quad \text{(same as tangential acceleration of point A)}$$

$$= 10 \text{ in./sec}^2 \quad \text{(directed along the belt)}$$

For point C on pulley P_3:

Normal acceleration:

$$\bar{A}_C^N = r_3\omega^2 \quad \text{(directed toward center of pulley)}$$

$$= 3\left(\frac{2}{3}\right)^2 = 4/3 \text{ in./sec}^2 \text{ (directed toward center of pulley)}$$

Tangential acceleration:

$$\bar{A}_C^T = \bar{A}_B \quad \text{(same as belt acceleration)}$$

$$= 10 \text{ in./sec}^2 \quad \text{(directed along the belt)}$$

Absolute acceleration:

$$\bar{A}_C = \bar{A}_C^N + \bar{A}_C^T \quad \text{(vectorial sum)}$$

$$\bar{A}_C = 10.1 \text{ in./sec}^2 \quad \text{(directed as shown)}$$

For point D on pulley P_2:

Normal acceleration:

$$\bar{A}_D^N = r_2\omega^2 \quad \text{(directed toward center of pulley)}$$

$$= 1\left(\frac{2}{1}\right)^2 = 4 \text{ in./sec}^2 \quad \text{(directed toward center of pulley)}$$

Tangential acceleration:

$$\bar{A}_D^T = \bar{A}_B \quad \text{(same as belt acceleration)}$$

$$= 10 \text{ in./sec}^2 \quad \text{(directed along the belt)}$$

Absolute acceleration:

$$\bar{A}_D = \bar{A}_D^N + \bar{A}_D^T \quad \text{(vectorial sum)}$$

$$= 10.8 \text{ in./sec}^2 \quad \text{(directed as shown)}$$

Figure 7.5 also shows the graphical solution.

7.5   PROPORTIONALITY OF ACCELERATIONS

It was shown earlier that if a body rotates about a fixed point, the normal
and tangential accelerations of a point located on that body at a distance r
from the center of rotation are given by

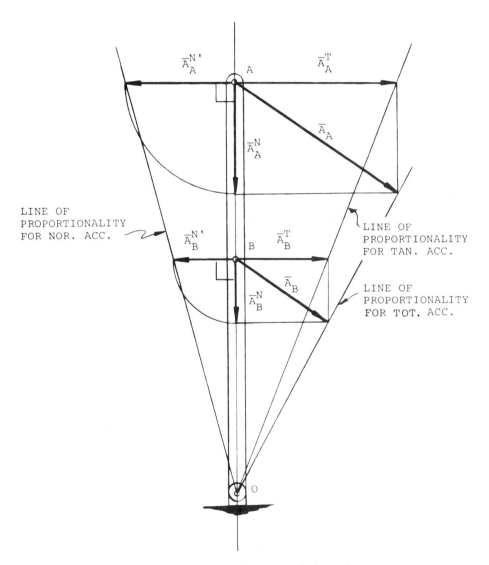

Figure 7.6   Proportionality of accelerations: link OBA.

$$\bar{A}^N = r\omega^2 \quad \text{(directed along r toward the center)}$$

$$\bar{A}^T = r\alpha \quad \text{(directed perpendicular to r)}$$

Therefore, if we consider points A and B on link OAB in Figure 7.6, it is easy to see that

$$\frac{A_B^T}{A_A^T} = \frac{r_B}{r_A}$$

and

$$\frac{A_B^N}{A_A^N} = \frac{r_B}{r_A}$$

which indicates that both normal and tangential components of any two points along the radial line OA are proportional. Similarly, since the acceleration components of the points are proportional, the sum of these components, or the total acceleration, must be proportional.

A graphical representation of the proportionality of accelerations $\bar{A}_B$ and $\bar{A}_A$ and their components can be seen in Figure 7.6, where the termini of these vectors lie on respective straight lines (termed lines of proportionality) that pass through the center of rotation O. Note that the proportionality line for the normal components $\bar{A}_B^N$ and $\bar{A}_A^N$ is a straight line drawn through the center O to touch the termini of these vectors when rotated 90° from the radial line.

7.6   RELATIVE ACCELERATION OF TWO POINTS ON A RIGID BODY

Just as in the velocity case, if the absolute accelerations of two point on a body are known, the relative acceleration between these points is the vectorial difference of their absolute accelerations. For example, if in link AC shown in Figure 7.7a the accelerations of both points A and C are known, then (Figure 7.7b)

$$\bar{A}_{A/C} = \bar{A}_A - \bar{A}_C \qquad (7.3)$$

or

$$\bar{A}_{C/A} = \bar{A}_C - \bar{A}_A \qquad (7.4)$$

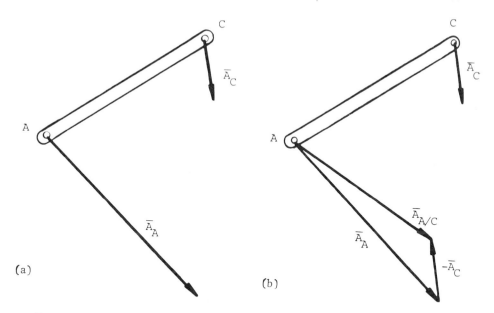

Figure 7.7   Relative acceleration of points on a rigid body.

Normal and Tangential Accelerations

Using the vector $\bar{A}_{A/C}$, we can resolve this vector into its normal and tangential components simply by dropping two perpendiculars from the terminus of this vector: one to the line AC and the other to a line normal to AC passing through point A. This construction yields $\bar{A}^N_{A/C}$ along AC and $\bar{A}^T_{A/C}$ along the line normal to AC (see Figure 7.8).

For the vector $\bar{A}_{C/A}$, a similar construction can be made at point C. Consequently, the components for $\bar{A}^N_{C/A}$ and $\bar{A}^T_{C/A}$ will have opposite directions to those of $\bar{A}^N_{A/C}$ and $\bar{A}^T_{A/C}$.

Normal and tangential acceleration relationships for a point relative to another point on a link are similar to those for a point that undergoes pure rotation. For example, in Figure 7.8, where the acceleration of point A relative to point C (which is assumed to have a nonzero acceleration) is known, the following relationships apply:

1.   Normal acceleration of A relative to C:

$$\bar{A}^N_{A/C} = AC \times \omega^2_{AC} \qquad \text{(directed from A to C)}$$

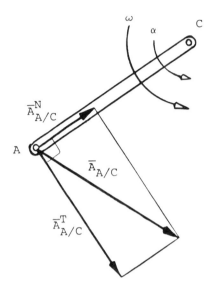

Figure 7.8   Relative acceleration.

2.   Tangential acceleration of A relative to C:

$$\bar{A}^T_{A/C} = AC \times \alpha_{AC} \quad \text{(directed perpendicular to AC}$$

3.   Absolute acceleration of A relative to C:

$$\bar{A}_{A/C} = \bar{A}^N_{A/C} + \bar{A}^T_{A/C} \quad \text{(vectorially)}$$

4.   Magnitude of acceleration of A relative to C:

$$|\bar{A}_{A/C}| = \sqrt{(A^N_{A/C})^2 + (A^T_{A/C})^2}$$

5.   Direction of absolute acceleration $\bar{A}_{A/C}$:

$$\phi = \tan^{-1} \frac{A^T_{A/C}}{A^N_{A/C}}$$

It should be evident that if the acceleration of point C were zero, Equation (7.4) would become

$$\bar{A}_{C/A} = \bar{A}_A$$

and all relative relationships would be reduced to absolute relationships, as in pure motion.

EXAMPLE 7.3

In link BC shown in Figure 7.9a, the angular velocity of point C relative to point B is 1 rad/min (counterclockwise) and the relative angular acceleration is 0.5 rad/min² (clockwise). If the absolute linear acceleration of point B is 5 in./min, as shown, determine the absolute linear acceleration of point C.

SOLUTION

The absolute linear acceleration of point C can be represented in terms of the acceleration of point B and the relative motion of point C to point B, by the vectorial relationship

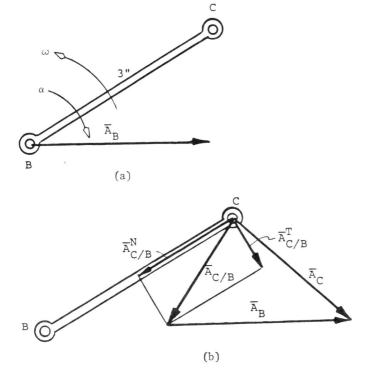

(a)

(b)

Figure 7.9   Example problem: (a) floating link; (b) acceleration diagram.

$$\bar{A}_C = \bar{A}_B + \bar{A}_{C/B} \tag{7.5}$$

where

$$\bar{A}_{C/B} = \bar{A}_{C/B}^N + \bar{A}_{C/B}^T \tag{7.6}$$

But

$$A_{C/B}^N = BC \times \omega^2$$

$$= 3(1.0)^2$$

$$= 3 \text{ in./min}^2$$

and

$$A_{C/B}^T = BC \times \alpha$$

$$= 3(0.5)$$

$$= 1.5 \text{ in./min}^2$$

Therefore, application of Equations (7.5) and (7.6) yields

$$\bar{A}_{C/B} = 3.4 \text{ in./min}^2 \quad \text{(directed as shown in Figure 7.9b)}$$

$$\bar{A}_C = 4.2 \text{ in./min}^2 \quad \text{(directed as shown in Figure 7.9b)}$$

7.7 ACCELERATION OF ANY POINT IN A FLOATING LINK

Method 1: Normal and Tangential Component Method

Consider floating link AB shown in Figure 7.10a. Suppose that the absolute accelerations of two points, A and B, are known and the absolute acceleration of a third point, C, is required.

The acceleration of point C can be found by applying the relationship

$$\bar{A}_C = \bar{A}_A + \bar{A}_{C/A}$$

where $\bar{A}_{C/A}$ is unknown but can be determined by proportionality with $\bar{A}_{B/A}$, which is obtainable from

$$\bar{A}_{B/A} = \bar{A}_B - \bar{A}_A$$

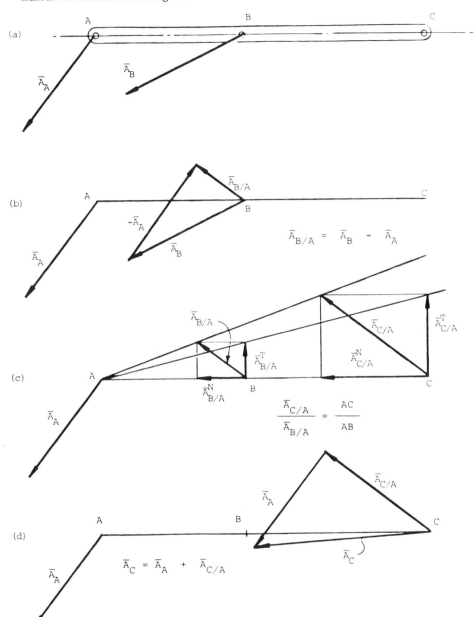

$$\overline{A}_{B/A} = \overline{A}_B - \overline{A}_A$$

$$\frac{\overline{A}_{C/A}}{\overline{A}_{B/A}} = \frac{AC}{AB}$$

$$\overline{A}_C = \overline{A}_A + \overline{A}_{C/A}$$

Figure 7.10   Method 1.

The proportionality between $\bar{A}_{C/B}$ and $\bar{A}_{B/A}$ is based on the fact that A, B, and C are points on the same link and therefore have the same angular velocity and acceleration. Hence

$$\frac{A^N_{C/A}}{A^N_{B/A}} = \frac{CA}{BA}$$

$$\frac{A^T_{C/A}}{A^T_{B/A}} = \frac{CA}{BA}$$

PROCEDURE

1. Determine the acceleration of point B relative to point A (Figure 7.10b), using the vector equation

$$\bar{A}_{B/A} = \bar{A}_B - \bar{A}_A$$

2. Resolve vector $A_{B/A}$ into its normal and tangential components, $\bar{A}^N_{B/A}$ along the link and $\bar{A}^T_{B/A}$ perpendicular to the link (Figure 7.10c).

3. Determine $\bar{A}_{C/A}$ (Figure 7.10c). Since $\bar{A}_{C/A}$ is proportional to $\bar{A}_{B/A}$, complete the following steps:

    a. From point A, draw a line proportionality for the tangential accelerations, touching the terminus of $\bar{A}^T_{B/A}$.

    b. From point C, construct a vector perpendicular to AB, terminating at the proportionality line in step (a). This vector defines $\bar{A}^T_{C/A}$.

    c. Again from point A, draw a second line of proportionality for the relative accelerations, touching the terminus of $\bar{A}_{B/A}$.

    d. From point C, construct a vector parallel to $\bar{A}_{B/A}$, terminating at the second line of proportionality in step (c). This vector defines $\bar{A}_{C/A}$.

4. Determine $\bar{A}_C$ (Figure 7.10d) using the vector equation

$$\bar{A}_C = \bar{A}_A + \bar{A}_{C/A}$$

Method 2: Orthogonal Component Method

Consider again floating link AB in Figure 7.11a, where the accelerations of points A and B are known and the acceleration of point C is required. In this

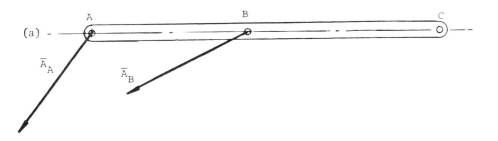

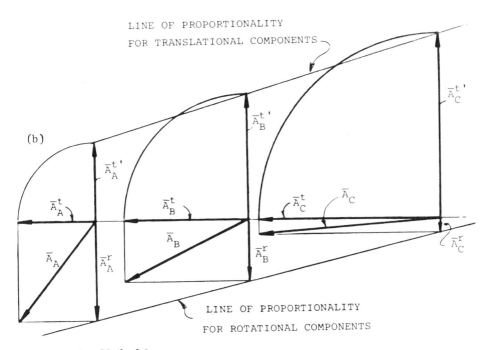

Figure 11   Method 2.

method we make use of the fact that since A, B, and C are on the rigid body, their orthogonal (rotational and translational) components of acceleration must be proportional to each other.

If we consider the rotational components for all points on the link centerline, the termini of these vectors must lie on a straight line. This line may be termed the line of proportionality for the rotational components of acceleration.

Similarly, if we consider the translational components, the termini of these vectors, when rotated through a common angle away from the link centerline, must lie on a straight line. This line may be termed the <u>line of proportionality for the translational components</u> of acceleration.

Therefore, to determine the acceleration of point C, we must first determine the rotational and translational components of acceleration at point C, using their respective lines of proportionality. Then, by adding these components, we obtain the required absolute acceleration of point C. Following is the procedure for construction.

PROCEDURE

1. Resolve vectors $\bar{A}_A$ and $\bar{A}_B$ into their translational and rotational components (along the link and perpendicular to the link).

2. Establish the line of proportionality for rotational components of acceleration. This is a straight line drawn to touch vectors $\bar{A}_A^r$ and $\bar{A}_B^r$.

3. Determine $\bar{A}_C^r$. This is a vector drawn from point C perpendicular to the link and terminating at the proportionality line.

4. Establish the line of proportionality for translational components of acceleration. This is obtained by first rotating vectors $\bar{A}_A^t$ and $\bar{A}_B^t$ in a direction perpendicular to the link, then joining the termini with a straight line. Note that the line of proportionality is used to obtain the magnitudes <u>only</u> (not directions) of the translational components of accelerations of all points along the link.

5. Determine the magnitude of $\bar{A}_C^t$. This is obtained by drawing an equivalent vector $\bar{A}_C^{t'}$, which extends from point C perpendicular to the link and terminates at the proportionality line. To obtain the true direction of $\bar{A}_C^t$, the equivalent vector $\bar{A}_C^{t'}$ must be rotated 90° to act in the direction consistent with $\bar{A}_A^t$ and $\bar{A}_B^t$ along the link.

6. Determine $\bar{A}_C$. This is obtained by summing the tangential and translational components $\bar{A}_C^r$ and $\bar{A}_C^t$ obtained above.

$$\bar{A}_C = \bar{A}_C^r + \bar{A}_C^t$$

Note also that if the directions of the given translational components of the two points oppose each other, this will indicate the the vectors are proportional with respect to a point that lies between the two points. Hence these vectors must be rotated to opposite sides of the link centerline to establish the line of proportionality. This line of proportionality must therefore intersect the line joining the two points.

Method 3: Instant Center of Acceleration Method

It has been established that the components of acceleration of a point on a rotating body are proportional to the distance of that point from the center of rotation. Therefore, if we consider any floating link, such as AB in Figure 7.12a, having the same conditions as in methods 1 and 2, by locating the center of acceleration of the two given points A and B, we can readily determine the acceleration of any other point, such as point C, by proportion.

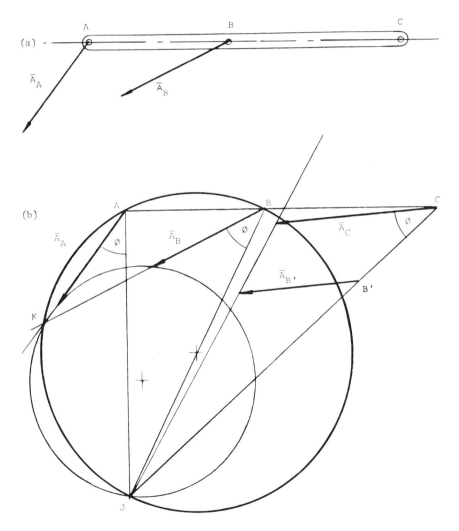

Figure 7.12   Method 3.

Here it should be noted that the center of acceleration or point of zero acceleration is the center of rotation. This should not be confused with the instant center used for velocity analysis. Although the instant center as defined for velocity analysis does have zero velocity, it may or may not have zero acceleration.

To determine the center of acceleration we employ the following construction, called the four-circle method.

The Four-Circle Method (Figure 7.12b)

1. Extend the lines of action of vectors A and B until they intersect at a point K.
2. Construct a circle to circumscribe the triangle formed by points A, B, and K.
3. Construct a second circle to pass through point K and the termini of vectors $\bar{A}_A$ and $\bar{A}_B$. This circle will intersect the first circle at another point, J. This point locates the center of acceleration of points A and B.
4. Connect points A and B to point J, using straight lines. Lines AJ and BJ are, therefore, the radii of rotation for points A and B.

Note the proportionality between the absolute acceleration vectors $\bar{A}_A$ and $\bar{A}_B$ and their respective radii of rotation. Also note the equality between phase angles JAK and JBK.

PROCEDURE (for locating point C)

Having determined the center of acceleration, complete the following steps:

1. Draw a straight line to connect point C and center J.
2. Rotate vector $\bar{A}_B$ to a point B' on radius CJ.
3. Using the rotated vector $\bar{A}_{B'}$ as a gauge, construct a proportionality line from point J to pass through the terminus of $\bar{A}_{B'}$.
4. From point C, construct a vector parallel to $\bar{A}_{B'}$ and terminating at the line of proportionality. This vector defines the required acceleration of point C ($\bar{A}_C$).

Note that although called the four-circle method, the method requires only two circles to locate the center of acceleration. The method takes its name from the four-circle theorem in geometry, which states that the four circles that circumscribe each side of a quadrilateral, and the intersection formed by the extension of two adjacent sides, intersect at a point.

Method 4: Relative Acceleration (or Polygon) Method

Consider again floating link AB where, as before, the accelerations of
points A and B are known and the acceleration of point C is required
(Figure 7.13a). This method employs the <u>acceleration image concept,</u>
which is similar to the velocity image concept considered earlier. Based
on the geometric similarities between points on the link and their relative
accelerations on the polygon, we can use proportions to determine the relative acceleration of any other point on the link, and hence the absolute
acceleration of that point.

PROCEDURE (Figure 7.13b)

1. Define a point o', called the pole, from which all absolute acceleration vectors originate.
2. Lay out given acceleration vectors $\bar{A}_A$ and $\bar{A}_B$ originating from
   the pole and label the respective termini a' and b'.
3. Draw a straight line to connect terminus a' to terminus b'. This
   line represents the magnitude of the relative acceleration between points B and A (or A and B) and, as in the velocity case,
   is called the acceleration image of link AB.
4. Locate point c on line a'b' in step 3 such that

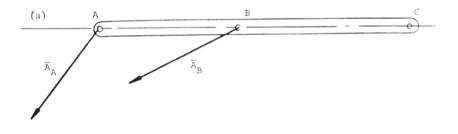

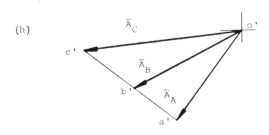

Figure 7.13   Method 4.

$$\frac{a'c'}{a'b'} = \frac{AC}{AB}$$

This point defines the terminus of vector $\bar{A}_C$, the required acceleration vector.

5.   Construct the vector $\bar{A}_C$ extending from the pole and terminating at point c' to determine its magnitude and direction.

7.8   CORIOLIS ACCELERATION

If one link slides radially on another link which is rotating, the sliding link (or slider) experiences an acceleration perpendicular to the radial line joining it to the center of rotation. Part of this acceleration is the effect of the changing distance of the slider from the center, and part is the effect of rotation of the radial sliding velocity vector. The tangential acceleration of the slider relative to the other link is called the <u>Coriolis acceleration</u>.

Consider rotating link AD in Figure 7.14, which turns at a constant velocity ω about a fixed axis A as slider S freely slides (radially outward) on it. B and C are two coincident points at a distance r from A. Point C is on link AD directly beneath point B, which is on the slider.

FIRST CHANGE (Figure 7.15)

Consider the relative velocity of point B to point C radially along the link as the slider rotates from B, C to B', C' (a change in the direction of $\bar{V}_{B/C}$ due

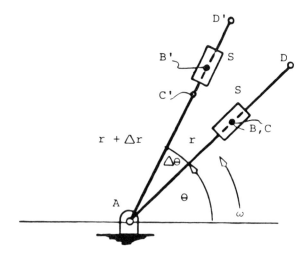

Figure 7.14   Rotating link with slider.

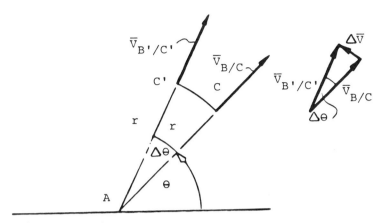

Figure 7.15   First change.

to rotation). Assuming that $\Delta\theta$ is small, the magnitude of $\Delta\bar{V}$ can be expressed as

$$\Delta V = V_{B/C}\,\Delta\theta$$

and

$$\frac{\Delta V}{\Delta T} = V_{B/C}\,\frac{\Delta\theta}{\Delta T}$$

Therefore,

$$A_B^{(1)} = V_{B/C}\,\omega$$

or

$$\bar{A}_B^{(1)} = V_{B/C}\,\omega \quad \text{(in the direction of } \Delta\bar{V})$$

SECOND CHANGE (Figure 7.16)

Consider the tangential velocity of slider S as it moves outward from the center (a change in $\bar{V}_B$ due to the change in slider distance from A). Here the magnitude of $\Delta\bar{V}$ is given by

$$\Delta V = V_{B'} - V_B$$

$$= \Delta r\,\omega$$

and

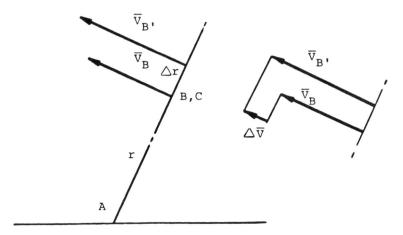

Figure 7.16   Second change.

$$\frac{\Delta V}{\Delta T} = \frac{\Delta r}{\Delta T} \omega$$

Therefore,

$$A_B^{(2)} = V_{B/C} \omega$$

or

$$\bar{A}_B^{(2)} = V_{B/C} \omega \quad \text{(in the direction of } \Delta \bar{V})$$

SUMMARY

Combining the results of the two changes, we obtain the Coriolis acceleration:

$$\bar{A}^{Cor} = \bar{A}_B^{(1)} + \bar{A}_B^{(2)}$$

or

$$\bar{A}^{Cor} = 2V_{B/C} \omega \quad \text{(in the direction of } \Delta \bar{V})$$

The Coriolis acceleration vector is always directed perpendicular to the rotating link and pointed as if it had been rotated about its tail through an angle 90° with vector $\bar{V}_{B/C}$ in the direction of ω, the angular velocity of the rotating link on which the sliding occurs.

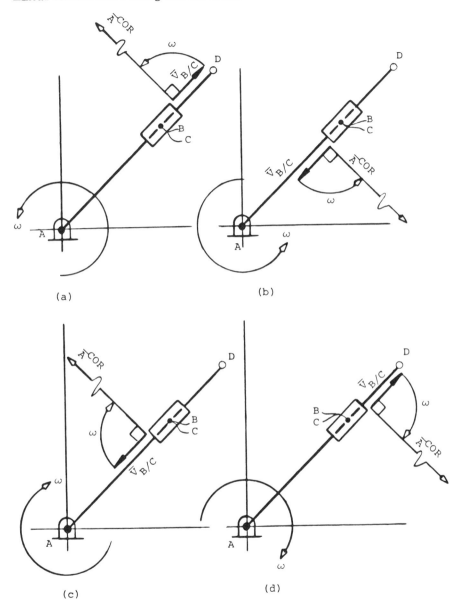

Figure 7.17   Coriolis acceleration for the four cases of Example 7.4:
(a) case 1; (b) case 2; (c) case 3; (d) case 4.

EXAMPLE 7.4 (Figure 7.17)

Consider a block B which slides on a rotation rod AD and has a velocity of 5 ft/sec relative to a point C on the rod ($V_{B/C}$ = 5 ft/sec). The angular velocity of the rod is 4 rad/sec. Determine the Coriolis acceleration in terms of both magnitude and direction for the following cases.

1. Block B moves outward and toward D as AD rotates counter-clockwise (Figure 7.17a).
2. Block B moves inward and toward A as AD rotates counter-clockwise (Figure 7.17b).
3. Block B moves inward and toward A as AD rotates clockwise (Figure 7.17c).
4. Block B moves outward and toward D as A rotates clockwise (Figure 7.17d).

SOLUTION

The magnitude of the Coriolis acceleration in each case is given by

$$A^{Cor} = 2V_{B/C}\,\omega_{AD}$$

$$= 2(5)(4)$$

$$= 40 \text{ ft/sec}^2$$

Figure 7.17 shows the direction of the Coriolis acceleration in each case. Note that the direction of $\bar{A}^{Cor}$ is obtained from rotating the vector $\bar{V}_{B/C}$ through an angle 90° in the same direction (clockwise or counterclockwise) as that of rotating link AD.

8

Effective Component of Acceleration Method

8.1 INTRODUCTION

The effective component of acceleration method is similar to that of the effective component of velocity method in that it is also based on the rigid body principle. However, it is somewhat more complex for bodies with combined motion since it involves the concepts of relative acceleration and relative velocity. It will be seen that the effective component of acceleration along a line connecting two points, A and B, on a rigid body is composed of two parts:

1. The acceleration component of point A relative to point B due to rotation of A about B
2. The acceleration component of the reference point B due to translation

8.2 ACCELERATION OF END POINTS ON A LINK

Consider link AB shown in Figure 8.1. Both the velocity and acceleration of point A are known in magnitude and direction, whereas only the line of action of the acceleration of point B (or $\bar{A}_B$) is known. This line of action is the line "b"-"b". Determine the complete acceleration of point B ($\bar{A}_B$).

The acceleration of point B is related to that of point A by the vector equation

$$\bar{A}_B = \bar{A}_A + \bar{A}_{B/A}$$

and similarly, the effective components of $\bar{A}_B$ and $\bar{A}_C$ along link AB are related by

$$\bar{A}_B^{AB} = \bar{A}_A^{AB} + \bar{A}_{B/A}^{AB}$$

To determine $\bar{A}_{B/A}^{AB}$, we must consider the motion of point B relative to point A. That is, we assume that point A is fixed while B rotates about it. Therefore, the only effective component of acceleration that B can have along link AB is the normal or radial acceleration ($\bar{A}_{B/A}^{N}$). In other words,

$$\bar{A}_{B/A}^{AB} = \bar{A}_{B/A}^{N}$$

where

$$A_{B/A}^{N} = \frac{V_{B/A}^2}{AB}$$

Therefore,

$$A_{B/A}^{AB} = \frac{V_{B/A}^2}{AB}$$

This means that we must find $V_{B/A}$ in order to find $\bar{A}_{B/A}^{AB}$. $V_{B/A}$ can be determined from a velocity polygon.

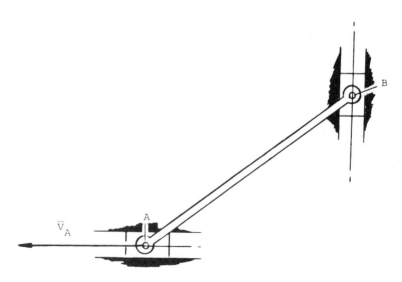

Figure 8.1   Linkage.

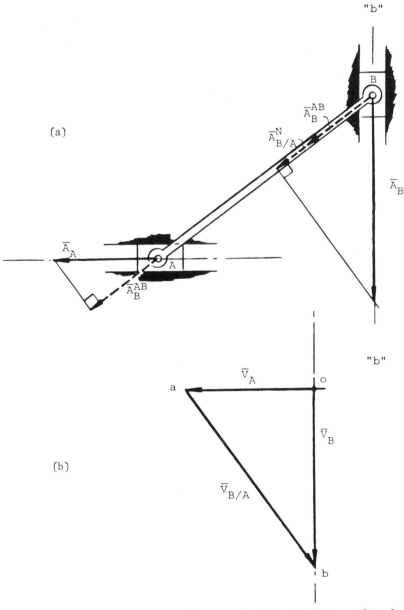

Figure 8.2   Acceleration of end points of a link: (a) linkage; (b) velocity polygon.

PROCEDURE (Figure 8.2a)

1. At point A, lay out the effective component of the given acceleration $\bar{A}_A$, that is, $\bar{A}_A^{AB}$. This is obtained, as in the velocity case, by dropping a perpendicular line from the terminus $\bar{A}_A$ to the extended line AB.

2. Find $\bar{A}_B^{AB}$ on link AB. This is obtained from the relationship

$$\bar{A}_B^{AB} = \bar{A}_A^{AB} + \bar{A}_{B/A}^{AB}$$

   where $\bar{A}_{B/A}^{AB}$ is the radial or normal component of acceleration of point B relative to point A.

3. Determine the effective component of $\bar{A}_B$ on link AB ($\bar{A}_B^{AB}$).
   a. Determine $V_{B/A}$. This may be obtained from a velocity polygon, as in Figure 8.2b.
   b. Calculate the magnitude of $\bar{A}_{B/A}^{AB}$.

$$A_{B/A}^{AB} = A_{B/A}^N = \frac{V_{B/A}^2}{AB}$$

   c. From point B, lay out vector $\bar{A}_B^{AB}$ (the summation of vectors $\bar{A}_{B/A}^{AB}$ and $\bar{A}_A^{AB}$) along link AB.

4. Knowing the path of acceleration $\bar{A}_B$ to be along line "b"-"b", this vector can now be defined by constructing a perpendicular line from the terminus of $\bar{A}_B^{AB}$ to meet line "b"-"b", similar to the procedure used in the velocity case.

8.3   SLIDER-CRANK ANALYSIS

Crank AB of the slider-crank mechanism shown in Figure 8.3 rotates with an angular velocity of 2 rad/sec (counterclockwise) and an angular acceleration of 2 rad/sec^2 (counterclockwise). Determine the acceleration of the slider.

   The acceleration of slider C is related to that of point B by the vector equation

$$\bar{A}_C = \bar{A}_B + \bar{A}_{C/B}$$

and similarly, the effective components of these accelerations along BC are related by

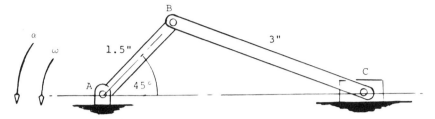

Figure 8.3   Slider-crank mechanism.

$$\bar{A}_C^{BC} = \bar{A}_B^{BC} + \bar{A}_{C/B}^{BC}$$

where

$$A_{C/B}^{BC} = A_{C/B}^{N}$$

and

$$A_{C/B}^{N} = \frac{V_{C/B}^{2}}{BC}$$

PROCEDURE (Figure 8.4b)

1. From point B, lay out acceleration vector $\bar{A}_B$ to a convenient scale using the vector equation

$$\bar{A}_B = \bar{A}_B^{N} + \bar{A}_B^{T}$$

    where

$$A_B^{N} = AB \times \omega^2$$

$$= 1.5(2)^2 = 6.0 \text{ in./sec}^2$$

$$A_B^{T} = AB \times \alpha$$

$$= 1.5(2) = 3 \text{ in./sec}^2$$

2. Construct the effective component of $\bar{A}_B$ on BC ($\bar{A}_B^{BC}$). Drop a perpendicular from the terminus of $\bar{A}_B$ to extended link BC.

3. Determine the effective component of $\bar{A}_C$ on link BC ($\bar{A}_C^{BC}$).

    a. Determine the magnitude $V_{C/B}$ from a velocity polygon (Figure 8.4a).

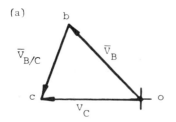

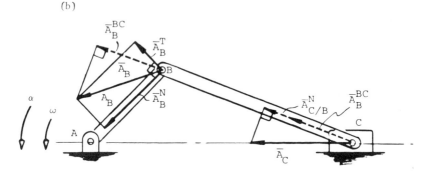

Figure 8.4   Acceleration of points on a slider–crank mechanism:
(a) acceleration diagram; (b) velocity polygon.

$$V_{C/B} = 2.3 \text{ in./sec}$$

b. Lay out $\bar{A}_{C/B}^{BC}$ from point C along link BC. The magnitude of
this vector is obtained

$$A_{C/B}^{BC} = \frac{(2.3)^2}{3} = 1.76 \text{ in./sec}^2$$

c. Lay out vector $\bar{A}_B^{BC}$ added to vector $\bar{A}_{C/B}^{BC}$ along BC.

4. Determine the absolute acceleration of point C ($\bar{A}_C$). Since the
path of the slider is a straight line, the direction of its acceleration must be along the same path. Therefore, to find the
magnitude of this acceleration, drop a perpendicular from the
effective component $\bar{A}_C^{BC}$ to intersect the line of action of $\bar{A}_C$.

The point of intersection defines the acceleration $\bar{A}_C$.

5. Scale the vector $\bar{A}_C$ to determine its magnitude.

$$A_C = 7.5 \text{ in./sec}^2$$

8.4   FOUR-BAR LINKAGE ANALYSIS

Crank AB of the four-bar linkage in Figure 8.5 has an angular velocity of
2 rad/sec (counterclockwise) and is accelerating at the rate of 1 rad/sec^2.
Determine the acceleration of point C.

In the example of Section 8.4, the acceleration of point C was related
to point B by the vector equation

$$\bar{A}_C = \bar{A}_B + \bar{A}_{C/B}$$

and the effective component of the acceleration of point C along link BC was
obtained from

$$\bar{A}_C^{BC} = \bar{A}_B^{BC} + \bar{A}_{C/B}^{BC}$$

In that example, since the direction of acceleration of point C was known,
only one effective component of that acceleration was required to define the
acceleration completely. In this example, however, the direction of acceleration of point C is not known. Therefore, another effective component is
needed to define $\bar{A}_C$. This leads us to consider the effective component of
point C along link CD, which is given by the equation

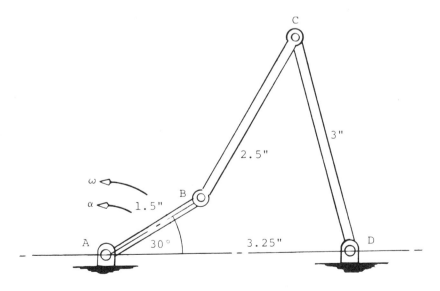

Figure 8.5   Four-bar mechanism.

$$\bar{A}_C^{-CD} = \bar{A}_D^{-CD} + \bar{A}_{C/D}^{-CD}$$

where

$$\bar{A}_D^{-CD} = 0 \quad \text{(since point D is fixed)}$$

$$\bar{A}_{C/D}^{-CD} = \bar{A}_{C/D}^N$$

where

$$\bar{A}_{C/D}^N = \frac{V_C^2}{CD} \quad \text{(directed from C to D)}$$

Therefore,

$$\bar{A}_C^{-CD} = \frac{V_C^2}{CD} \quad \text{(directed from C to D)}$$

With the two effective components $\bar{A}_C^{-BC}$ and $\bar{A}_C^{-CD}$ determined, the required acceleration of point C can readily be found.

PROCEDURE

1. From point B (Figure 8.6b), lay out acceleration vector $\bar{A}_B$ to a convenient scale, applying the relationship

$$\bar{A}_B = \bar{A}_B^N + \bar{A}_B^T \quad \text{(vectorial sum)}$$

where

$$A_B^N = AB \times \omega^2$$

$$= 1.5(4) = 6 \text{ in./sec}^2$$

$$A_B^T = AB \times \alpha$$

$$= 1.5(1) = 1.5 \text{ in./sec}^2$$

2. Construct the effective component of $\bar{A}_B$ on BC ($\bar{A}_B^{-BC}$). Drop a perpendicular from the terminus of $\bar{A}_B$ to extended link BC.
3. Find the effective component of $\bar{A}_C$ on link BC ($\bar{A}_C^{-BC}$).

$$\bar{A}_C^{-BC} = \bar{A}_B^{-BC} + \bar{A}_{C/B}^{-BC}$$

where

$$A^{BC}_{C/B} = A^{N}_{C/B} = \frac{V^2_{C/B}}{BC}$$

a. Therefore, construct a velocity polygon (Figure 8.6a) to find $V_{C/B}$.

$$V_{C/B} = 4.15 \text{ in./sec}$$

b. Determine $\bar{A}^{BC}_{C/B}$.

$$A^{BC}_{C/B} = \frac{4.15^2}{2.5} = 6.8 \text{ in./sec}^2$$

Therefore,

$$\bar{A}^{BC}_{C/B} = 6.8 \text{ in./sec}^2 \quad \text{(directed along BC toward B)}$$

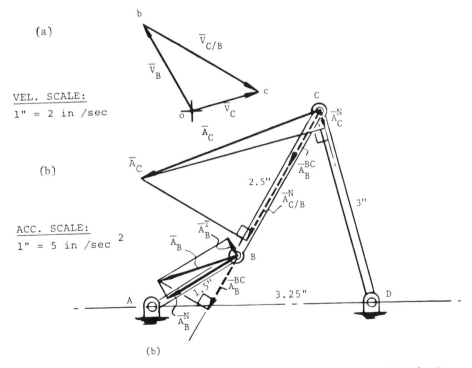

(a)

VEL. SCALE:
1" = 2 in /sec

(b)

ACC. SCALE:
1" = 5 in /sec ^{2}

(b)

Figure 8.6   Accelerations of points on a four-bar mechanism: (a) velocity polygon; (b) acceleration diagram.

c. From point C, add vector $\bar{A}^{BC}_{C/B}$ to vector $\bar{A}^{BC}_{B}$ along link BC
to obtain $\bar{A}^{BC}_{C}$.

4. Determine the effective component of $\bar{A}_C$ along link CD ($\bar{A}^{CD}_{C}$).

$$\bar{A}^{CD}_{C} = \bar{A}^{N}_{C} \quad \text{(radial acceleration of point C)}$$

where

$$A^{N}_{C} = \frac{V^2_C}{CD} \quad \text{(V_C from velocity polygon)}$$

$$= \frac{2.15^2}{3} = 1.54 \text{ in./sec}^2$$

Therefore,

$$\bar{A}^{CD}_{C} = 1.54 \text{ in./sec}^2 \quad \text{(directed from C to D)}$$

5. From point C, lay out vector $\bar{A}^{CD}_{C}$ on link CD directed toward the fixed axis D.

6. With the two effective components of point C now known, the required acceleration $\bar{A}_C$ is obtained by projecting perpendiculars from the termini of these two components until they intersect. This point of intersection defines the magnitude and direction of required vector $\bar{A}_C$.

7. Scale the magnitude of vector $\bar{A}_C$.

$$A_C = 14.5 \text{ in./sec}^2$$

Therefore,

$$\bar{A}_C = 14.5 \text{ in./sec}^2 \quad \text{(directed as shown)}$$

9

Relative Acceleration Method

9.1 INTRODUCTION

The relative acceleration method is probably the fastest and most common among the graphical methods used for acceleration analysis. This is merely an extension of the relative velocity method used for velocity analysis and utilizes an acceleration polygon that is very similar to the velocity polygon. The method is based on the following principles:

1. All motions are considered instantaneous.

2. The instantaneous acceleration of a point A relative to another point B on a rigid link is obtained from the vectorial relationship

$$\bar{A}_A = \bar{A}_B + \bar{A}_{A/B}$$

3. The instantaneous motion of a point may be considered one of pure rotation in which the acceleration can be resolved into two rectangular components: one normal and one tangential to the path of rotation. Thus the expression in step 2 can be represented graphically as

$$\bar{A}_A^N + \bar{A}_A^T = \bar{A}_B^N + \bar{A}_B^T + \bar{A}_{A/B}^N + \bar{A}_{A/B}^T$$

4. The absolute as well as relative velocities of various points in the mechanism are known. This requirement makes it most desirable to use the relative velocity or instant center method to determine the velocities involved.

9.2   THE ACCELERATION POLYGON

Let us consider the two points A and B on link AB shown in Figure 9.1. As we saw earlier, the acceleration of point B relative to point A is given by the vector expression

$$\bar{A}_B = \bar{A}_A + \bar{A}_{B/A}$$

Just as in the velocity case, an acceleration polygon can be developed to represent the vector expression. Furthermore, the construction procedure and convention employed for this development are the same in both cases.

> PROCEDURE
>
> 1. Define a point o', called the pole, as the origin for the construction of the polygon. This is the point of zero acceleration. All absolute acceleration vectors originate from this point. By "absolute acceleration" we mean the real or true acceleration of a body as observed from a stationary frame of reference such as the earth.
> 2. Since the accelerations $\bar{A}_A$ and $\bar{A}_B$ are absolute accelerations, draw vectors $\bar{A}_A$ and $\bar{A}_B$ from point o' and define their respective termini as a' and b'.
> 3. From point a', the terminus of $\bar{A}_{A'}$, draw a third vector to terminate at point b', the terminus of $\bar{A}_{B'}$, thereby closing the polygon. This vector defines the relative acceleration $\bar{A}_{B/A}$.

9.3   ACCELERATION POLYGON CONVENTION

The convention used to develop the acceleration polygon of Figure 9.1 can be summarized as follows:

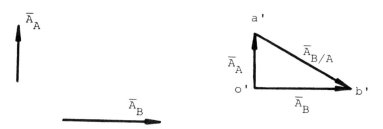

Figure 9.1 Acceleration polygon.

o'a' represents the acceleration vector $\bar{A}_A$.
o'b' represents the acceleration vector $\bar{A}_B$.
a'b' represents the acceleration vector $\bar{A}_{B/A}$.

As in the velocity polygon case (Section 6.4), note that the letter to which the vector is directed indicates the acceleration under consideration. For example, in the vectors o'a' and o'b', the arrows point toward a' and b', respectively. Hence the vectors represent the accelerations of points A and B, respectively. Similarly, in the vector a'b', the arrow points away from a' toward b'; hence the vector represents the acceleration of point B relative to point A ($\bar{A}_{B/A}$). If the arrow were reversed, pointing toward a' and away from b', the vector b'a' would represent the acceleration of A relative to B ($\bar{A}_{A/B}$).

9.4 GENERALIZED PROCEDURE

The construction procedure for developing the acceleration polygon of a mechanism will now be generalized using the four-bar mechanism as a model. Consider the four-bar mechanism ABCD shown in Figure 9.2. In this mechanism, AB, the driven member, has a clockwise angular ~~acceleration~~ _velocity_ ω_{AB} and a counterclockwise angular acceleration α_{AB}. It is required to find the acceleration of point C on the follower CD ($\bar{A}_C$).

PROCEDURE

1. Determine the velocities of all points on the mechanism, including those with combined motion. This may be done using the instant center method, the effective component method, or as in the present case, the relative velocity method (Figure 9.3).

2. Define a starting point o', called the polar origin. All absolute acceleration vectors originate from the polar origin. By "absolute acceleration" we mean the real or true acceleration of the point as observed from a fixed frame of reference such as the earth.

3. Lay out the known components of absolute acceleration of the drive member, starting with the normal acceleration, which has the magnitude $A_B^N = AB \times \omega_{AB}^2$ and is directed parallel to AB. Then add the tangential acceleration, which has the magnitude $A_B^T = AB \times \alpha_{AB}$ and is directed perpendicular to the normal acceleration, or $\bar{A}_B^N \perp \bar{A}_B^T$. The summation of these two components defines the absolute acceleration of point B ($\bar{A}_B = \bar{A}_B^N + \bar{A}_B^T$) (Figure 9.4). If an angular acceleration α_{AB} is not given, $\bar{A}_B^T$ does not exist ($\bar{A}_B^T = 0$), and $\bar{A}_B^N$ becomes the absolute acceleration of point B. Define the terminus of $\bar{A}_B$ as b'.

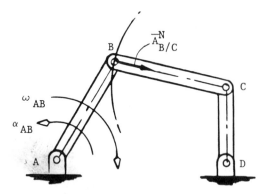

Figure 9.2   Four-bar mechanism.

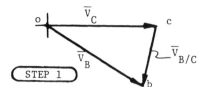

Figure 9.3   Velocity polygon.

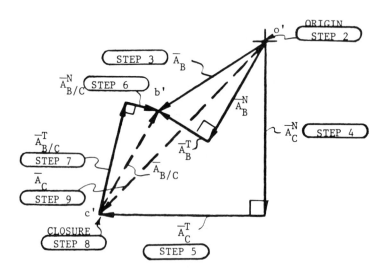

Figure 9.4   Acceleration polygon.

4. Select another point on the mechanism that has absolute motion and is rigidly connected to point B (call it point C) and lay out its normal acceleration component $\bar{A}_C^N$, noting that this vector is defined by the magnitude

$$A_C^N = \frac{V_{C/D}^2}{CD}$$

and is directed parallel to link CD. The normal components of acceleration are readily determined from the velocity data and the link orientation.

5. Through the terminus of vector $\bar{A}_C^N$, construct a perpendicular to represent the tangential acceleration ($\bar{A}_C^T$), whose direction is known but those magnitude is as yet undefined. This line contains point c', the terminus of vector $\bar{A}_C$.

6. Consider next the relationship between point B, whose acceleration is completely known, and point C, whose acceleration is only partially known. This means that we must consider the acceleration of point B relative to point C.

$$\bar{A}_{B/C} = \bar{A}_{B/C}^N + \bar{A}_{B/C}^T$$

where $\bar{A}_{B/C}^N$ can be completely determined from the velocity analysis and $\bar{A}_{B/C}^T$ is known in direction only. Therefore, lay out vector $\bar{A}_{B/C}^N$ knowing that:

a. Its magnitude is given by

$$A_{B/C}^N = \frac{V_{B/C}^2}{BC}$$

b. Its direction is obtained by assuming that point C on link is fixed while point B rotates about it. $\bar{A}_{B/C}^N$ therefore lies on the link and is directed toward the center of rotation assumed.

c. Also, in accordance with the polygon convention, relative acceleration vectors normally do not originate at the pole o' but extend between the termini of the absolute acceleration vectors to which they relate. For example, the notation $\bar{A}_{B/C}$ means that polygon vector is directed from c' to b', and $\bar{A}_{C/B}$ means that polygon vector is directed from b' to c', where b' and c' are the termini of absolute vectors on the acceleration polygon, and B and C are the corresponding points in the linkage.

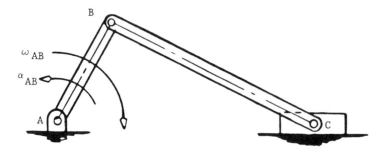

Figure 9.5   Slider-crank mechanism.

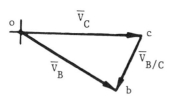

Figure 9.6   Velocity polygon—step 1.

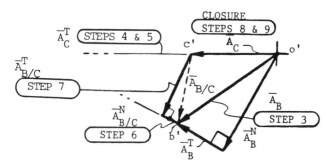

Figure 9.7   Acceleration polygon.

7. Through the tail* of $\bar{A}^N_{B/C}$, construct a perpendicular line of undefined length to represent the acceleration $\bar{A}^T_{B/C}$, whose direction is known but whose magnitude is yet to be determined. This line also contains point c', the terminus of $\bar{A}_C$.

8. Now since c' lies on both $\bar{A}^T_{B/C}$ (step 7) and $\bar{A}^T_C$ (step 5), it follows that the intersection of these two lines will define that point. Therefore, extend $\bar{A}^T_{B/C}$ and $\bar{A}^T_C$ until they intersect and label the point of intersection c'.

9. Finally, lay out a vector from origin o' to terminus c' to define the required absolute acceleration $\bar{A}_C$ and a vector from terminus c' to point b' to define the relative acceleration $\bar{A}_{B/C}$.

10. A quick check of the completed polygon should reveal the balanced vector equation

$$\bar{A}_B = \bar{A}_C + \bar{A}_{B/C}$$

or

$$\bar{A}^N_B + \bar{A}^T_B = \bar{A}^N_C + \bar{A}^T_C + \bar{A}^N_{B/C} + \bar{A}^T_{B/C}$$

<u>Special Cases</u>

Slider-Crank Mechanism

The slider-crank mechanism shown in Figures 9.5 to 9.7 may be considered a special case of the four-bar linkage where the follower link—for example, CD in Figure 9.2—is infinitely long. This means that the velocity of point C must be a straight line. Therefore, in step 4 (Figure 9.7) the normal acceleration of the slider $\bar{A}^N_C$ is zero, and

$$\bar{A}_C = \bar{A}^T_C$$

In effect, the tangential component of acceleration $\bar{A}^T_C$ becomes the absolute acceleration of C and originates at o'.

---

*Note that had the normal relative acceleration $\bar{A}^N_{C/B}$ been chosen instead of $\bar{A}^N_{B/C}$, the perpendicular drawn to represent $\bar{A}^T_{C/B}$ would have been constructed through the terminus of $\bar{A}^N_{C/B}$. This is because, by polygon convention, the normal acceleration vector $\bar{A}^N_{C/B}$ must head toward c' and away from b'.

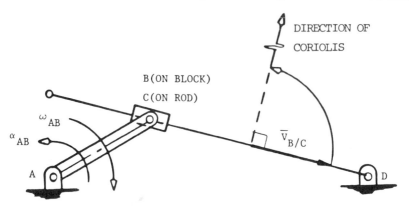

Figure 9.8   Quick-return mechanism.

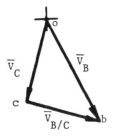

Figure 9.9   Velocity polygon—step 1.

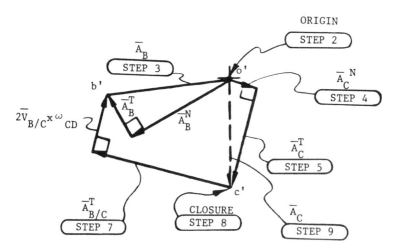

Figure 9.10   Acceleration polygon.

Quick-Return Mechanism

Another special case is the quick-return mechanism shown in Figures 9.8 to 9.10. Here point B slides on link CD, which is itself rotating. This means that in addition to the normal acceleration of point B relative to a coincident point C on CD, we must consider an additional component of acceleration: the Coriolis acceleration.

Also, since the path of B relative to C is a straight line, the acceleration of B relative to C can only be tangential. That is, B can have <u>no</u> <u>normal component</u> of acceleration relative to C; or

$$\bar{A}^{N}_{B/C} = 0$$

Had the path of B on CD been a curve, there would have existed a normal acceleration component of B relative to C ($\bar{A}^{N}_{B/C}$) as will be seen later (see Section 9.7).

To determine the Coriolis acceleration, we note from earlier discussion that the magnitude of this vector is given by

$$A^{Cor} = 2V_{B/C}\omega_{CD}$$

where $V_{B/C}$ and ω_{CD} are obtained from the velocity analysis. Also, we note that in defining the direction of this vector, we always consider (1) <u>the</u> <u>linear velocity of the sliding body relative to that of the rotating body</u>, and (2) <u>the angular velocity of the same body on which the sliding occurs</u>.

In this example, since body B slides on link CD, we consider the velocity of B relative to C or $\bar{V}_{B/C}$ (<u>not</u> $\bar{V}_{C/B}$) and the angular velocity of link CD or ω_{CD} (<u>not</u> ω_{AB}). Accordingly, in step 6 (Figure 9.10), connect the vector $\bar{A}^{Cor}$ to terminus b' on the polygon, observing the following rules:

1. This vector has the orientation of vector $\bar{V}_{B/C}$ when rotated 90° about its tail in the direction of ω_{CD} (counterclockwise in this case).

2. Since point <u>B relative to point C</u> on the link is being considered, the polygon vector must go from c' to b' (note the <u>reversed</u> <u>letter sequence</u>).

EXAMPLE 9.1 (see Figure 9.11)

Let AB = 1.5 in., BC = 1.5 in., CD = 1 in., ω_{AB} = 1 rad/sec (clockwise), and α_{AB} = 0.5 rad/sec² (counterclockwise). Determine $\bar{A}_{C}$.

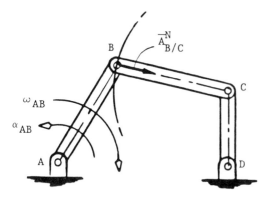

Figure 9.11   Four-bar mechanism.

SOLUTION

1. Determine the velocities. Construct a velocity polygon (Figure 9.12) and obtain

$$V_B = AB \times \omega_{AB} = 1.5(1) = 1.5 \text{ in./sec}$$

$$V_C = 1.45 \text{ in./sec}$$

$$V_{B/C} = 0.8 \text{ in./sec}$$

2. Define polar origin o' for the acceleration polygon (Figure 9.13).
3. Lay out the acceleration of the drive point, $\bar{A}_B$.

$$A_B^N = AB \times \omega_{AB}^2 = 1.5(1^2) = 1.5 \text{ in./sec}^2$$

$$A_B^T = AB \times \alpha_{AB} = 1.5(0.5) = 0.75 \text{ in./sec}^2$$

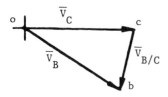

SCALE: 1 in. = $\dfrac{1 \text{ in.}}{\text{sec.}}$

Figure 9.12   Velocity polygon.

$$\bar{A}_B = \bar{A}_B^N + \bar{A}_B^T$$

The vector originates at o' and terminates at b'.

4. Lay out the normal acceleration of the driven point, $\bar{A}_C^N$. The magnitude of this vector is given as

$$A_C^N = \frac{V_C^2}{CD} = \frac{1.45^2}{1.0} = 2.13 \text{ in./sec}^2$$

The vector is directed parallel to link CD, originating from o'.

5. Add the tangential acceleration $\bar{A}_C^T$ (undefined length).

$$\bar{A}_C^T \perp \bar{A}_C^N$$

6. Lay out the normal acceleration of the drive point relative to the driven point, $\bar{A}_{B/C}^N$.

$$\bar{A}_{B/C}^N = \frac{V_{B/C}^2}{BC} = \frac{0.8^2}{1.5} = 0.426 \text{ in./sec}^2$$

The vector is directed parallel to link BC, pointing to b' (or away from c').

7. Add the tangential acceleration $\bar{A}_{B/C}^T$ (undefined length).

$$\bar{A}_{B/C}^T \perp \bar{A}_{B/C}^N$$

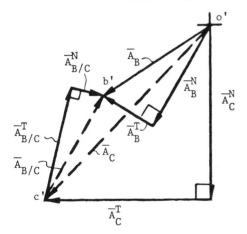

SCALE: 1 in. = $\dfrac{1 \text{ in}}{\text{sec.}}2$

Figure 9.13   Acceleration polygon.

8. Determine the magnitude of tangential accelerations $\bar{A}_C^T$ and $\bar{A}_{B/C}^T$. Extend undefined lines for these acceleration components in steps 6 and 7, until the polygon is closed. $\bar{A}_C^T$ intersects $\bar{A}_{B/C}^T$ at c'.

9. Determine the required acceleration $\bar{A}_C$. Connect point c' to point o' and measure o'c', or magnitude A_C.

$$\bar{A}_C = 3.2 \text{ in./sec}^2 \quad \text{(directed as shown in Figure 9.13)}$$

9.5   MECHANISM WITH EXPANDED FLOATING LINK

It was shown earlier that if the accelerations of two points on a link are known, the acceleration of a third point on that link can be found by proportion. In the following example it will be seen that the polygon construction procedure can also be used to determine the acceleration of the third point.

EXAMPLE 9.2

Consider the mechanism ABCE shown in Figure 9.14a, where the crank AB rotates with an angular velocity of 2 rad/sec (clockwise). It is required to find the acceleration of point E on the expanded floating link BEC.

SOLUTION

1. Develop the velocity polygon for the mechanism as shown in Figure 9.14b. From this polygon the absolute and relative velocity magnitudes are obtained as follows:

$$V_B = 2.5 \text{ in./sec}$$

$$V_C = 2.8 \text{ in./sec}$$

$$V_{E/B} = 1.0 \text{ in./sec}$$

$$V_{E/C} = 1.5 \text{ in./sec}$$

2. Develop the acceleration polygon for the basic mechanism ABC (Figure 9.14c) (ignoring point E for the time being), following the construction procedure discussed earlier. From this polygon the accelerations of points B and C are obtained.

$$\bar{A}_B = 5 \text{ in./sec}^2 \quad \text{(directed as shown)}$$

$$\bar{A}_C = 3.9 \text{ in./sec}^2 \quad \text{(directed as shown)}$$

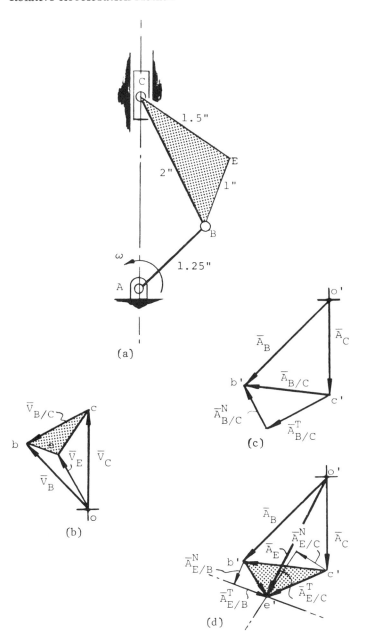

Figure 9.14   Mechanism with expanded floating link: (a) mechanism;
(b) velocity polygon; (c) basic acceleration polygon (for ABC); (d) complete
acceleration polygon.

3. To determine the acceleration of point E ($\bar{A}_E$), complete the
    following steps (Figure 9.14d).
    a. Lay out the vector for the normal acceleration of E relative
       to B ($\bar{A}_{E/B}^N$). The magnitude of this vector is obtained from

$$A_{E/B}^N = \frac{V_{E/B}^2}{EB} = \frac{(1)^2}{1}$$

$$= 1 \text{ in./sec}^2$$

   This vector is directed from point E toward point B on the
   mechanism, but on the polygon, heads from point b' (which
   is defined) to point e' (which is undefined), <u>still maintaining
   its basic orientation.</u>
    b. Through the terminus of vector $\bar{A}_{E/B}^N$, draw a line of unde-
       fined length perpendicular to this vector to indicate the direc-
       tion of $\bar{A}_{E/B}^T$. This line contains point e', the terminus of

       vector $\bar{A}_E$.
    c. From point c' on the polygon, lay out the vector for the nor-
       mal acceleration of E relative to C. The magnitude of this
       acceleration is obtain from

$$A_{E/C}^N = \frac{V_{E/C}^2}{EC} = \frac{(1.5)^2}{1.5}$$

$$= 1.5 \text{ in./sec}^2$$

   This vector is directed from point E to point C on the mechanism but on the polygon heads from point c' (which is defined) toward point e' (which is undefined), <u>still maintaining
   its basic orientation.</u>
    d. Through the terminus of vector $\bar{A}_{E/C}^N$, draw a line of unde-
       fined length perpendicular to this vector to indicate the direc-
       tion of $\bar{A}_{E/C}^T$. This line contains point e', the terminus of $\bar{A}_E$.
    e. Since point e' is to be found on both lines $\bar{A}_{E/B}^T$ [step (c)] and
       $\bar{A}_{E/C}^T$ [step (d)], it must be located at the intersection of

       these lines. This point e' defines the terminus of the vector
       $\bar{A}_E$ drawn from the pole o'. Therefore,

$$\bar{A}_E = 5.6 \text{ in./sec}^2 \quad \text{(directed as shown)}$$

Acceleration Image

As in the relative velocity case, it should be observed from the polygon that the triangle formed by points b', e', c' is similar to link BEC, and for this reason it is often referred to as the acceleration image of link BEC. Consequently, the acceleration of point E could have been determined more directly by constructing the acceleration image on link b'c' of the polygon for the basic mechanism ABC. (Note that the letters used to designate both the links and the image must run in the same order and direction.)

9.6   COMPOUND MECHANISM

Let ABCDEF represent a compound mechanism in the form of a shaper, where crank AB rotates with a constant angular velocity of 1 rad/sec (counterclockwise) (Figure 9.15a). Determine the acceleration of point E on the slider.

PROCEDURE

1. Develop the velocity polygon (Figure 9.15b) and determine the velocities as follows:

$$V_B = AB \times \omega = 1.5(1) = 1.5 \text{ in./sec}$$

$$V_C = 1.2 \text{ in./sec}$$

$$V_{B/C} = 1.0 \text{ in./sec}$$

$$V_D = \frac{OD}{OC} \times V_C = \frac{6}{4}(1.2) = 1.8 \text{ in./sec}$$

$$V_E = 2.1 \text{ in./sec}$$

$$V_{D/E} = 0.75 \text{ in./sec}$$

2. Define polar origin o' for the acceleration polygon (Figure 9.15c).
3. Lay out the acceleration of the driver at point B ($\bar{A}_B$).

$$A_B^N = AB \times \omega^2 = 1.5(1)^2 = 1.5 \text{ in./sec}^2$$

$$A_B^T = AB \times \alpha = 0$$

$$\bar{A}_B = \bar{A}_B^N + \bar{A}_B^T = 1.5 \text{ in./sec}^2 \quad \text{(directed as shown)}$$

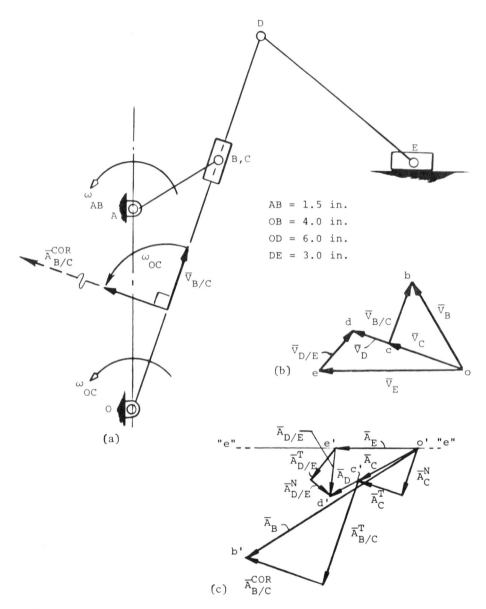

AB = 1.5 in.
OB = 4.0 in.
OD = 6.0 in.
DE = 3.0 in.

Figure 9.15   Acceleration analysis of a shaper mechanism: (a) shaper mechanism; (b) velocity polygon; (c) acceleration polygon.

4. Lay out the normal acceleration of point-driven point C ($\bar{A}_C^N$).

$$A_C^N = \frac{V_C^2}{OC}$$

$$A_C^N = \frac{(1.2)^2}{4}$$

$$\bar{A}_C^N = 0.36 \text{ in./sec}^2 \quad \text{(directed as shown)}$$

5. Add the tangential acceleration of point C (undefined length) to $\bar{A}_C^N$.

$$\bar{A}_C^T \perp \bar{A}_C^N$$

6. Determine the relative acceleration between drive point B and related point C ($\bar{A}_{B/C}$).

$$\bar{A}_{B/C} = \bar{A}_{B/C}^N + \bar{A}_{B/C}^T + \bar{A}_{B/C}^{Cor}$$

$$A_{B/C}^N = 0 \quad \text{(since the path of the slider is a straight line)}$$

and

$$\bar{A}_{B/C}^{Cor} = 2V_{B/C}\,\omega_{OC} \quad \text{(properly directed)}$$

or

$$A_{B/C}^{Cor} = 2(1)\left(\frac{1.2}{4}\right) = 0.6 \text{ in./sec}^2$$

a. Lay out the vector $\bar{A}_{B/C}^{Cor}$ in accordance with convention. That is, the vector must have the same orientation as vector $\bar{V}_{B/C}$ when rotated 90° in the direction of ω_{OC}.

b. Add the tangential relative acceleration $\bar{A}_{B/C}^T$ (undefined length) to $\bar{A}_{B/C}^{Cor}$.

$$\bar{A}_{B/C}^T \perp \bar{A}_{B/C}^{Cor}$$

7. Extend the tangential accelerations in steps 4 and 5 until the polygon is closed. The acceleration of point C on the polygon is defined at the intersection of $\bar{A}_C^T$ and $\bar{A}_{B/C}^T$, that is, at point c'.

8. Draw vector $\bar{A}_C$ extending from the pole o' to c'. Note:

$$\bar{A}_C = \bar{A}_C^N + \bar{A}_C^T$$

and

$$\bar{A}_{B/C} = \bar{A}_{B/C}^{Cor} + \bar{A}_{B/C}^T$$

$$A_C = 0.5 \text{ in./sec}^2$$

9. Determine magnitude of the acceleration of point D (A_D). This is obtained from the proportion

$$\frac{A_D}{A_C} = \frac{OD}{OC}$$

$$A_D = \frac{6}{4}(0.5) = 0.75 \text{ in./sec}^2$$

10. Define the terminus of vector $\bar{A}_D$ as point d' on polygon.
11. Determine the acceleration of point E relative to point D.

$$\bar{A}_{D/E} = \bar{A}_{D/E}^N + \bar{A}_{D/E}^T$$

where

$$\bar{A}_{D/E}^N = \frac{V_{D/E}^2}{DE} \quad \text{(properly directed)}$$

$$= \frac{(0.75)^2}{3} = 0.1875 \text{ in./sec}^2 \quad \text{(properly directed)}$$

a. Lay out vector $\bar{A}_{D/E}^N$. This vector must be pointed toward d' on the polygon and be parallel to DE on the mechanism.
b. Add vector $\bar{A}_{D/E}^T$ to vector $\bar{A}_{D/E}^N$.

$$\bar{A}_{D/E}^T \perp \bar{A}_{D/E}^N$$

12. Determine the acceleration of point E ($\bar{A}_E$). Note that since the slider path is a straight line, there is no normal acceleration ($\bar{A}_E^N = 0$), so $\bar{A}_E$ is equal to the tangential acceleration $\bar{A}_E^T$, which exists along the same path. Therefore:

a. Lay out the direction of vector $\bar{A}_E$ by drawing a line "e"-"e" through pole o' parallel to the slider path.
b. Extend vector $\bar{A}^T_{D/E}$ in step 11(b) until it intersects line "e"-"e" (direction line). This point of intersection defines the magnitude of $\bar{A}_E$.

13. Define the point of intersection in step 12(b) as e'.
14. Measure line o'-e', the vector $\bar{A}_E$.

$$\bar{A}_E = 0.625 \text{ in./sec}^2 \quad \text{(directed as shown)}$$

9.7 CAM-FOLLOWER MECHANISM

Consider the cam-follower mechanism shown in Figure 9.16a. The cam (2) rotates counterclockwise at a constant angular velocity of 2 rad/sec. Find the acceleration of the follower (4).

At first glance it would appear that to determine the acceleration of 4 it would be necessary first to determine the motion of the roller (3). However, since the path that 3 traces on 2 is generally not easily recognizable, because of the two curved surfaces in contact, a more direct approach is to do the following:

1. Assume that the roller does not turn. Then the acceleration of contact point C is the same as that of any point in 4.
2. Construct a pitch line to represent the locus of the center of the roller as it rolls on the cam.
3. Designate the new point of contact between 2 and 4 as P_2 or P_4.
4. Proceed with the analysis based on the assumption that the follower 4 actually slides on the expanded cam defined by the pitch line.

Figure 9.16b shows the velocity polygon for the mechanism. Note that since 4 rides on 2, the relative velocity $\bar{V}_{P4/P2}$ is considered for Coriolis acceleration. This velocity has a direction tangent to the curvature of the cam at the contact point.

Figure 9.16c shows the acceleration polygon for the mechanism. Here it should be noted that the acceleration of P4 relative to P2 (or $\bar{A}_{P4/P2}$) consists of three components:

$$\bar{A}_{P4/P2} = \bar{A}^N_{P4/P2} + \bar{A}^T_{P4/P2} + \bar{A}^{Cor}_{P4/P2}$$

This equation, incidentally, is unlike those of earlier examples where $\bar{A}^N_{P4/P2}$ did not exist, because the sliding paths in those cases were straight

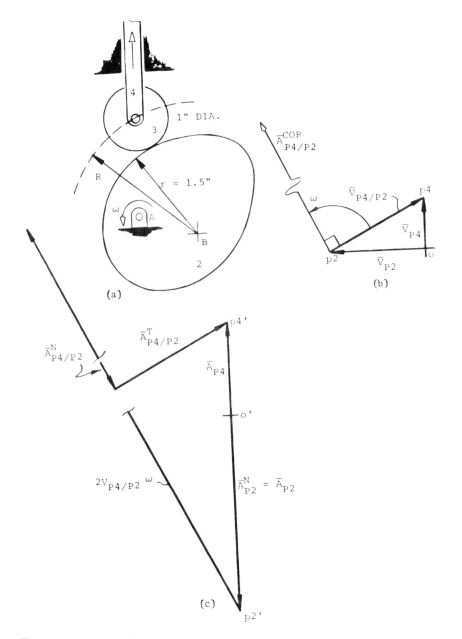

Figure 9.16   Acceleration analysis of a cam-follower mechanism: (a) cam-follower mechanism; (b) velocity polygon; (c) acceleration polygon.

lines. In the present case, since the path of P4 on P2 is a curve (or 4 is forced to move in a curved path as it rides on 2), there must exist a normal acceleration, $\bar{A}^N_{P4/P2}$, whose magnitude is given by

$$A^N_{P4/P2} = \frac{V_{P4/P2}}{r + 0.5}$$

where $r + 0.5$ or R is the radius of curvature of the path.

For the Coriolis acceleration $\bar{A}^{Cor}$, we consider, as before, the motion of P4 relative to P2, computing its magnitude from

$$A^{Cor}_{P4/P2} = 2V_{P4/P2}\,\omega$$

and defining its direction by rotating the vector $\bar{V}_{P4/P2}$ through an angle of 90° in the same angular direction (counterclockwise) as ω_2.

Following is a summary of calculations and results:

$$V_{P2} = 1.5(2) = 3 \text{ in./sec}$$

$$V_{P4} = 1.6 \text{ in./sec} \quad \text{(from the velocity polygon)}$$

$$V_{P4/P2} = 3.35 \text{ in./sec} \quad \text{(from the velocity polygon)}$$

$$A^N_{P2} = 1.5(2)^2 = 6 \text{ in./sec}^2$$

$$A^{Cor}_{P4/P2} = 2(3.35)(2) = 13.4 \text{ in./sec}^2$$

$$A^N_{P4/P2} = \frac{(3.4)^2}{2.0} = 5.6 \text{ in./sec}^2$$

$$A^T_{P4/P2} = 4.0 \text{ in./sec}^2 \quad \text{(from the acceleration polygon)}$$

$$A_{P4} = 2.8 \text{ in./sec}^2 \quad \text{(from the acceleration polygon)}$$

9.8   SUMMARY

The generalized procedure for constructing the acceleration polygon may be summarized as follows:

1.   Proceed from the "known" to the "unknown." That is:
    a. Lay out those absolute vectors whose magnitude and direction are known.

    b. Lay out those components of absolute and relative vectors that are known (magnitude and direction) or can be determined. These include normal accelerations and Coriolis acceleration.

    c. Add to the components in step (b) their corresponding tangential accelerations (directions only), and extend these to close the polygon.

2. All absolute vectors on the acceleration polygon originate from pole o', while relative vectors extend between the termini of the absolute vectors.

3. a. A vector that originates from b' and terminates at c' represents the relative acceleration of point C to that of point B on the link.

    b. The choice between $\bar{A}_{B/C}$ and $\bar{A}_{C/B}$ makes no difference in the polygon configuration or the results, except that these vectors have opposite senses.

4. Two undefined vectors such as $\bar{A}^T_{B/C}$ and $\bar{A}^T_C$ will contain a common point c' on the polygon, and will define that point where they intersect.

10
Velocity-Difference Method

10.1 INTRODUCTION

The velocity-difference method of determining accelerations of points in a mechanism is probably the most straightforward of all the graphical methods used in kinematic analysis. Applicable to any type of mechanism—pin connected, rolling contact, or sliding contact—the method does not rely on the use of sophisticated formulas, but instead employs the simple relationship

$$A = \frac{\Delta V}{\Delta T}$$

based on the fundamental definition, which states: The acceleration (A) of a point is the rate of change of velocity (ΔV) of that point over the time interval (ΔT) in which the change occurs. In terms of mechanism analysis, if the linear acceleration is required for a point in a mechanism at any given position of the mechanism, this acceleration may be found by first considering a small change in position ($\Delta\theta$) of the point, then determining the change in velocity (ΔV) resulting from the change in position, and finally dividing the change of velocity by the time interval (ΔT) during which the change has taken place.

In applying the velocity-difference method, experience has shown that, generally, for reasonably accurate results, a position change of the mechanism based on a crank angular displacement of 1/10 rad is acceptable. This displacement is normally measured such that the given crank position is centrally located between its initial and final positions. That is, the initial and final crank positions are each 1/20 rad on either side of the given position.

The associated time interval ΔT depends on the angular velocity of the crank and its angular displacement. If, for example, the crank of a mechanism turns at a constant speed of 5 rad/sec, ΔT is the time it takes

for the crank to move through 1/10 rad. In our earlier discussion it was
noted that for constant angular speed, ΔT is obtained from

$$\omega = \frac{\Delta\theta}{\Delta T}$$

or

$$\Delta T = \frac{\Delta\theta}{\omega}$$

$$= \frac{1}{5 \text{ rad/sec}} \left(\frac{\text{rad}}{10}\right)$$

$$= 1/50 \text{ sec}$$

If, on the other hand, the crank turns with a uniform acceleration,
which means that the angular velocity is changing, then again from our
earlier discussion, ΔT is obtained from

$$\alpha = \frac{\Delta\omega}{\Delta T}$$

or

$$\Delta T = \frac{\Delta\omega}{\alpha}$$

where $\Delta\omega$ is the change in angular velocity between the initial and final
positions of the mechanism. This uniform or constant acceleration case
will be discussed in more detail in Section 10.4.

The following two sections illustrate the procedure for applying the
velocity–difference technique to mechanisms where the crank arm rotates
with constant angular velocity.

10.2   SLIDER–CRANK MECHANISM ANALYSIS

Consider slider–crank mechanism ABC shown in Figure 10.1, where crank
arm AB is rotating at a constant angular velocity of 1 rad/sec (counterclockwise). Let it be required to find the acceleration of point C for the
position shown.

PROCEDURE

1.   Lay out the given mechanism in the position $AB_i C_i$ (see Figure
     10.2), where AB_i indicates the initial position of the crank, dis-
     placed 1/20 rad (or 1/2 of 1/10 rad) clockwise from the given

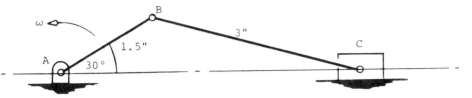

Figure 10.1   Slider-crank mechanism.

position. Note that the angular displacement 1/10 rad can be accurately laid out by applying the well-known relationship

$$S = R \, \Delta\theta$$

where S is the length of an arc that subtends angle $\Delta\theta$ and R is the radius of the arc. Hence the angle 1/20 rad is obtained by laying out a segment of a circle having some convenient radius, say 6 in., in which case the arc length is

$$S = 6\left(\frac{1}{20}\right) = 0.3 \text{ in.}$$

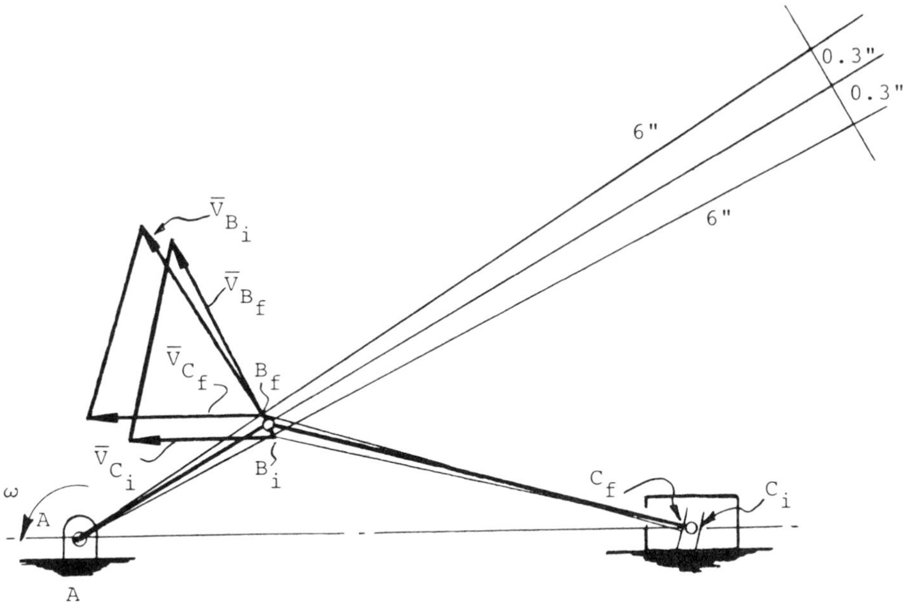

Figure 10.2   Slider-crank mechanism.

2. Determine the linear velocity of point C_i using any of the methods studied earlier. The relative velocity method, used in this example, yields

$$\bar{V}_{C_i} = 1.0 \text{ in./sec} \quad \text{(directed as shown in Figure 10.2)}$$

3. Lay out the given mechanism in position AB_fC_f (see Figure 10.2), where AB_f indicates the final position of the crank, displaced 1/20 rad counterclockwise from the given position.
4. Determine the linear velocity from point C_f as in step 2. This velocity has been determined to be

$$\bar{V}_{C_f} = 1.17 \text{ in./sec} \quad \text{(directed as shown in Figure 10.2)}$$

5. Determine the velocity difference ΔV_C, which is given by

$$\Delta\bar{V}_C = \bar{V}_{C_f} - \bar{V}_{C_i} \quad \text{(vectorial difference)}$$

This equation, which is represented graphically in Figure 10.2, yields

$$\Delta\bar{V}_C = 0.17 \text{ in./sec} \quad \text{(directed to the left)}$$

Note that the vectorial difference in this example yields the same results as the algebraic difference since $\bar{V}_{C_f}$ and $\bar{V}_{C_i}$ both have the same line of action.

6. Determine the acceleration of point C. The magnitude of this vector is given by

$$A_C = \frac{\Delta V_C}{\Delta T}$$

where

$$\Delta T = \frac{\Delta\theta}{\omega}$$

$$= 0.1 \text{ rad/rad/sec} = 0.1 \text{ sec}$$

Therefore,

$$A_C = \frac{0.17}{0.1} \text{ in./sec}^2$$

and the vector $\bar{A}_C = 1.70 \text{ in./sec}^2$ (directed as shown in

Figure 10.2). Note that the direction of acceleration is always the same as that for $\Delta \bar{V}_C$.

10.3   QUICK-RETURN MECHANISM ANALYSIS

Now consider the quick-return mechanism ABCD in Figure 10.3. Crank AB turns with a constant angular speed of 30 rad/sec (counterclockwise) and the linear acceleration of point C on the rod is required.

PROCEDURE

We follow the same procedure as that given in Section 10.2.

1. Lay out the mechanism AB_iC_iD (Figure 10.4), showing the initial position of the mechanism rotated back 1/20 rad from the given position.
2. Determine the linear velocity of point C ($\bar{V}_{C_i}$) for this position (see Figure 10.4).

$$\bar{V}_{C_i} = 72 \text{ in./sec} \quad \text{(directed as shown)}$$

3. Lay out the mechanism AB_fC_fD (Figure 10.4), showing the final position of the mechanism rotated forward 1/20 rad from the given position.

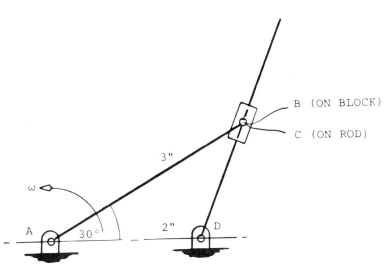

B (ON BLOCK)

C (ON ROD)

3"

ω

A   30°   2"   D

Figure 10.3   Quick-return mechanism.

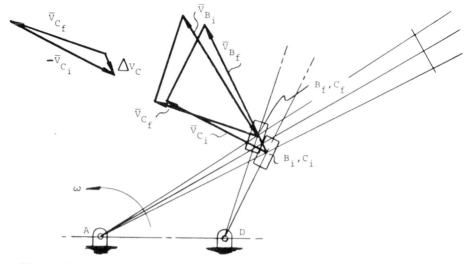

Figure 10.4   Quick-return mechanism.

4. Determine the linear velocity of point C ($\bar{V}_{C_f}$) for this position
    (see Figure 10.4),

$$\bar{V}_{C_f} = 69.5 \text{ in./sec} \quad \text{(directed as shown)}$$

5. Determine the velocity difference $\Delta\bar{V}_C$ from the relationship

$$\Delta\bar{V}_C = \bar{V}_{C_f} - \bar{V}_{C_i} \quad \text{(vectorial difference)}$$

$$= 15.0 \text{ in./sec} \quad \text{(directed as shown in Figure 10.4)}$$

6. Determine the linear acceleration $\bar{A}_C$. The magnitude of this
    vector is given by

$$A_C = \frac{\Delta V_C}{\Delta T}$$

$$= \Delta V_C \frac{\omega}{\Delta\theta}$$

$$= 15.0\left(\frac{30}{0.1}\right)$$

$$= 4500 \text{ in./sec}^2$$

Therefore,

$$\bar{A}_C = 4500 \text{ in./sec}^2 \quad \text{(directed like } \Delta\bar{V}_C \text{ as shown)}$$

7. To obtain the angular acceleration of CD (α_{CD}), apply the relationship

$$\alpha_{CD} = \frac{\Delta\omega_{CD}}{\Delta T}$$

where

$$\Delta\omega_{CD} = \omega_{CD_f} - \omega_{CD_i}$$

$$= \frac{V_{C_f}}{CD_f} - \frac{V_{C_i}}{CD_i}$$

$$= \frac{69.5}{1.7} - \frac{72}{1.52}$$

$$= 40.9 - 47.4 = -6.5 \text{ rad/sec}$$

$$\Delta T = \frac{\Delta\theta}{\omega}$$

$$= \frac{1}{10}\left(\frac{1}{30}\right) = 1/300 \text{ sec}$$

Therefore,

$$= \frac{-6.5}{1/300}$$

$$= -1950 \text{ rad/sec}^2$$

Note that in this analysis there was no need to determine the Coriolis acceleration, which is ordinarily the case in the relative acceleration method. In this respect, the velocity–difference method offers additional simplification to the solution of a problem that can otherwise be more complicated.

10.4   FOUR–BAR MECHANISM ANALYSIS

Consider the four-bar mechanism ABCD, where crank AB turns with a constant angular acceleration of 1 rad/sec^2 and for the position shown has an angular speed of 2 rad/sec (Figure 10.5). It is required to find the linear acceleration of point C on the follower.

The procedure for solving this problem is basically the same as that used in Sections 10.2 and 10.3, with the exception that in this case there is

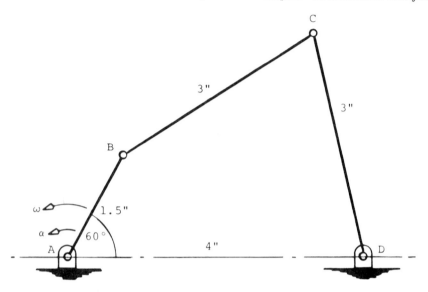

Figure 10.5   Four-bar mechanism.

an angular acceleration, and therefore the magnitudes of the linear velocities of $\bar{V}_{B_i}$ and $\bar{V}_{B_f}$ are not the same. These velocity magnitudes are calculated from

$$V_{B_i} = AB\omega_i \tag{10.1}$$

and

$$V_{B_f} = AB\omega_f \tag{10.2}$$

where ω_i is the initial angular velocity of AB_i obtained from

$$\omega^2 = \omega_i^2 + 2\alpha\,\Delta\theta \tag{10.3}$$

and ω_i is the final angular velocity of AB_f, obtained from

$$\omega^2 = \omega_f^2 - 2\alpha\,\Delta\theta \tag{10.4}$$

Also, since the angular acceleration is constant, the angular speed is changing. ΔT must be expressed in terms of the angular acceleration α and the changing speed ω. That is,

$$\Delta T = \frac{\Delta \omega}{\alpha} \tag{10.5}$$

where $\omega_f = \omega_i$.

PROCEDURE

1. Figure 10.6 shows the mechanisms AB_iC_iD in the initial position and AB_fC_fD in the final position.

2. Determine ω_i. This is found by applying Equation (10.3).

$$\omega^2 = \omega_i^2 + 2\alpha \, \Delta\theta$$

$$2^2 = \omega_i^2 + 2(1)\left(\frac{1}{20}\right)$$

$$\omega_i^2 = 4 - 0.1 = 3.9$$

$$\omega_i = 1.97 \text{ rad/sec}$$

3. Determine $\bar{V}_{B_i}$. From Equation (10.1), the magnitude of this vector is computed as

$$V_{B_i} = 1.5(1.9)$$

$$= 2.95 \text{ in./sec}$$

Therefore,

$$\bar{V}_{B_i} = 2.45 \text{ in./sec} \quad \text{(directed as shown)}$$

4. Determine ω_f. Since α is constant, ω_f is easily determined by recognizing that its increment from ω at the given crank position is equal to that of the same ω from ω_i, due to the same angular displacement of $1/20$ rad in both cases. Therefore,

$$\omega_f = 2 + (2 - 1.97)$$

$$= 2.03 \text{ rad/sec}$$

5. Determine $\bar{V}_{B_f}$. The magnitude of this vector is computed as

$$V_{B_f} = 1.5(2.03)$$

$$= 3.04 \text{ in./sec}$$

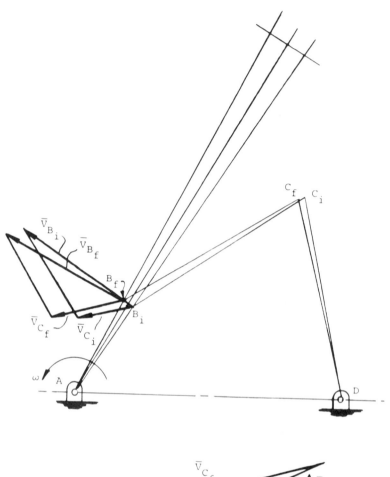

Figure 10.6   Four-bar mechanism.

Therefore,

$$\bar{V}_{B_f} = 3.04 \text{ in./sec} \quad \text{(directed as shown)}$$

6.   Determine $\bar{V}_{C_f}$ using the mechanism AB_fC_fD, as before.

$$\bar{V}_{C_f} = 1.7 \text{ in./sec} \quad \text{(directed as shown)}$$

7. Determine $\bar{V}_{C_i}$ using the mechanism AB_iC_iD, as before.

$$\bar{V}_{C_i} = 1.25 \text{ in./sec} \quad \text{(directed as shown)}$$

8. Determine $\Delta\bar{V}_C$ from

$$\Delta\bar{V}_C = \bar{V}_{C_f} - \bar{V}_{C_i} \quad \text{(vectorial difference)}$$

$$= 0.46 \text{ in./sec} \quad \text{(directed as shown)}$$

9. Determine $\bar{A}_C$. The magnitude of this vector is computed as

$$A_C = \frac{\Delta V_C}{\Delta T}$$

where ΔT, from Equation (10.5), is given by

$$\Delta T = \frac{\Delta\omega}{\alpha}$$

$$= \frac{2.03 - 1.97}{1}$$

$$= 0.06 \text{ sec}$$

Therefore,

$$A_C = \frac{0.46}{0.06}$$

$$= 7.7 \text{ in./sec}^2$$

Hence

$$\bar{A}_C = 7.7 \text{ in./sec}^2 \quad \text{(directed like } \Delta\bar{V}_C \text{ as shown)}$$

Despite the simplicity of the velocity-difference method, it should be understood that because of the need to measure small changes in displacements as well as velocities, a high degree of drawing precision is required if sufficiently accurate results are to be achieved. In some cases, this requirement can impose practical limitations such that the results obtained may not be reliable.

11

Graphical Calculus Method

11.1   GRAPHICAL DIFFERENTIATION

A common method for obtaining velocity and acceleration curves for a
mechanism consists of constructing a time-displacement curve and then by
graphical differentiation developing the velocity and acceleration curves.
Graphical differentiation consists of obtaining the slopes of various tangents
along one curve and plotting these values as ordinates to establish a second
curve. The method, as applied to the displacement and velocity curves, is
based on the following principles.

1.   The instantaneous velocity of a moving point can be represented
      graphically as the slope of the displacement curve at that instant, or

$$v = \frac{\Delta s}{\Delta t} \quad \text{where} \quad \Delta t \text{ is very small}$$

2.   The instantaneous acceleration can be represented as the slope of the
      velocity curve at that instant, or

$$a = \frac{\Delta v}{\Delta t} \quad \text{where} \quad \Delta t \text{ is very small}$$

   Graphical differentiation can be an effective tool for the designer,
particularly when used during the preliminary stages of a mechanism design
since it affords an overall picture of the velocity and acceleration throughout
the motion cycle. For example, from motion curves such as the s-t, v-t,
and a-t curves, one can readily determine:

   The maximum absolute values of displacement, velocity, and accel-
      eration

Where and when the maximum values of angular displacement occur
Whether there is any abrupt change in displacement, velocity, and
acceleration during the cycle

Also, there are many instances where it is necessary to differentiate a
curve for which an equation is difficult to obtain. In such instances, graphical differentiation is the most convenient approach.

Tangent Method

Consider the displacement-time curve shown in Figure 11.1. Suppose that
it is required to develop the velocity-time curve from this curve.
    Since the instantaneous velocity at any point is represented by the
slope of the displacement curve at that point, a tangent drawn through any
point of the curve defines the velocity at that point. Therefore, at point A,
the velocity v_A is obtained from

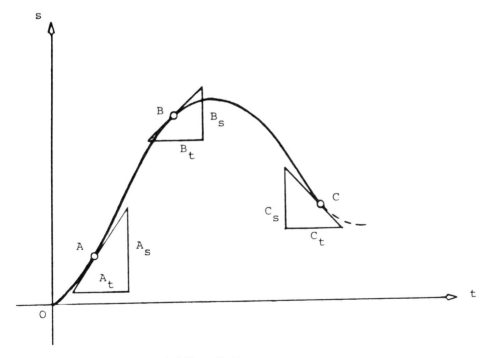

Figure 11.1   Graphical differentiation.

v_A = slope of tangent at A

$$= \frac{A_s k_s}{A_t k_t}$$

where

k_s = displacement scale, in./in.

k_t = time scale, sec/in.

Similarly,

$$v_B = \text{slope of tangent at B} = \frac{B_s(k_s)}{B_t(k_t)}$$

$$v_C = \text{slope of tangent at C} = -\frac{C_s(k_s)}{C_t(k_t)}$$

Here the negative sign indicates a negative slope of the curve and hence a negative velocity. Also, when the tangent to the curve is a horizontal line, this simply means that the velocity at the point of tangency is zero.

In addition, it should be clear that by making the time intervals A_t, B_t, C_t, and so on, equal, the velocities v_A, v_B, v_C, and so on, will be proportional to A_s, B_s, C_s, and so on. Hence, once the velocity v_A has been found, the velocities v_B, v_C, and so on, are readily obtained as follows:

$$v_B = \frac{B_s}{A_s} v_A$$

$$v_C = -\frac{C_s}{A_s} v_A$$

In this manner, the velocities of other points can be obtained to enable a smooth continuous curve to be drawn.

EXAMPLE 11.1

The curve shown in Figure 11.2a represents a typical displacement–time curve for a cam mechanism. It is desired to obtain the velocity–time curve from this displacement curve, and subsequently, the acceleration–time curve from the velocity curve.

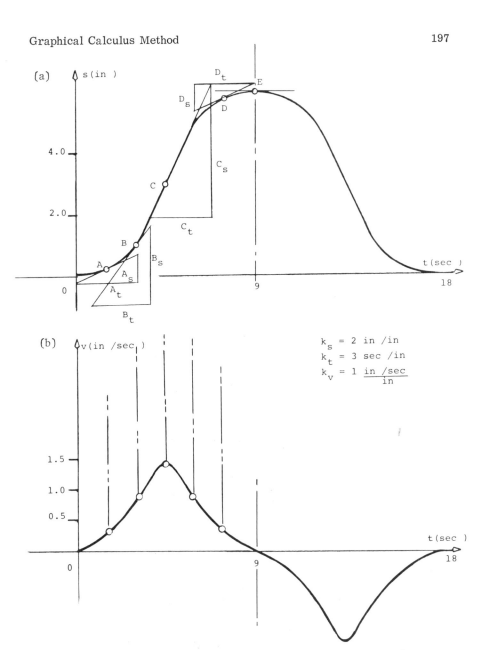

Figure 11.2   Graphical differentiation—tangent method: (a) displacement curve; (b) velocity curve.

SOLUTION

1. Draw tangents to various points on the given displacement-time curve: points A, B, C, D, and E.

2. Using the tangents as hypotenuses, construct right triangles at each point, making all the bases equal, that is,

$$A_t = B_t = C_t \quad \text{etc.}$$

3. Calculate the velocities at points A, B, C, D, and E as follows:

$$k_s = 2 \text{ in./in.}$$

$$k_t = 3 \text{ sec/in.}$$

$$k_v = 1 \text{ (in./sec)/in.}$$

$$v_A = \frac{A_s k_s}{A_t k_t} = \left(\frac{0.5}{1.0}\right)\frac{2}{3} = 0.33 \text{ in./sec}$$

$$v_B = \frac{B_s}{A_s} v_A = \left(\frac{1.3}{0.5}\right) 0.33 = 0.86 \text{ in./sec}$$

$$v_C = \frac{C_s}{A_s} v_A = \left(\frac{2.20}{0.5}\right) 0.33 = 1.45 \text{ in./sec}$$

$$v_D = \frac{D_s}{A_s} v_A = \left(\frac{0.5}{0.5}\right) 0.33 = 0.33 \text{ in./sec}$$

$$v_E = \frac{E_s}{A_b} v_A = \left(\frac{0}{0.5}\right) 0.33 = 0 \text{ in./sec}$$

Note that the remaining points on the velocity curve can be located by inspection based on the symmetrical shape of the displacement curve.

4. Draw the velocity-time axis using the same time scale as that used for the displacement-time axis (Figure 11.2b).

5. Plot the velocities calculated in step 3 on the velocity-time axis.

6. Draw a smooth curve connecting the plotted points.

7. To obtain the acceleration curve, repeat steps 1 through 6, replacing the displacement curve with the velocity curve just found (Figure 11.3a). The required acceleration curve is shown in Figure 11.3b.

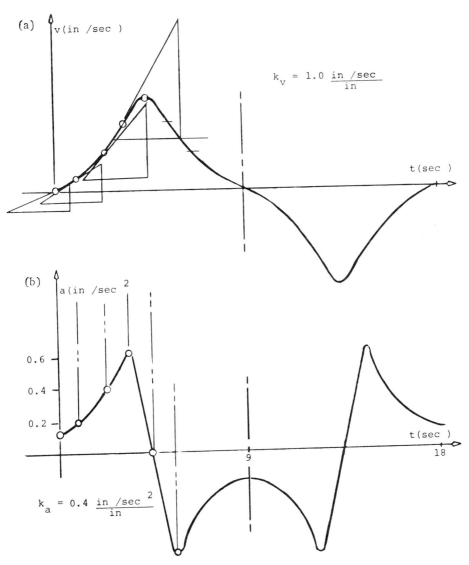

Figure 11.3   Graphical differentiation—tangent method: (a) velocity curve;
(b) acceleration curve.

Polar Method

An alternative graphical differentiation method commonly used is the <u>polar</u>
<u>method</u>. This method has been found to produce greater accuracy than the

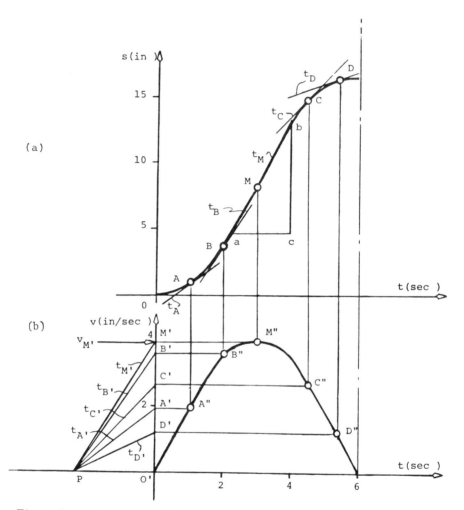

Figure 11.4   Graphical differentiation—polar method: (a) displacement curve; (b) velocity curve.

tangent method mainly because it requires less graphical construction. The method is best described by considering the displacement–time curve given in Figure 11.4a. Let it be required to develop the velocity and acceleration diagrams from the given curve.

PROCEDURE

1. Lay out the axes for the velocity-time graph directly below those for the displacement curve, maintaining the same time scale in both cases (see Figure 11.4b).

2. Through the point of steepest slope on the displacement curve (point M), draw a tangent line to define the point of maximum velocity (t_M). Now, since by definition, velocity equals slope, the maximum velocity at point M is given by

$$v_{max} = \frac{\Delta s}{\Delta t}$$

where Δs is the change of displacement over any interval Δt, or

$$v_{max} = \left(\frac{bc}{ac}\right)\frac{k_s}{k_t}$$

where

$\quad k_s \equiv$ displacement scale, in./in.

$\quad k_t \equiv$ time scale, sec/in.

3. Define a point P (called the pole) on the time axis of the velocity-time curve, at any convenient distance left of the origin, and from it draw line $t_{M'}$ parallel to tangent line t_M to intersect the velocity axis. This line locates a point M' on the velocity axis which represents the point of maximum velocity, or

$$v_{M'} = v_{max}$$

4. Plot the velocity of point M at the intersection of the horizontal line drawn from M' on the velocity curve and a vertical line drawn from M on the displacement curve. Label this point M".

5. Use the value obtained for v_{max} to establish the scale for the velocity axis of the velocity-time graph. To do this, compare the actual measurement on the diagram with the computed maximum velocity and then by ratio determine the velocity value that is represented by 1 in. on the diagram. In this example we find that

$$v_{max} = \left(\frac{bc}{ac}\right)\frac{k_s}{k_t}$$

where

$$k_s = 5 \text{ in./in.}$$

$$k_t = 2 \text{ sec/in.}$$

Therefore,

$$v_{max} = \left(\frac{1.6}{1.0}\right)\frac{5}{2} = 4 \text{ in./sec}$$

(a)

(b)

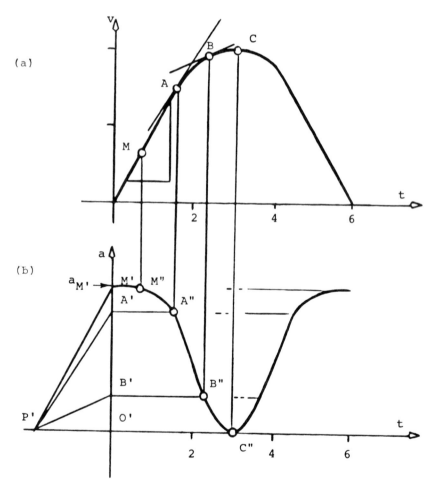

Figure 11.5   Graphical differentiation—polar method: (a) velocity curve; (b) acceleration curve.

Hence 2 in. on the velocity curve represents 4 in./sec, or 1 in. represents 2 in./sec, or

$$k_v = 2 \ (in./sec)/in.$$

6.   To establish the remaining velocity points at ordinates A, B, C,

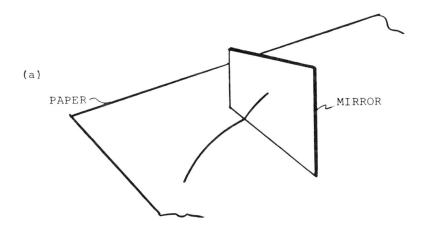

(a)

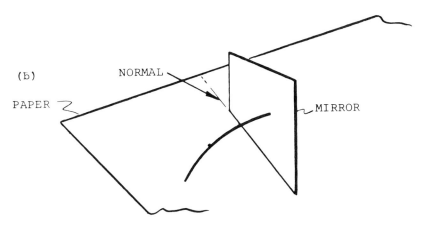

(b)

Figure 11.6   Use of a mirror in constructing tangents: (a) incorrect mirror position; (b) correct mirror position.

> D, and so on, follow the procedure outlined in steps 3 and 4.
> Figure 11.5a shows the completed velocity diagram.

7. After the velocity diagram has been completed, the acceleration
    diagram can be developed by following the procedure outlined
    above. Figure 11.5b shows the required curve.

Note on Accuracy

It should be emphasized that accuracy of the graphical differentiation methods depends primarily on:

1. How accurately the tangents can be drawn
2. The number of intervals chosen
3. The ability to fit a smooth curve to a given set of points

For greater accuracy in constructing tangents, any straight edge having a
reflective surface, such as the rectangular mirror shown in Figure 11.6,
may be used. The straight edge, held vertical to the paper, is placed across
the curve and rotated such that the visible curve and its reflection form a
continuous curve. In this position, the straight edge is then normal to the
curve. Therefore, a line drawn perpendicular to the normal and touching
the curve is the required tangent.

Accuracy is increased with the number of intervals into which the
curve is divided. The greater the number of intervals chosen, the greater
the accuracy that is achieved. Finally, the ability to fit a smooth curve to a
given set of points, particularly if double differentiation is required, greatly
minimizes the risk of compounding errors from one curve to the other.

11.2   GRAPHICAL INTEGRATION

Just as it is possible to go from the displacement curve to the velocity curve
and then to the acceleration curve using graphical differentiation, it is also
possible to reverse the process, going from acceleration to velocity and
then to displacement. The process by which this is achieved is called
graphical integration.

The process of integration can be thought of as the procedure for
obtaining the area under a given curve. As applied to the motion of a mechanism, the integration of the acceleration curve gives the velocity curve,
and the integration of the acceleration curve gives the displacement curve.
This is based on the motion laws, which state that

$$\Delta v = \bar{a} \, \Delta t \quad \text{or} \quad v_F - v_O = \bar{a}(t_F - t_O)$$

and

$$\Delta s = \bar{v} \, \Delta t \quad \text{or} \quad s_F - s_O = \bar{v}(t_F - t_O)$$

where the subscripts F and O denote final and original conditions, and $\bar{a}$ and $\bar{v}$ denote average acceleration and velocity. Hence the following rules apply.

1. To derive the velocity-time curve from the acceleration-time curve, the change in velocity between any two times equals the the area under the acceleration-time curve between the same two times.
2. To derive the displacement-time curve from the velocity-time curve, the change in displacement between any two times equals the area under the velocity-time curve between the same two times.

Mid-ordinate Method

There are several graphical methods available to determine the area under the curve. One of the simplest is the mid-ordinate method. The following example will illustrate the procedure.

EXAMPLE 11.2

The curve shown in Figure 11.7a represents a typical acceleration curve for a cam mechanism. It is desired to obtain the velocity-time curve.

SOLUTION

1. Divide the given curve into an equal number of sections, S_1, S_2, S_3, and so on, as shown in Figure 11.7a, and construct mean ordinates $\bar{a}_1$, $\bar{a}_2$, $\bar{a}_3$, and so on, for each section to intersect the curve at 1, 2, 3, and so on.
2. Through points 1, 2, 3, and so on, draw horizontal lines extending between the boundaries of alternate sides of the curve. For example, in section S_1, triangles A and B are on alternate sides of the curve. From this it is easily seen that, provided that the slope of the curve within the section remains fairly constant, the alternate triangles are approximately equal. This means that in section S_1, the area of triangle A can be considered approximately equal to that of triangle B; similarly, in section S_2, the area of triangle C is equal to that of triangle D; and so on. Therefore, the area under the curve in section S_1 can be approximated by the rectangle $\bar{a}_1 \, \Delta t_1$; the area under section S_2 can be approximated by the rectangle $\bar{a}_2 \, \Delta t_2$; and so on.
3. Since the velocity equals the area under the curve, the velocities v_1, v_2, and v_3 may be found as follows:

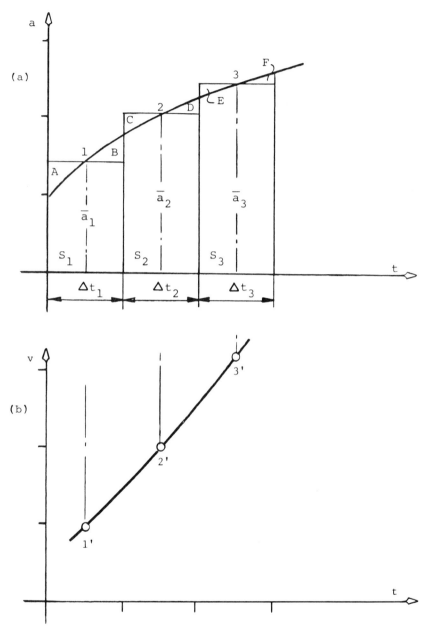

Figure 11.7   Graphical integration—mid-ordinate method: (a) acceleration curve; (b) velocity curve.

$$v_1 = \bar{a}_1 \, \Delta t_1$$

$$v_2 = v_1 + \bar{a}_2 \, \Delta t_2$$

$$v_3 = v_2 + \bar{a}_3 \, \Delta t_3 \quad \text{etc.}$$

For example, in Figure 11.7a,

$$v_1 = 1.4(1) = 1.4 \text{ in./sec}$$

$$v_2 = 1.4 + 2(1) = 3.4 \text{ in./sec}$$

$$v_3 = 3.4 + 2.4(1) = 5.8 \text{ in./sec}$$

The velocity curve is shown in Figure 11.7b.

4. After the required velocity curve has been completed, repeat steps 1 to 3 to develop the required displacement curve.

Polar Method

Another graphical integration method that is commonly used is the polar method, which is relatively fast and easy to apply. To illustrate the method, let us consider the acceleration curve shown in Figure 11.8a. From this curve we will develop the velocity curve and, in turn, use the velocity curve to develop the displacement curve.

PROCEDURE

1. Erect coordinates BB, CC, DD, and so on, to divide the curve into an equal number of line intervals, as shown in Figure 11.8a. In this case, five intervals have been chosen with ordinates $\bar{a}_1$ to $\bar{a}_5$ defined at the midpoints of the intervals. Also, since the horizontal axis is the time axis, each interval is defined at Δt_1, Δt_2, Δt_3, and so on. Note that all time scales for acceleration, velocity, and displacement curves must be the same, as with graphical differentiation.

2. From points 1, 2, 3, and so on, project horizontals to meet a line parallel to the a axis at points 1', 2', 3', and so on.

3. From a point P', located on the time axis, at any convenient distance left of the origin, draw connecting straight lines to points 1', 2', 3', and so on.

4. From the origin of the velocity axis (Figure 11.8b) draw line ob' parallel to P'1' to meet ordinate BB at b', line b'c' parallel to P'2' to meet ordinate CC at c', line c'd' parallel to P'3' to meet ordinate DD at d, and so on, until the complete velocity curve is obtained. From this velocity curve, the velocities at points 1, 2, 3, and so on, are given by

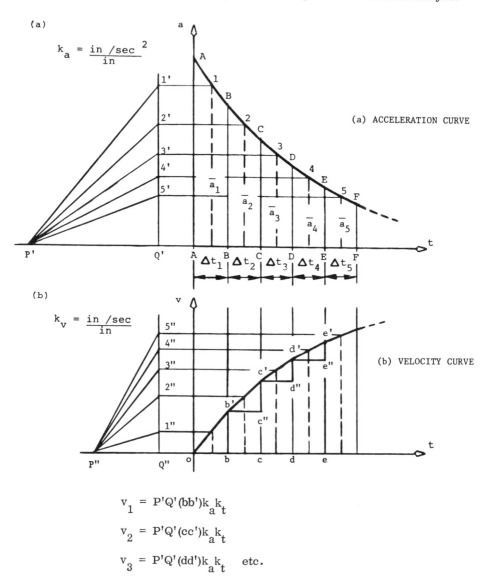

(a) ACCELERATION CURVE

(b) VELOCITY CURVE

$$v_1 = P'Q'(bb')k_a k_t$$

$$v_2 = P'Q'(cc')k_a k_t$$

$$v_3 = P'Q'(dd')k_a k_t \quad \text{etc.}$$

where b, c, d, and so on, are the intercepts of the first, second, third, and so on, ordinates with the time axis, and k_a [(in./sec^2)/ in.] and k_t (sec/in.) are the acceleration and time scales, respectively.

The relationships above can be verified by applying the properties of similar triangles to triangles P'Q'1' and obb',

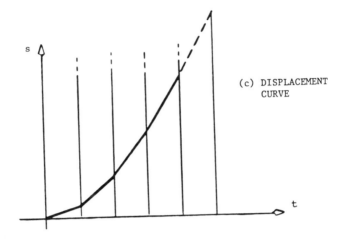

(c)

$$k_s = \frac{in}{in}$$

(c) DISPLACEMENT
    CURVE

Figure 11.8   Graphical integration—polar method: (a, facing page)
acceleration curve; (b, facing page) velocity curve; (c) displacement curve.

where bb' is the vertical leg of the velocity curve segment ob.
From these triangles, we obtain

$$\frac{Q'1'}{bb'} = \frac{P'Q'}{ob} \qquad (11.1)$$

But

$$Q'1' = \bar{a}_1$$

and

$$ob = \Delta t_1$$

Therefore, Equation (11.1) becomes

$$\frac{\bar{a}_1}{bb'} = \frac{P'Q'}{\Delta t_1}$$

from which follows

$$\bar{a}_1 \Delta t_1 = P'Q'(bb') \quad \text{(all parameters in inches)}$$

But

$$\bar{a}_1 \, \Delta t_1 (k_a k_t) = v_1 - v_0$$

where

$$v_0 = 0$$

Therefore,

$$v_1 = P'Q'(bb')k_a k_t$$

Similarly, from triangles P'Q'2' and b'c''c', where c''c' is the vertical leg of the velocity curve segment b'c'', we obtain

$$\frac{Q'2'}{c''c'} = \frac{P'Q'}{b'c''}$$

from which follow

$$\bar{a}_2 \, \Delta t_2 = P'Q'(c''c')$$

$$v_2 - v_1 = P'Q'(c''c')k_a k_t$$

$$v_2 = v_1 + P'Q'(c''c')k_a k_t$$

$$v_2 = P'Q'(bb')k_a k_t + P'Q'(c''c')k_a k_t$$

$$= P'Q'(bb' + c''c')k_a k_t$$

$$= P'Q'(cc')k_a k_t$$

and from triangles P'Q'3' and c'd''d',

$$\frac{Q'3'}{d''d'} = \frac{P'Q'}{c'd''}$$

from which follow

$$\bar{a}_3 \, \Delta t_3 = P'Q'(d''d')$$

$$v_3 - v_2 = P'Q'(d''d')k_a k_t$$

$$v_3 = v_2 + P'Q'(d''d')k_a k_t$$

$$v_3 = P'Q'(cc')k_a k_t + P'Q'(d''d')k_a k_t$$

$$= P'Q'(cc' + d''d')k_a k_t$$

$$= P'Q'(dd')k_a k_t$$

To obtain the displacement diagram, repeat steps 1 to 4, replacing the acceleration curve with the velocity curve. Figure 11.8c shows the required displacement diagram.

Note on Accuracy

As in the case of the graphical differentiation methods, accuracy of the graphical integration method increases with the number or size of the intervals chosen. The smaller the interval chosen, the more likely the slope of the curve within that interval (or section) will remain constant, and hence the closer the approximation of equality between the alternate triangles. It is therefore desirable to make the intervals or sections as small as possible for greater accuracy. Also, if double differentiation is required, accuracy will depend on the ability to fit a smooth curve to a set of given points.

12

Special Methods

12.1 COMPLETE GRAPHICAL ANALYSIS METHOD

In the graphical methods for acceleration analysis presented so far, one of the required calculations to be performed was that used to determine the normal acceleration components for points in the mechanism. In this section we demonstrate a complete graphical method where, by proper choice of link, velocity, and acceleration scales, both the magnitude and direction of all normal acceleration components can be determined without the need for calculations.

To illustrate this method, let us consider link AB, shown in Figure 12.1a, which rotates at an angular velocity of ω rad/sec. First, we obtain the magnitude of the velocity of point B relative to A ($V_{B/A}$) by multiplying the length of the link by the angular velocity,

$$V_{B/A} = AB \times \omega_{AB} \tag{12.1}$$

We then lay out the vector perpendicular to the link as shown. Here it is useful to recall that the magnitude of $\bar{V}_{B/A}$ is related to that of $\bar{A}^N_{B/A}$ by the expression

$$A^N_{B/A} = \frac{V^2_{B/A}}{BA} \tag{12.2}$$

Rearranging this equation, we obtain the relationship

$$\frac{A^N_{B/A}}{V_{B/A}} = \frac{V_{B/A}}{BA} \tag{12.3}$$

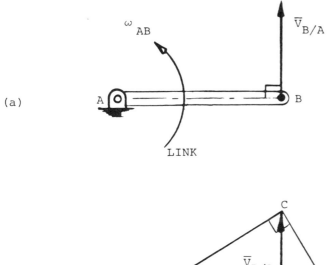

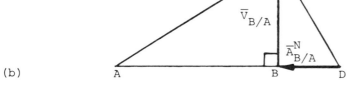

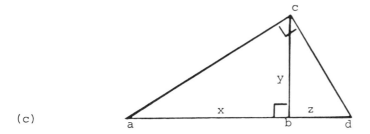

Figure 12.1   Normal acceleration construction.

which can be represented as

$$\frac{BD}{BC} = \frac{BC}{BA} \qquad (12.4)$$

using two similar triangles, ACB and CDB, as shown in Figure 12.1b, where

$$BD = A^N_{B/A}$$

and

$$BC = V_{B/A}$$

Here is can be seen that if the scales of link AB, velocity $V_{B/A}$, and acceleration $A^N_{B/A}$ in Equation (12.3) are properly chosen, in accordance with the relationship expressed in Equation (12.4), the length of line BD will accurately represent the required normal component.

Let the scales be defined as follows:

k_s = space scale, ft/in.

k_v = velocity scale, (ft/sec)/in.

k_a = acceleration scale, (ft/sec^2)/in.

where the inch units in the denominators represent actual measurements of the drawing in inches. For example, if the space scale is given as k_s = 5 in./in., a link length of 5 in. will be represented by a line 1 in. long on the drawing. Similarly, if the velocity scale is given as k_v = 10 (ft/sec)/in., a velocity of 20 ft/sec will be represented by a line 2 in. long on the drawing; and for an acceleration scale of 50 (ft/sec^2)/in. an acceleration of 150 ft/sec^2 will be represented by a line 3 in. long.

The required relationship of the scales is obtained by considering triangle abc in Figure 12.1c, where sides x, y, and z are proportional, respectively, to sides AB, BC, and BD of triangle ABC in Figure 12.1b. In other words, if x represents the link length, y the velocity, and z the acceleration, we can write

$$xk_s = BA$$

or

$$x = \frac{BA}{k_s} \tag{12.5}$$

$$yk_v = BC$$

or

$$y = \frac{BC}{k_v} \tag{12.6}$$

$$zk_a = BD$$

or

$$z = \frac{BD}{k_a} \tag{12.7}$$

and from similarity,

$$\frac{z}{y} = \frac{y}{x} \tag{12.8}$$

Substitution of Equations (12.5) to (12.7) into Equation (12.8) yields

$$\frac{BD/k_a}{BC/k_v} = \frac{BC/k_v}{BA/k_s} \tag{12.9}$$

or

$$\frac{A_{B/A}^N/k_a}{V_{B/A}/k_v} = \frac{V_{B/A}/k_v}{BA/k_s} \tag{12.10}$$

from which

$$A_{B/A}^N = \frac{V_{B/A}^2}{BA} \frac{k_a k_s}{k_v^2} \tag{12.11}$$

Thus the scales must be related by the equation

$$\frac{k_a k_s}{k_v^2} = 1 \tag{12.12}$$

or

$$k_a k_s = k_v^2 \tag{12.13}$$

This means that any two scales may be chosen arbitrarily, but the third must be chosen from Equation (12.12).

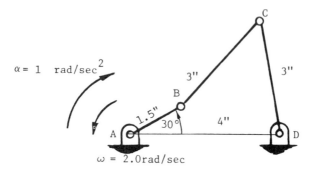

Figure 12.2   Four-bar mechanism.

EXAMPLE 12.1

For the four-bar mechanism ABCD shown in Figure 12.2, make a complete
graphical acceleration analysis given that crank AB is rotating with an
angular velocity of 2 rad/sec (counterclockwise) and an angular acceleration
of 1 rad/sec² (clockwise).

SOLUTION

1.   Select a space scale k_S = 2 in./in. or 1 in. (space scale) = 2 in.
     and velocity scale k_V = 2 (in./sec)/in. or 1 in. (velocity scale) =
     2 in./sec, and from the relationship

$$\frac{k_S k_a}{k_V^2} = 1$$

     obtain

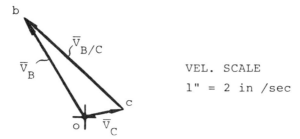

VEL. SCALE

1" = 2 in /sec

Figure 12.3   Velocity polygon.

$$k_a = \frac{(2/1)^2}{2/1} = 2 \ (\text{in.}/\text{sec}^2)/\text{in.}$$

or 1 in. (acceleration scale) = 2 in./sec^2.

2. Construct the velocity polygon (Figure 12.3) using the k_v scale. This requires the calculation of the velocity magnitude $V_{B/A}$ as a first step, then the layout of the vectors in accordance with the velocity polygon procedure already discussed.

3. Transfer vectors $\bar{V}_B$, $\bar{V}_C$, and $\bar{V}_{B/C}$ from the velocity polygon to points B and C on the linkage, maintaining the same orientation. $\bar{V}_B$ is drawn from B perpendicular to AB; $\bar{V}_C$ is drawn from C perpendicular to CD; and $\bar{V}_{B/C}$ is drawn from B perpendicular to BC.

4. Determine $\bar{A}_B^N$, $\bar{A}_C^N$, and $\bar{A}_{B/C}^N$ using the construction outlined above. For example, in Figure 12.4, $\bar{A}_B^N$ is obtained by construct-a right-angle triangle Abx in which $\angle$ b is 90°, Ax is the extended link AB, and the line Bb, perpendicular to AB, is the velocity magnitude of point B. The line segment Bx then represents the required value of $\bar{A}_B^N$.

5. Construct the acceleration polygon using the k_a scale (1 in. = 2 in./sec^2), starting with the calculation of A_B^T, then following the acceleration polygon procedure already discussed. Figure

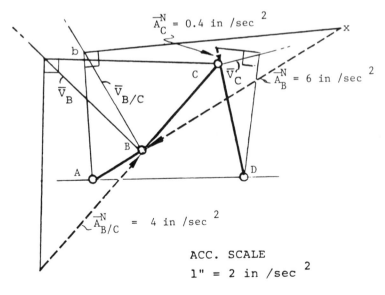

Figure 12.4   Normal acceleration construction

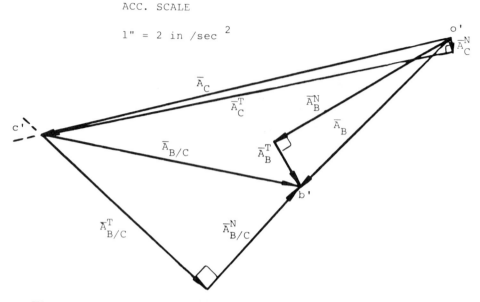

Figure 12.5   Acceleration polygon.

12.5 shows the acceleration polygon, from which the required acceleration $\bar{A}_C$ is determined to be

$$\bar{A}_C = 12.2 \text{ in./sec}^2 \quad \text{(as directed)}$$

In the example above, it is to be noted that only two calculations were necessary to complete the analysis after the scales were determined. These were to determine V_B and A_B^T, the velocity and acceleration magnitudes of the first link. If this link were to rotate with a constant angular velocity, calculation of A_B^T would not have been necessary, since the value of this acceleration would be zero. Thus, for a complete graphical analysis, the maximum number of calculations necessary is two.

12.2   EQUIVALENT LINKAGE METHOD

Determining the acceleration of points on many higher-paired mechanisms, such as those with rolling and sliding contacts, can become rather involved if a point-to-point analysis is attempted. This is because of the need to know the curvature of the path traced by a point on one link relative to the

MECHANISM   EQUIVALENT LINKAGE

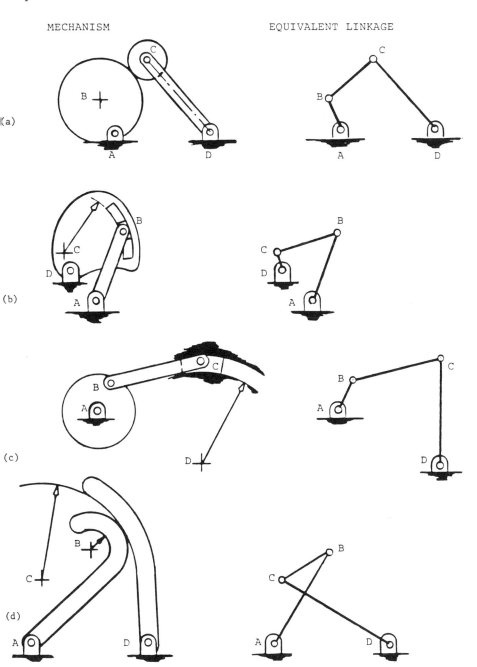

Figure 12.6   Kinematically equivalent four-bar linkages.

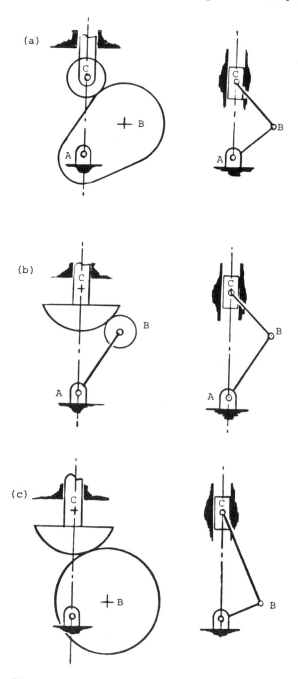

Figure 12.7   Kinematically equivalent four-bar linkages.

other and to apply the Coriolis law. If no easily recognized path is found, the analysis can be difficult.

To simplify this problem, the use of equivalent linkages has been found most effective. In application, an equivalent linkage replaces a higher-paired contact with appropriate lower pairs that will produce the correct values of velocities and accelerations for the instantaneous phase under consideration. An equivalent linkage may then be defined as one that produces identical motion as the part being analyzed for a given position or phase.

Figures 12.6 and 12.7 show several mechanisms with their equivalent linkages depicted by dashed lines. Note that the rolling and sliding surfaces have been replaced by pin joints as part of a more simplified four-bar linkage. Note also that in each case, the floating link of the equivalent linkage is drawn along the common normal of the two contacting surfaces and connects the centers of curvature of the surfaces.

Although an equivalent linkage is generally valid only for a given instant or phase and does not ordinarily apply to a complete cycle, there are some instances where the equivalent linkages of some higher-paired mechanisms will duplicate the input/output motion of those mechanisms throughout their motion cycle. Some examples are shown in Figures 12.6 and 12.7.

EXAMPLE 12.2

Consider the cam mechanism shown in Figure 12.8a. The cam (2) rotates counterclockwise at a constant angular velocity of 2 rad/sec. Find the acceleration of the follower (4) using the equivalent linkage method.

SOLUTION

The equivalent mechanism for the cam mechanism given is the simple slider-crank ABC shown in Figure 12.8b, for which the velocity and acceleration diagrams are readily obtained, as shown in Figure 12.8c and d. Applying the velocity polygon construction procedure, we obtain

$$V_B = 0.9(2) = 1.8 \text{ in./sec}$$

$$V_C = 1.45 \text{ in./sec} \quad \text{(from the velocity polygon)}$$

$$V_{C/B} = 0.6 \text{ in./sec} \quad \text{(from the velocity polygon)}$$

Applying the acceleration construction procedure, we obtain

$$A_B^N = 0.9(2)^2 = 3.6 \text{ in./sec}^2$$

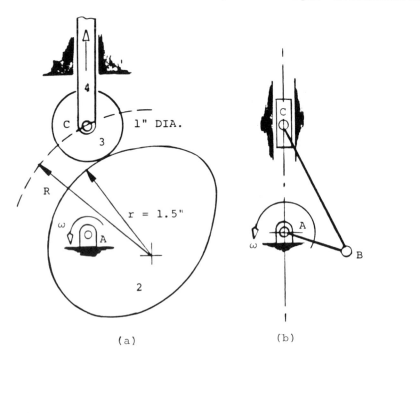

(a)   (b)

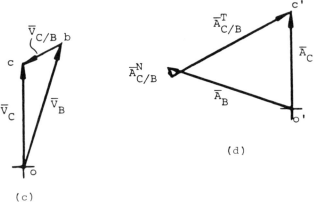

(c)

(d)

Fig. 12.8   Acceleration analysis of a cam-follower mechanism: (a) cam-follower mechanism; (b) equivalent linkage; (c) velocity polygon; (d) acceleration polygon.

$$A^T_B = 0$$

$$A^N_{C/B} = \frac{(0.6)^2}{2.0} = 0.18 \text{ in./sec}^2$$

$$A^T_{C/B} = 3.8 \text{ in./sec}^2 \quad \text{(from the acceleration polygon)}$$

$$\bar{A}_C = 2.8 \text{ in./sec}^2 \quad \text{(directed as shown)}$$

Note that the resulting acceleration of point C ($\bar{A}_C$) is exactly the same as that obtained for point P4 (the same point) in Example 12.1, using an alternative method.

12.3 SLIDER-CRANK ACCELERATION: PARALLELOGRAM METHOD

Despite the wide use of the relative acceleration method in solving linkage problems, it is not uncommon for one to experience some confusion in correctly applying this method to the slider-crank mechanism. The confusion most often encountered arises from uncertainties or oversight as to the proper location of the relative radial acceleration vector in the acceleration polygon. As a result, many errors are made.

To avoid such confusion or oversight, the graphical method presented here makes use of simple parallelogram constructions which serve as guides in laying out the vectors. The method is not only simple to apply, but also saves time.

Scope

In a typical slider-crank mechanism, as shown in Figure 12.9a, crank AB has angular rotation ω and angular deceleration α about point A. Angle θ_O is the instantaneous angular position of the crank with respect to the line of action AC, and θ_C is the angle between the slider arm BC and line AC. It is required to determine slider acceleration $\bar{A}_C$ at any angle θ_O.

Construction Development

1. Define o' as the point of zero acceleration coincident with pivot B on the mechanism, and draw vector o'b' (see the acceleration diagram in Figure 12.9c) to represent acceleration, $\bar{A}_B$, as determined from

$$\bar{A}_B = \bar{A}^N_B + \bar{A}^T_B \tag{12.14}$$

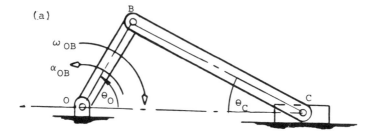

(a)

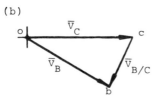

(b)

KEY POINT:

$\bar{A}^N_{B/C}$ (or $\bar{A}^N_{C/B}$ if used) is always located on the 'reflected' connecting rod.

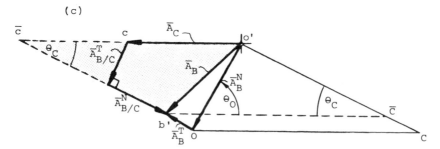

(c)

Figure 12.9   Graphical construction procedure: (a) slider–crank mechanism; (b) velocity polygon; (c) acceleration polygon.

where the magnitudes of $\bar{A}_B^N$ and $\bar{A}_B^T$ are given by

$$A_B^N = OB\omega^2$$

$$A_B^T = OB\alpha$$

2. With o'b' as the diagonal and θ_C as the subtended angle, construct a parallelogram o'$\bar{C}$b'$\bar{c}$ such that o'$\bar{C}$ is equal and parallel to $\bar{c}$b', and b'$\bar{C}$ is equal and parallel to o'$\bar{c}$. (Note the geometric identity between triangular sections o'b'$\bar{c}$ and o'b'$\bar{C}$. Vector o'b' may be considered the axis of asymmetry to the parallelogram.)

3. Determine $\bar{V}_{B/C}$ from the vectorial relationship

$$\bar{V}_{B/C} = \bar{V}_B - \bar{V}_C \tag{12.15}$$

where

$$\bar{V}_B = OB\omega \quad \text{(directed perpendicular to OB)} \tag{12.16}$$

and $\bar{V}_C$ is known in direction only (along the slider path). $\bar{V}_{B/C}$ can therefore be obtained from a velocity diagram, as follows. From a point o (see the velocity diagram in Figure 12.9b), draw line ob scaled to represent $\bar{V}_B$ perpendicular to the instantaneous position of crank OB and pointing in the direction of the motion. Then draw a line parallel to the direction of the slider motion to represent velocity $\bar{V}_C$. Finally, draw a line from point b perpendicular to arm BC so that it intersects the $\bar{V}_C$ line at point c. The velocity magnitude $V_{B/C}$ is determined by measuring line bc and converting to the appropriate velocity according to the chosen scale.

4. Compute the acceleration magnitude $A_{B/C}^N$ from

$$A_{B/C}^N = \frac{V_{B/C}^2}{BC} \tag{12.17}$$

Next, mark a segment of line b'$\bar{c}$ on the parallelogram to represent vector $\bar{A}_{B/C}^N$ scaled the same as o'b' and heading toward b'.

(This requirement is in keeping with the polygon convention, in which the acceleration vector relative to a point on a link must be be directed toward the corresponding point on the acceleration polygon. Alternatively, if $\bar{A}_{C/B}^N$ is considered, this vector must point from b' to c' on the polygon.)

5. From the tail of vector $\bar{A}^N_{B/C}$, draw a perpendicular line to inter-
    sect line o'c̄ at point c'. Point c' defines the acceleration, $\bar{A}_C$,
    along line o'c̄, and the perpendicular drawn from $\bar{A}^N_{B/C}$ represents
    the tangential acceleration, $\bar{A}^T_{B/C}$. The value of $\bar{A}_C$ may be
    checked with the vectorial relationship

$$\bar{A}_C = \bar{A}^N_B + \bar{A}^T_B - \bar{A}^N_{B/C} - \bar{A}^T_{B/C} \tag{12.18}$$

(Note that for a uniform velocity of crank OB, triangle o'b'c̄ is
identical to the configuration of the given mechanism OBC, in
which case $\bar{C}$ and C are coincident points. Note also that $A^N_{B/C}$ always lies on the "reflected" slider arm.)

As an example, consider a slider-crank mechanism where OB = 1.5 in.,
BC = 3 in., ω = 1 rad/sec, and α = 0. Determine the slider acceleration, $\bar{A}_C$,
for θ_O = 45° (Figure 12.10), 135° (Figure 12.11), 225° (Figure 12.12), and
315° (Figure 12.13).

PROCEDURE

1. From Equation (12.14), acceleration A_B = 1.5 in./sec and line
    o'b' is drawn 1.5 in. in length to represent this vector.

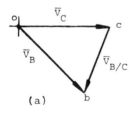

(a)

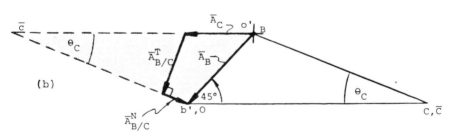

(b)

Figure 12.10   Crank angle at 45°: (a) velocity polygon; (b) acceleration
polygon.

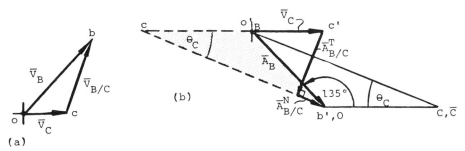

Figure 12.11   Crank angle at 135°: (a) velocity polygon; (b) acceleration polygon.

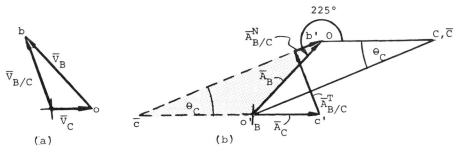

Figure 12.12   Crank angle at 225°: (a) velocity polygon; (b) acceleration polygon.

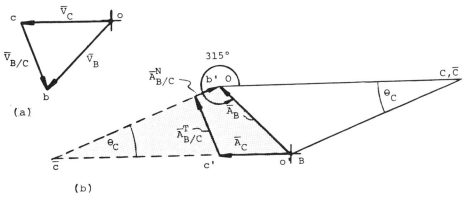

Figure 12.13   Crank angle at 315°: (a) velocity polygon; (b) acceleration polygon.

2.   Construct the parallelogram $o'\bar{C}b'\bar{c}$, using $o'b'$ as the diagonal.

3.   From Equation (12.16), $V_B = 1.5$ in./sec. With line ob drawn
     1.5 in. long to represent vector $\bar{V}_B$, a velocity polygon is con-
     structed from which $V_{B/C} = 1.12$ in./sec.

4.   From Equation (12.17), $A_{B/C}^N = 0.42$ in. and a 0.42-in. segment
     of line $b'\bar{c}$ is denoted on the parallelogram to represent $\bar{A}_{B/C}^N$.

5.   A perpendicular line is drawn from the tail of vector $\bar{A}_{B/C}^N$ to
     intercept line $o'\bar{c}$ at point c'. The length of line $o'c'$ is measured
     to be 1.05 in., so that $\bar{A}_C = 1.05$ in./sec^2 (to the left). Use of
     Equation (12.18) confirms this value as being accurate. Similarly,
     $\bar{A}_C$ for the crank position $\theta_O = 135°$, 225°, and 315° are found to
     be 1.05 in./sec^2 (to the right), 1.05 in./sec^2 (to the right), and
     1.05 in./sec^2 (to the left), respectively.

III
ANALYTICAL TECHNIQUES

Analytical techniques or mathematical methods for velocity and accelerations determination can be a powerful tool in the analysis and design of a mechanism. Compared to graphical techniques, analytical techniques offer two major advantages. They are faster and more accurate, provided that a valid expression has been derived for the mechanism. Graphical methods typically involve numerous diagrams that must be constructed for each shift in the position of the mechanism. The use of mathematics, on the other hand, allows general expressions for velocity and acceleration to be developed in terms of link geometry for all positions. Once a general expression is obtained for a mechanism, the velocity and acceleration values for points on that mechanism can readily be determined. This expression also reveals how various parameters such as lengths and angular positions of the links affect motion characteristics. This information is most important in the synthesis of a mechanism where a specific output motion is desired.

The main disadvantage of analytical techniques is that the mathematical analysis required to obtain general velocity and acceleration expressions for a mechanism are typically complex, lengthy, and error-prone. Conventional mathematical methods require both complex numbers and calculus, and generally, these methods are too theoretical to provide the quick insight the designer needs to develop a practical linkage.

Fortunately, the mathematical analysis can be simplified using the simplified vector method presented in Chapters 14 through 17. Unlike the conventional methods, this method does not rely on calculus, but combines trigonometry, complex numbers, and principles of relative motion to obtain required motion relationships. Plane trigonometry and complex numbers are used to express motion relationships in concise vectorial forms, and relative motion principles help to simplify equation development.

Besides the simplified vector method, two alternative mathematical methods are demonstrated in this part. These are the modified vector method, which is basically a variation of the simplified vector method, and the calculus method, which is conventional. In both cases it is clear that the application of calculus is essential.

13

Complex Algebra

13.1 INTRODUCTION

In mechanism analysis, a convenient way to describe the position of velocity or acceleration of a link is by the use of a complex number. A complex number consists of two parts, normally written in the form

$a + ib$

where a is the real part and b is the imaginary part, denoted by the letter i, which has a value of $\sqrt{-1}$.

Two complex numbers that differ only in the sign of their imaginary parts, such as a + ib and a - ib, are termed conjugates. Thus a - ib is the conjugate of a + ib; and conversely, a + ib is the conjugate of a - ib.

13.2 COMPLEX VECTOR OPERATIONS

Computations involving complex numbers follow the rules for algebra, with the additional requirement that all powers of i be reduced to the lowest terms by applying the following properties:

$i^2 = -1$

$i^3 = -i$

$i^4 = +1$

$i^5 = +i$ etc.

For example, if $\bar{r}_1 = a + ib$ and $\bar{r}_2 = c + id$, then

$$\bar{r}_1 + \bar{r}_2 = (a + ib) + (c + id)$$
$$= (a + c) + i(b + d)$$

$$\bar{r}_1 - \bar{r}_2 = (a + ib) - (c + id)$$
$$= (a + c) - i(b + d)$$

$$\bar{r}_1 \times \bar{r}_2 = (a + ib)(c + id)$$
$$= (ac - bd) + i(bc + ad)$$

$$\bar{r}_1/\bar{r}_2 = \frac{a + ib}{c + id}$$

$$= \frac{(a + ib)(c - id)}{(c + id)(c - id)}$$

$$= \frac{ac + bd}{c^2 + d^2} + \frac{i(bc - ad)}{c^2 + d^2} = \frac{(ac + bd) + i(bc - ad)}{c^2 + d^2}$$

Note that in the division case, the quotient is conveniently found by multiplying both numerator and denominator by the conjugate of the denominator.

13.3   GEOMETRIC REPRESENTATION
OF A COMPLEX VECTOR

A complex number is represented geometrically by a vector in the complex plane defined by two mutually perpendicular axes: the real and imaginary axes. For example, in Figure 13.1 the complex number $(a + ib)$ is shown to be represented by the vector $\bar{R}_1$, which extends from the origin o of the complex axes to a point P in the plane such that the horizontal component along the positive real axis represents the real part of the number, a, and the vertical component along the positive imaginary axis represents the imaginary part, b. Thus the magnitude of $\bar{R}_1$ is given by

$$R_1 = |\bar{R}_1| = \sqrt{a^2 + b^2}$$

and the direction is given by

$$\theta_1 = \tan^{-1} \frac{b}{a} \quad (0° < \theta_1 < 90°)$$

Note that the angle θ_1 is conventionally defined as the angular displacement of the vector from the positive real axis and is positive when measured counterclockwise.

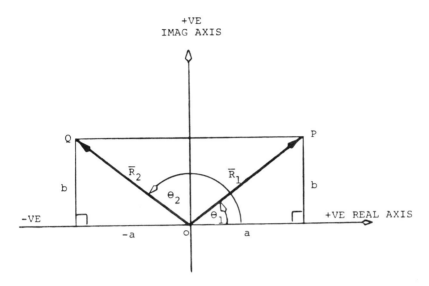

Figure 13.1   Graphical representation of complex numbers.

Similarly, the complex number $(-a + ib)$ is represented by a vector $\bar{R}_2$ as shown in Figure 13.1. Here the horizontal distance $-a$ along the negative real axis represents the real part of the number, and the vertical distance b along the positive imaginary axis represents the imaginary part. Therefore, the magnitude and direction of vector $\bar{R}$ are obtained as follows:

$$R_2 = |\bar{R}_2| = \sqrt{(-a)^2 + b^2}$$

$$\theta_2 = \tan^{-1} \frac{b}{-a} \quad (90° < \theta_2 < 180°)$$

Note that care should be taken in interpreting the value of θ_2 since in the range 0 to 360°, there are two values of θ_2 for each value of $\tan \theta_2$. To avoid ambiguity, it is usually advisable first to determine the quadrant in which the vector lies, using a simple sketch, before deciding which of the two angles applies.

In general, to represent a complex quantity, say $\bar{R} = a + ib$, as a vector:

1.   Find the absolute value of the components

$$|\bar{R}| = (a^2 + b^2)^{\frac{1}{2}}$$

2.   Find the direction (θ) from

$$\theta = \tan^{-1} \frac{\text{imaginary}}{\text{real}}$$

or

$$\theta = \tan^{-1} \frac{b}{a}$$

noting that:

If a is positive and b positive, θ is in first quadrant.
If a is negative and b positive, θ is in second quadrant.
If a is negative and b negative, θ is in third quadrant.
If a is positive and b negative, θ is in fourth quadrant.

13.4   COMPLEX FORMS

The vector $\bar{R}$ may be expressed in one of four complex forms: (1) rectangular, (2) trigonometric, (3) exponential, and (4) polar.
   In the rectangular form,

$$\bar{R} = a + ib$$

where, as before, a and b are, respectively, the real and imaginary components of $\bar{R}$.
   In the trigonometric form,

$$\bar{R} = R \cos \theta + iR \sin \theta$$

where $R \cos \theta = a$ and $R \sin \theta = b$, as shown in Figure 13.2. Alternatively,

$$\bar{R} = R (\cos \theta + i \sin \theta)$$

where $(\cos \theta + i \sin \theta)$ is the trigonometric form of the unit vector that defines the direction of $\bar{R}$.
   In the exponential form,

$$\bar{R} = Re^{i\theta}$$

where $e^{i\theta}$ is the exponential equivalent of the unit vector $(\cos \theta + i \sin \theta)$ given above.
   In the polar form,

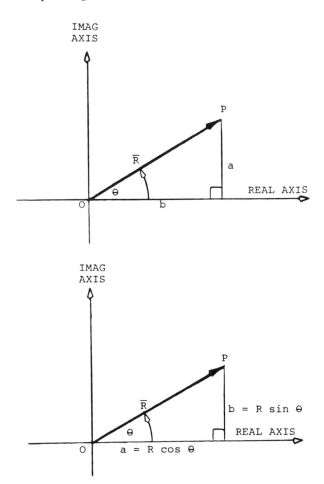

Figure 13.2   Graphical representation of the complex vector.

$$\bar{R} = R \angle \theta$$

where $\angle \theta$ defines the angular position of $\bar{R}$.

13.5   THE UNIT VECTOR

A unit vector is a vector that has a magnitude of unity. If $\bar{V}$ is a vector with magnitude $V \neq 0$, then $\bar{V}/V$ is a unit vector having the same direction of $\bar{V}$. Defining this unit vector as $\bar{V}$, we can write

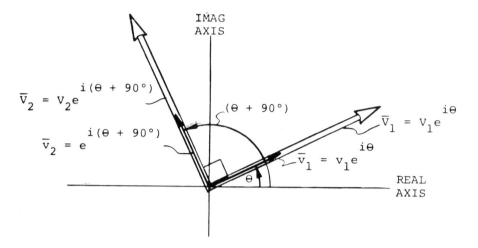

Figure 13.3   Unit vector.

$$\frac{\bar{V}}{V} = \bar{v}$$

or

$$\bar{V} = V\bar{v}$$

which means that any vector $\bar{V}$ can be represented by the magnitude of $\bar{V}$ multiplied by the unit vector $\bar{v}$ in the direction of $\bar{V}$.

The unit vector may be expressed in exponential form, such as $e^{i\theta}$ and $e^{i(\theta+90°)}$, where θ and $(\theta + 90°)$ are the position angles of the vectors, measured from the real axis (see Figure B.1 in Appendix B) and i is equal to $\sqrt{-1}$. However, for the purpose of computation, it is more convenient to convert these unit vectors to their equivalent trigonometric forms, as follows:

$$e^{i\theta} = \cos\theta + i\sin\theta$$

and

$$e^{i(\theta + 90°)} = \cos(\theta + 90°) + i\sin(\theta + 90°)$$

Thus, from Figure 13.3,

$$\bar{V}_1 = V_1 e^{i\theta}$$
$$\bar{V}_1 = V_1(\cos\theta + i\sin\theta)$$

and

$$\bar{V}_2 = V_2 e^{i(\theta + 90°)}$$

$$\bar{V}_2 = V_2[\cos(\theta + 90°) + i \sin(\theta + 90°)]$$

13.6 LINKAGE APPLICATION

In linkage analysis it is convenient to express vector quantities such as displacement velocity and acceleration as complex numbers which can be written in any of the complex forms discussed previously. Consider the case of a link AB that rotates counterclockwise with angular velocity ω and angular acceleration α. The displacement or change in position of the point B can be described by a position factor $\bar{R}$ directed from the origin of the complex axes to that point (Figure 13.4a). That is, the displacement of link AB is given by

$$\bar{R} = Re^{i\theta}$$

where R is the link length and $e^{i\theta}$ is the exponential form of the unit reactor which defines the instantaneous angular position of the link.

From previous velocity studies we know that the magnitude of velocity of point B can be obtained from the relationship

$$V_B = R\omega$$

and that the vector describing this velocity acts perpendicularly to the link in the same sense as θ. This velocity vector $\bar{V}_B$, shown in Figure 13.4b, can therefore be expressed in the complex form

$$\bar{V}_B = R\omega e^{i(\theta + 90°)}$$

where $e^{i(\theta + 90°)}$ is the unit vector used to indicate the direction of the velocity.

Similarly, the tangential, acceleration, shown in Figure 13.4c may be written as

$$\bar{A}_B^T = R\alpha e^{i(\theta + 90°)}$$

where $R\alpha$ is the magnitude of the acceleration and $e^{i(\theta + 90°)}$ is the unit vector defining the direction. The normal acceleration, shown in Figure 13.4d, may be written as

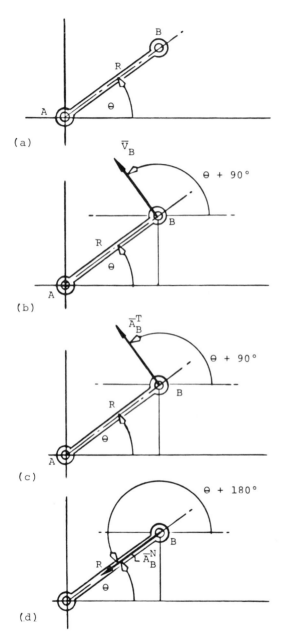

Figure 13.4   Graphical representations: (a) displacement; (b) velocity;
(c) tangential acceleration; (d) normal acceleration.

$$\bar{A}_B^N = R\omega^2 e^{i(\theta+180°)} = -R\omega^2 e^{i\theta}$$

where $R\omega^2$ is the magnitude of the acceleration and $-e^{i\theta}$ is the unit vector defining the direction.

EXAMPLE 13.1

Link AB of length R in Figure 13.5a rotates counterclockwise at an angular velocity ω.

   a. Develop a complex expression for the velocity V_B.
   b. Evaluate the expression found in part (a) given that R = 1.5 in.,
       ω = 1 rad/sec, and θ = 120°.

SOLUTION

Proceed as follows:

   1. Using point B as the intersection of the real and imaginary axes
       of the complex plane (Figure 13.5b), draw line Bo, perpendicular
       to AB, to represent $\bar{V}_B$.
   2. Drop a perpendicular from point b or terminus of $\bar{V}_B$ to intersect
       the imaginary axis at point o (Figure 13.5c).

From this construction we obtain the vector equation

$$\bar{V}_B = \bar{ob} + \bar{Bo} \quad \text{(vectorial summation)}$$

where

$$ob = -V_B \sin\theta \quad \text{(real)}$$

$$Bo = iV_B \cos\theta \quad \text{(imaginary)}$$

Therefore,

$$\bar{V}_B = -V_B \sin\theta + iV_B \cos\theta$$

$$= -V_B(\sin\theta - i\cos\theta)$$

$$= -R\omega(\sin\theta - i\cos\theta)$$

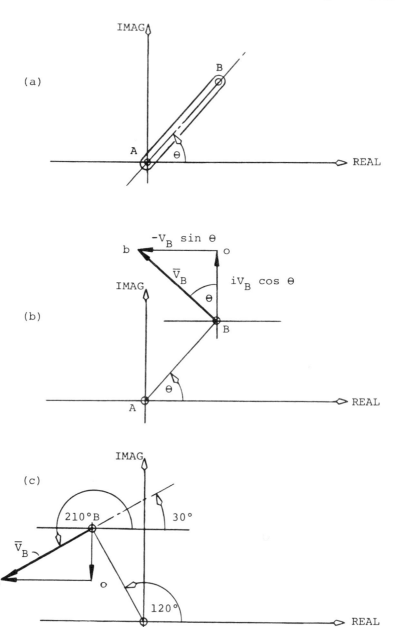

Figure 13.5   Velocity representation for Example 13.1.

$$\bar{V}_B = -R\omega(\sin\theta - i\cos\theta)$$

$$= -(1.5)(1)(\sin 120° - i\cos 120°)$$

$$= -1.5[0.866 - i(-0.5)]$$

$$= -1.5(0.866 + i0.5)$$

$$= -1.5\sqrt{(0.866)^2 + 0.5^2}$$

$$= -1.5(1) \text{ at } \tan^{-1}\frac{0.5}{0.866}$$

$$= -1.5 \text{ in./sec at } 30°$$

$$= 1.5 \text{ in./sec at } 210°$$

14

Four-Bar Mechanism Analysis: Simplified Vector Method

14.1 INTRODUCTION

The four-bar mechanism is often considered the most basic of all kinematic mechanisms. Consisting of four rigid links (AB, BC, CD, and AD) and four turning pairs (A, B, C, and D) as shown in Figure 14.1, this mechanism can be arranged in three basic configurations and, accordingly, is classified as follows:

> The <u>crank-rocker mechanism</u>, where drive crank AB is capable of complete rotation while the follower CD oscillates. Here the drive crank must be the shortest link.
>
> The <u>drag link mechanism</u>, where both drive crank AB and follower CD are capable of complete rotation. Here the frame AD must be the shortest link.

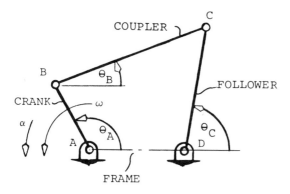

Figure 14.1   Four-bar mechanism.

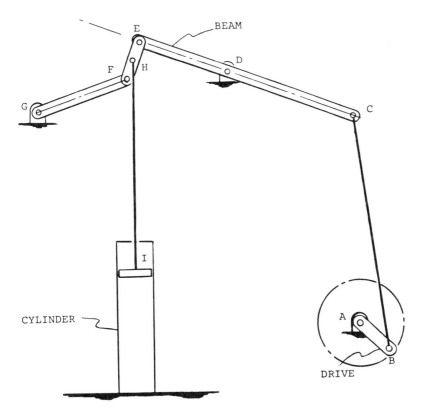

Figure 14.2   Beam pump mechanism.

The double rocker mechanism, where both drive crank AB and follower CD oscillate and neither is capable of complete rotation. Here the coupler BC must be the shortest link.

The four-bar mechanism is important in many ways:

1. It is the simplest possible plane linkage that can provide virtually any type of output motion.
2. Variations and combinations of this linkage make possible an almost limitless variety of mechanisms. Typical examples can be seen in some common applications shown in Figures 14.2 to 14.4.
3. Analysis of many complex direct contact mechanisms can be simplified by replacing the mechanism with an equivalent four-bar linkage.

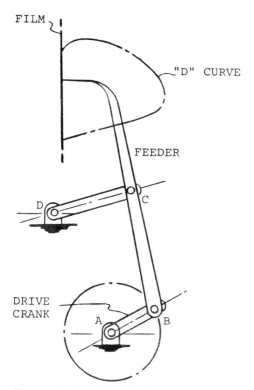

Figure 14.3   Film feeder mechanism.

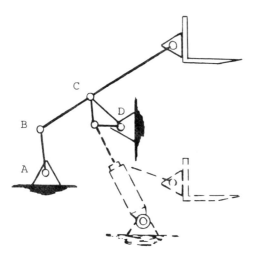

Figure 14.4   Forklift truck mechanism. (From A. S. Hall, 1961.)

For these reasons, the motion characteristics of the four-bar linkage are studied more than those of other kinematic mechanism.

The simplified mathematical method presented here quickly determines linear velocity and acceleration relationships of key points in the four-bar linkage for any given position of the drive crank during its motion cycle.

14.2   SCOPE AND ASSUMPTIONS

The mechanism ABCD in Figure 14.1 represents a typical four-bar linkage mechanism in which links AB, BC, CD, and AD have known lengths, and θ_A, θ_B, and θ_C are their respective angular positions measured from base link AD. Crank AB rotates with an angular velocity ω and an angular acceleration α about pivot point A.

In this analysis, equations for computing linear velocities and accelerations of the mechanism for a given instantaneous crank angle θ_A will be determined. Note that all angular displacements, velocities, and accelerations are considered positive for counterclockwise rotation and negative for clockwise rotation.

14.3   GEOMETRIC RELATIONSHIPS

First, we determine the angular relationships of θ_B and θ_C by constructing a diagonal BD, as in Figure 14.5, to form two triangles, ABD and BCD, where

$$ABD = \phi_B$$

$$ADB = \phi_D$$

$$CBD = \gamma_B$$

$$CDB = \gamma_D$$

Using basic trigonometric relationships, we can show that

$$\theta_B = \gamma_B - \phi_D \tag{14.1}$$

and

$$\theta_C = 180° - \phi_D - \gamma_D \tag{14.2}$$

where

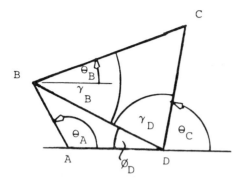

Figure 14.5   Geometric relationships.

$$\gamma_B = \cos^{-1} \frac{BC^2 + BD^2 - CD^2}{2(BD)(CD)}$$

$$\phi_D = \sin^{-1}\left(\frac{AB}{BD} \sin \theta_A\right)$$

$$\gamma_D = \cos^{-1} \frac{BD^2 + CD^2 - BC^2}{2(BD)(CD)}$$

$$BD = [AB^2 + AD^2 - 2(AB)(AD) \cos \theta_A]^{\frac{1}{2}}$$

14.4   VELOCITY ANALYSIS

To determine the velocities $\bar{V}_B$, $\bar{V}_C$, and $\bar{V}_{C/B}$, we construct the velocity polygon Bcb for the mechanism, as in Figure 14.6, letting:

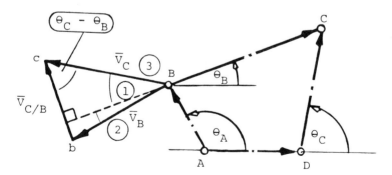

Figure 14.6   Velocity polygon.

Bb represent magnitude of the linear velocity of point B (V_B)
Bc represent magnitude of the linear velocity of point C (V_C)
bc represent magnitude of the relative velocity of point C with respect
to point B ($V_{C/B}$)

Then from this polygon, we determine the angular relationships of the velocities as follows:

$$③ = \theta_C + 90° - \theta_A$$

$$① = 180° + \theta_B - \theta_A - ③ = 180° + \theta_B - \theta_A - (\theta_C + 90° - \theta_A)$$

$$= 180° + \theta_B - \theta_A - \theta_C - 90° + \theta_A$$

$$= 90° + \theta_B - \theta_C$$

$$② = 90° - ① - ③ = 90° - (90° + \theta_B - \theta_C) - (\theta_C + 90° - \theta_A)$$

$$= 90° - 90° - \theta_B + \theta_C - \theta_C - 90° + \theta_A$$

$$= \theta_A - \theta_B - 90°$$

$$\angle c = 90° - ① = 90° - (90° + \theta_B - \theta_C)$$

$$= 90° - 90° - \theta_B + \theta_C$$

$$= \theta_C - \theta_B$$

$$\angle b = 90° - ② = 90° - (\theta_A - \theta_B - 90°)$$

$$= 90° - \theta_A + \theta_B + 90°$$

$$= 180° - \theta_A + \theta_B$$

$$\underline{\angle bBc} = ① + ② = (90° + \theta_B - \theta_C) + (\theta_A - \theta_B - 90°)$$

$$= 90° + \theta_B - \theta_C + \theta_A - \theta_B - 90°$$

$$= \theta_A - \theta_C$$

V_B, V_C, and $V_{C/B}$ can now be found by applying the rule of sines from
trigonometry.

$$\frac{V_B}{\sin(\theta_C - \theta_B)} = \frac{V_C}{\sin(180° - \theta_A + \theta_B)} = \frac{V_{C/B}}{\sin(\theta_A - \theta_C)} \qquad (14.3)$$

Noting that

$$\sin(180° - \theta_A + \theta_B) = \sin[180° - (\theta_A - \theta_B)]$$
$$= \sin(\theta_A - \theta_B)$$

Equation 14.3 can be written

$$\frac{V_B}{\sin(\theta_C + \theta_B)} = \frac{V_C}{\sin(\theta_A - \theta_B)} = \frac{V_{C/B}}{\sin(\theta_A - \theta_C)}$$

or

$$\frac{V_B}{\sin(\theta_B - \theta_C)} = \frac{V_C}{\sin(\theta_B - \theta_A)} = \frac{V_{C/B}}{\sin(\theta_C - \theta_A)}$$

Also, since

$$V_B = \omega AB \tag{14.4}$$

then

$$V_C = \omega AB \frac{\sin(\theta_B - \theta_A)}{\sin(\theta_B - \theta_C)} \tag{14.5}$$

and

$$V_{C/B} = \omega AB \frac{\sin(\theta_C - \theta_A)}{\sin(\theta_B - \theta_C)} \tag{14.6}$$

Since $\bar{V}_B$, $\bar{V}_C$, and $\bar{V}_{C/B}$ are oriented, respectively, at angles $(\theta_A + 90°)$, $(\theta_C + 90°)$, and $(\theta_B + 90°)$ from the real or reference axis, we can write the required vectorial expressions as follows:

$$\bar{V}_B = V_B e^{i(\theta_A + 90°)} \tag{14.7}$$

$$\bar{V}_C = V_C e^{i(\theta_C + 90°)} \tag{14.8}$$

$$\bar{V}_{C/B} = V_{C/B} e^{i(\theta_B + 90°)} \tag{14.9}$$

where $e^{i(\theta_A+90°)}$, $e^{i(\theta_C+90°)}$, and $e^{i(\theta_B+90°)}$ are unit vectors, used to define the direction of $\bar{V}_B$, $\bar{V}_C$, and $\bar{V}_{C/B}$, respectively. (Note that $e^{i\theta} = \cos\theta + i\sin\theta$, where θ is the position angle.)

14.5   ACCELERATION ANALYSIS

To find the linear acceleration of point C ($\bar{A}_C$), we apply the relative motion theory, which states that

$$\bar{A}_C = \bar{A}_B + \bar{A}_{C/B} \tag{14.10}$$

Expanding this equation into its normal and tangential component form, we have

$$\bar{A}_C^N + \bar{A}_C^T = \bar{A}_B^N + \bar{A}_B^T + \bar{A}_{C/B}^N + \bar{A}_{C/B}^T \tag{14.11}$$

where

$$\bar{A}_C^N = -\frac{V_C^2}{CD}e^{i\theta_C}$$

$$\bar{A}_C^T = A_C^T e^{i(\theta_C+90°)}$$

$$\bar{A}_B^N = -\omega^2 ABe^{i\theta_A}$$

$$\bar{A}_B^T = \alpha ABe^{i(\theta_A+90°)}$$

$$\bar{A}_{C/B}^N = -\frac{V_{C/B}^2}{BC}e^{i\theta_B}$$

$$\bar{A}_{C/B}^T = A_{C/B}^T e^{i(\theta_B+90°)}$$

or

$$-\frac{V_C^2}{CD}e^{i\theta_C} + A_C^T e^{i(\theta_C+90°)} = -\omega^2 ABe^{i\theta_A} + \alpha ABe^{i(\theta_A+90°)}$$

$$-\frac{V_{C/B}^2}{BC}e^{i\theta_B} + A_{C/B}^T e^{i(\theta_B+90°)} \tag{14.12}$$

Examination of this complex equation indicates that the only unknown quantities are the magnitudes A_C^T and $A_{C/B}^T$. All other quantities (magnitudes and directions) are either known or can readily be determined from the problem data. Note that the directions of $\bar{A}_C^T$ and $\bar{A}_{C/B}^T$, although not precisely known, are assumed positive (or counterclockwise) for convenience only. If, in reality, any of the directions is reversed, the numerical solution of the equation will automatically produce a negative sign for the unknown quantity.

To solve Equation (14.12), we equate the real and imaginary parts as follows:

Real:

$$-\frac{V_C^2}{CD}\cos\theta_C + A_C^T\cos(\theta_C + 90°) = -\omega^2 AB\cos\theta_A + \alpha AB\cos(\theta_A + 90°)$$

$$-\frac{V_{C/B}^2}{BC}\cos\theta_B + A_{C/B}^T\cos(\theta_B + 90°)$$

$$(14.13)$$

Imaginary:

$$-\frac{V_C^2}{CD}\sin\theta_C + a_C^T\sin(\theta_C + 90°) = -\omega^2 AB\sin\theta_A + \alpha AB\sin(\theta_A + 90°)$$

$$-\frac{V_{C/B}^2}{BC}\sin\theta_B + A_{C/B}^T\sin(\theta_B + 90°)$$

$$(14.14)$$

The solution of these simultaneous equations yields

$$A_C^T = \frac{C_1 B_2 - C_2 B_1}{A_1 B_2 - A_2 B_1} \tag{14.15}$$

$$A_{C/B}^T = \frac{A_1 C_2 - A_2 C_1}{A_1 B_2 - A_2 B_1} \tag{14.16}$$

where

$$A_1 = \cos(\theta_C + 90°) \tag{14.17}$$

$$A_2 = \sin(\theta_C + 90°) \tag{14.18}$$

$$B_1 = -\cos(\theta_B + 90°) \tag{14.19}$$

$$B_2 = -\sin(\theta_B + 90°) \tag{14.20}$$

$$C_1 = \frac{V_C^2}{CD} \cos \theta_C - \omega^2 AB \cos \theta_A + \alpha AB \cos (\theta_A + 90°) - \frac{V_{C/B}^2}{BC} \cos \theta_B$$

$$(14.21)$$

$$C_2 = \frac{V_C^2}{CD} \sin \theta_C - \omega^2 AB \sin \theta_A + \alpha AB \sin (\theta_A + 90°) - \frac{V_{C/B}^2}{BC} \sin \theta_B \quad (14.22)$$

Given that the values V_C, $V_{C/B}$, A_C^T, and $A_{C/B}^T$ have been determined, general equations for computing the linear accelerations are:

$$\bar{A}_B = -\omega^2 ABe^{i\theta_A} + \alpha ABe^{i(\theta_A + 90°)} \tag{14.23}$$

$$\bar{A}_C = -\frac{V_C^2}{CD} e^{i\theta_C} + A_C^T e^{i(\theta_C + 90°)} \tag{14.24}$$

$$\bar{A}_{C/B} = -\frac{V_{C/B}^2}{BC} e^{i\theta_B} + A_{C/B}^T e^{i(\theta_B + 90°)} \tag{14.25}$$

EXAMPLE 14.1: Crank Rocker

Considering the four-bar linkage in Figure 14.1, let AB = 1.5 in., BC = 3 in., CD = 3 in., AD = 4 in., $\theta_A = 30°$, $\omega = 2$ rad/sec (counterclockwise), and $\alpha = 1$ rad/sec². It is required to find $\bar{V}_C$, $\bar{V}_{C/B}$, $\bar{A}_C$, and $\bar{A}_{C/B}$.

SOLUTION

1. Determine θ_B and θ_C using Equations (14.1) and (14.2).

$$\theta_B = 62.14° - 15.51° = 46.63°$$

$$\theta_C = 180° - 15.51° - 62.14° = 102.35°$$

2. Determine V_B, V_C, and $V_{C/B}$ using Equations (14.4), (14.5), and (14.6).

$$V_B = 2(1.5) = 3.00 \text{ in./sec}$$

$$V_C = 2(1.5) \frac{\sin(46.6° - 30°)}{\sin(46.6° - 102.3°)} = -1.04 \text{ in./sec}$$

$$V_{C/B} = 2(1.5) \frac{\sin(102.3° - 30.0°)}{\sin(46.6° - 102.3°)} = -3.46 \text{ in./sec}$$

3. Determine $\bar{V}_B$, $\bar{V}_C$, and $\bar{V}_{C/B}$ using Equations (14.7), (14.8), and (14.9).

$$\bar{V}_B = 3.0e^{i(120°)} = 3.0(\cos 120° + i \sin 120°)$$
$$= 3.0 \text{ in./sec } \underline{/120°}$$

$$\bar{V}_C = -1.04e^{i(192.3°)} = -1.04(\cos 192.3° + i \sin 192.3°)$$
$$= 1.04 \text{ in./sec } \underline{/12.3°}$$

$$\bar{V}_{C/B} = -3.46e^{i(136.4°)} = -3.46(\cos 136.6° + i \sin 136.6°)$$
$$= 3.46 \text{ in./sec } \underline{/-43.4°}$$

4. Determine the constants A_1, A_2, B_1, B_2, C_1, and C_2 using Equations (14.17) through (14.22).

$$A_1 = \cos(102.35 + 90°)$$
$$= -0.98$$
$$A_2 = \sin(102.35 + 90°)$$
$$= -0.21$$

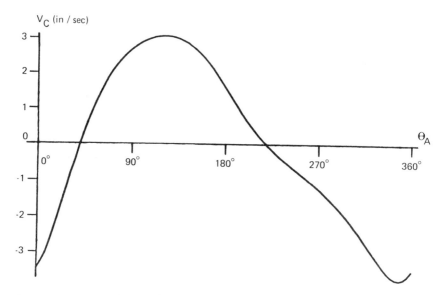

Figure 14.7   Velocity of point C versus crank angle.

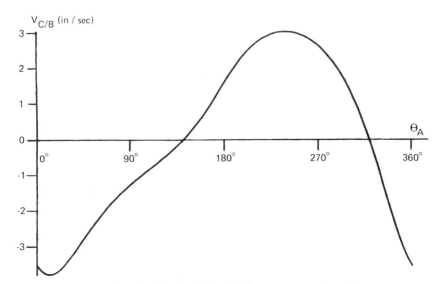

Figure 14.8   Velocity of C relative to B versus crank angle.

$$B_1 = -\cos(46.6° + 90°)$$
$$= 0.73$$
$$B_2 = -\sin(46.6° + 90°)$$
$$= -0.69$$
$$C_1 = -\frac{(1.04)^2}{3.0}\cos 102.35° - 2^2(1.5)\cos 30°$$
$$+ 1.0(1.5)\cos(30° + 90°) - \frac{(-3.46)^2}{3.0}\cos 46.6° = -8.76$$
$$C_2 = -\frac{1.04^2}{3.0}\sin 102.35° - 2^2(1.5)\sin 30°$$
$$+ 1.0(1.5)\sin(30° + 90°) - \frac{(-3.46)^2}{3.0}\sin 46.6° = -4.25$$

5.   Determine A_C^T and $A_{C/B}^T$ using Equations (14.15) and (14.16).

$$A_C^T = \frac{(-8.76)(0.69) - (-4.25)(0.73)}{(-0.98)(-0.69) - (-0.21)(0.73)}$$
$$= 11.02 \text{ in.}/\text{sec}^2$$

$$A_{C/B}^{T} = \frac{(-0.98)(-4.25) - (-0.21)(-8.76)}{(-0.98)(-0.69) - (-0.21)(0.73)}$$

$$= 2.76 \text{ in.}/\text{sec}^2$$

6. Substitute the values found for A_C^T and $A_{C/B}^T$ in Equations (14.24) and solve for $\bar{A}_C$ and $\bar{A}_{C/B}$.

$$\bar{A}_C = -\frac{(-1.04)^2}{3.0} e^{i(102.3°)} + 11.02 e^{i(192.3°)}$$

$$= 0.36 \, \underline{/-77.6°} + 11.02 \, \underline{/-167.6°}$$

$$= 11.03 \text{ in.}/\text{sec}^2 \, \underline{/-165.7°}$$

$$\bar{A}_{C/B} = -\frac{(-3.46)^2}{3.0} e^{i(46.63°)} + 2.76 e^{i(136.63°)}$$

$$= 3.99 \, \underline{/-133.4°} + 2.76 \, \underline{/136.6°}$$

$$= 4.85 \text{ in.}/\text{sec}^2 \, \underline{/-168.04°}$$

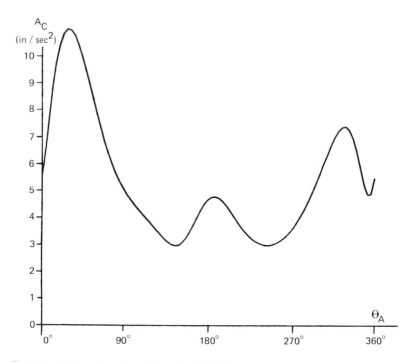

Figure 14.9   Acceleration of point C versus crank angle.

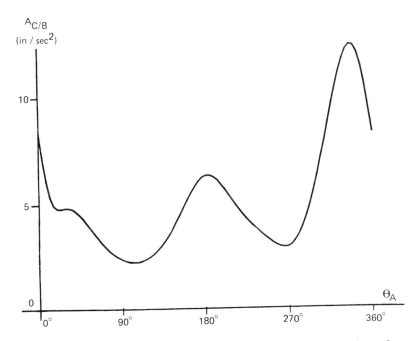

Figure 14.10   Acceleration of C relative to B versus crank angle.

See Figures 14.7 to 14.10 and Table 14.1 for velocity and acceleration profiles of the complete crank cycle (θ_A = 15 to 360°).

EXAMPLE 14.2: Drag Link

Considering the four-bar linkage in Figure 14.1, let AB = 2.0 in., BC = 1.5 in., CD = 2.5 in., AD = 1.0 in., θ_A = 240°, ω = 1 rad/sec (counter-clockwise), and α = 0.5 rad/sec^2. It is required to find $\bar{V}_C$, $\bar{V}_{C/B}$, $\bar{A}_C$, and $\bar{A}_{C/B}$.

SOLUTION

1.   Determine θ_B and θ_C using Equations (14.1) and (14.2).

$$\theta_B = 67.8° - (-40.9°) = 108.7°$$

$$\theta_C = 180° - (-40.9°) - 33.7° = 187.14°$$

TABLE 14.1   Velocity and Acceleration Profiles for Example Problem

AB = 1.50   BC = 3.00   CD = 3.00   AD = 3.00   OMEGA(A) = 4.00   ALPHA(A) = 2.00   = 1.00

θ(A)	θ(B)	θ(C)	V(B)	ARG	V(C)	ARG	V(CB)	ARG	A(B)	ARG	A(C)	ARG	A(CB)	ARG
0.	65.	115.	3.00	90.0	3.60	24.6	3.60	-24.6	6.18	166.0	5.55	-104.2	8.29	-56.0
15.	56.	107.	3.00	105.0	2.53	16.8	3.86	-34.1	6.18	-179.0	9.49	-150.2	5.05	-114.0
30.	47.	102.	3.00	120.0	1.04	12.3	3.46	-43.4	6.18	-164.0	11.03	-165.8	4.85	-168.0
45.	39.	102.	3.00	135.0	.37	-168.5	2.81	-51.2	6.18	-149.0	9.84	-168.2	4.48	164.9
60.	33.	104.	3.00	150.0	1.46	-166.1	2.20	-57.5	6.18	-134.0	7.84	-160.9	3.63	148.8
75.	28.	109.	3.00	165.0	2.23	-161.4	1.68	-62.3	6.18	-119.0	6.15	-145.8	2.85	137.0
90.	24.	115.	3.00	180.0	2.74	-155.2	1.26	-66.0	6.18	-104.0	5.04	-125.4	2.37	127.0
105.	21.	122.	3.00	-165.0	3.03	-147.9	.90	-68.6	6.18	-89.0	4.33	-102.7	2.23	118.3
120.	20.	130.	3.00	-150.0	3.15	-140.1	.55	-70.5	6.18	-74.0	3.73	-77.9	2.48	111.9
135.	19.	138.	3.00	-135.0	3.07	-132.3	.16	-71.4	6.18	-59.0	3.16	-46.9	3.17	108.8
150.	19.	145.	3.00	-120.0	2.80	-124.9	.32	108.8	6.18	-44.0	3.03	-4.4	4.31	109.3
165.	20.	151.	3.00	-105.0	2.30	-118.5	.93	110.4	6.18	-29.0	3.85	34.1	5.62	113.3
180.	24.	156.	3.00	-90.0	1.64	-113.6	1.64	113.6	6.18	-14.0	4.73	55.6	6.34	121.6
195.	29.	160.	3.00	-75.0	.93	-110.4	2.30	118.5	6.18	1.0	4.69	66.1	5.99	135.7
210.	35.	161.	3.00	-60.0	.32	-108.8	2.80	124.9	6.18	16.0	3.99	70.7	5.07	156.0
225.	42.	161.	3.00	-45.0	.16	71.4	3.07	132.3	6.18	31.0	3.33	71.2	4.23	-179.6
240.	50.	160.	3.00	-30.0	.55	70.5	3.15	140.1	6.18	46.0	3.02	68.6	3.59	-152.9
255.	58.	159.	3.00	-15.0	.90	68.6	3.03	147.9	6.18	61.0	3.12	63.7	3.07	-121.8
270.	65.	156.	3.00	.0	1.26	66.0	2.74	155.2	6.18	76.0	3.60	57.5	2.99	-81.6
285.	71.	152.	3.00	15.0	1.68	62.3	2.23	161.4	6.18	91.0	4.47	50.1	4.05	-42.8
300.	76.	147.	3.00	30.0	2.20	57.5	1.46	166.1	6.18	106.0	5.68	41.1	6.38	-20.3
315.	78.	141.	3.00	45.0	2.81	51.2	.37	168.5	6.18	121.0	6.96	28.9	9.47	-11.8
330.	78.	133.	3.00	60.0	3.46	43.4	1.04	-12.3	6.18	136.0	7.39	10.7	12.07	-14.0
345.	73.	124.	3.00	75.0	3.86	34.1	2.53	-16.8	6.18	151.0	5.79	-25.0	11.97	-27.1
360.	65.	115.	3.00	90.0	3.60	24.6	3.60	-24.6	6.18	166.0	5.55	-104.2	8.29	-56.0

2. Determine V_B, V_C, and $V_{C/B}$ using Equations (14.4), (14.5), and (14.6).

$$V_B = 1(2.0) = 2.0 \text{ in./sec}$$

$$V_C = 1(2.0) \frac{\sin(108.7° - 240°)}{\sin(108.7° - 187.14°)} = 1.53 \text{ in./sec}$$

$$V_{C/B} = 1(2.0) \frac{\sin(187.14° - 240°)}{\sin(108.7° - 187.14°)} = 1.63 \text{ in./sec}$$

3. Determine $\bar{V}_B$, $\bar{V}_C$, and $\bar{V}_{C/B}$ using Equations (14.7), (14.8), and (14.9).

$$\bar{V}_B = 2.0\,e^{i(330°)} = 2.0 \text{ in./sec } \underline{/-30°}$$

$$\bar{V}_C = 1.53\,e^{i(267.1°)} = 1.53 \text{ in./sec } \underline{/-82.9}$$

$$\bar{V}_{C/B} = 1.63\,e^{i(198.7°)} = 1.63 \text{ in./sec } \underline{/-161.3}$$

4. Determine the constants A_1, A_2, B_1, B_2, C_1, and C_2 using Equations (14.17) through (14.22).

$$A_1 = \cos(187.1° + 90°)$$
$$= 0.12$$

$$A_2 = \sin(187.1° + 90°)$$
$$= -0.99$$

$$B_1 = -\cos(108.7 + 90°)$$
$$= 0.95$$

$$B_2 = -\sin(108.7 + 90°)$$
$$= 0.32$$

$$C_1 = \frac{1.53^2}{2.5} \cos 187.1° + (1)^2(2.0) \cos 240°$$
$$+ 0.5(2.0) \cos(240° + 90°) - \frac{1.63^2}{1.5} \cos 108.7° = 1.49$$

$$C_2 = \frac{1.53^2}{2.5} \sin 187.1° + (1)^2 (2.0) \sin 240°$$

$$+ 0.5(2.0) \sin (240° + 90°) - \frac{1.63^2}{1.5} \sin 108.7° = -0.56$$

5. Determine A_C^T and $A_{C/B}^T$ using Equations (14.15) and (14.16).

$$A_C^T = \frac{(1.49)(0.32) - (-0.56)(0.95)}{(0.12)(0.32) - (-0.99)(0.95)}$$

$$= 1.03 \text{ in./sec}^2$$

$$A_{C/B}^T = \frac{(0.12)(-0.56) - (-0.99)(1.49)}{(0.12)(0.32) - (-0.99)(0.95)}$$

$$= 1.45 \text{ in./sec}^2$$

6. Substitute the values found for A_C^T and $A_{C/B}^T$ in Equations (14.24) and (14.25) and solve for $\bar{A}_C$ and $\bar{A}_{C/B}$.

$$\bar{A}_C = -\frac{1.53^2}{2.5} e^{i(187.14°)} + 1.03 e^{i(277.14°)}$$

$$= 0.94 \underline{/7.15°} + 1.03 \underline{/-82.9°}$$

$$= 1.39 \text{ in./sec}^2 \underline{/-40.4°}$$

$$\bar{A}_{C/B} = -\frac{163^2}{1.5} e^{i(108.7°)} + 1.45 e^{i(198.7°)}$$

$$= 1.76 \underline{/-71.3°} + 1.45 \underline{/-161.3°}$$

$$= 2.28 \text{ in./sec}^2 \underline{/-110.7°}$$

Note that the four-bar mechanism may be presented in a crossed-phase configuration as shown in Figure 14.11, where, with crank AB in the first quadrant and rotating counterclockwise, link BC crosses the fixed link AD. In this configuration, the oscillation of the follower is below the fixed link and the equations for θ_B and θ_C do not apply directly. However, it is useful to note that this mechanism is the mirror image of the one in Figure 14.1 when the crank AB is in the fourth quadrant and its rotation is in the clockwise direction. Hence the equations derived above are applicable to the velocity and acceleration analysis, provided that the reflected crank angle and reversed rotation sense are considered.

Alternatively, if we elect to analyze the mechanism in the crossed phase as presented, Equations (14.1) and (14.2) must be modified as follows:

(a)

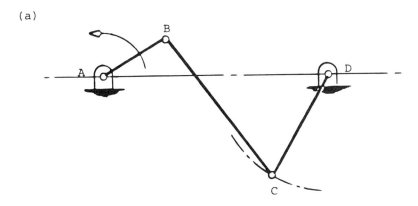

(b)

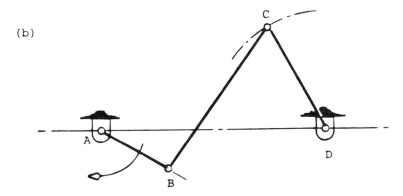

Figure 14.11   Four-bar mechanism: crossed-link configurations.

$$\theta_B = -\gamma_B - \phi_D$$

$$\theta_C = 180° + \gamma_D - \phi_D$$

All other relationships remain unchanged.

15

Slider-Crank Mechanism Analysis: Simplified Vector Method

15.1 INTRODUCTION

The slider-crank mechanism is probably the most common kinematic element to be found in most machines. A basic variation of the four-bar linkage, in which the follower crank is replaced by a sliding block, this mechanism is capable of converting rotary motion to linear motion, and vice versa.

The mechanism is generally found in three basic arrangements:

1. The <u>central</u> or <u>in-line type</u>, as in ABC of Figure 15.1, where the slider path of travel essentially passes through the crank pin or point A.

2. The <u>positive-offset type</u>, as in ABCD of Figure 15.2, where the slider path is offset a distance CD (or eccentricity e) above a reference line AD drawn parallel to the path and passing through point A.

3. The <u>negative-offset type</u>, as in ABCD of Figure 15.3, where the slider path is offset a distance CD below the reference line AD, drawn parallel to the path and passing through point A.

Among the countless applications of this mechanism are piston engines, pumps, compressors, saws, and other forms of reciprocating machinery. Some examples of these applications can be seen in Figures 15.4 to 15.7. Also, like the four-bar linkage, this mechanism is generally used in combination with other basic mechanisms to produce a wide variety of output motions as well as a model to simplify the analysis of more complex machines. For these reasons, the slider crank is one of the most frequently analyzed mechanisms. In this analysis, the linear velocity and acceleration relationships will be developed for any angular position of the drive crank.

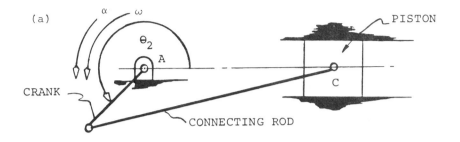

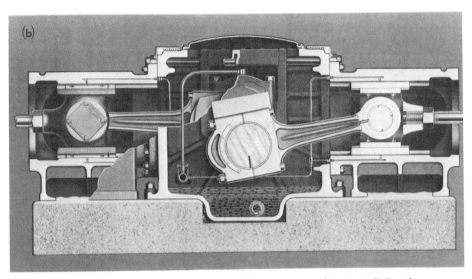

Figure 15.1   Slider-crank mechanism. (Courtesy of Ingersoll Rand, Woodcliff Lake, N.J.)

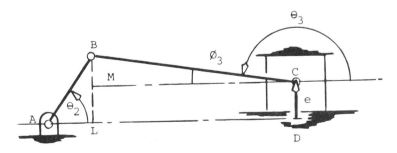

Figure 15.2   Slider crank with positive offset.

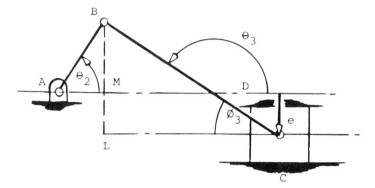

Figure 15.3   Slider crank with negative offset.

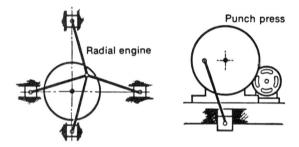

Figure 15.4   Typical slider-crank applications: machinery.

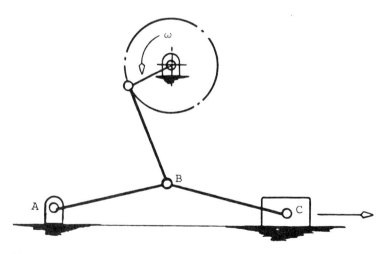

Figure 15.5   Toggle mechanism.

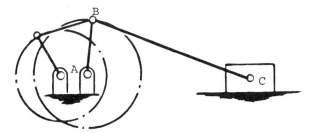

Figure 15.6   Drag-link quick-return mechanism.

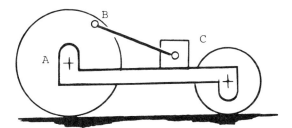

Figure 15.7   Steam locomotive.

15.2   SCOPE AND ASSUMPTIONS

The mechanism ABC in Figure 15.1 represents a typical slider-crank mechanism in which AB, the length of the drive crank 2, and BC, the length of the connecting rod 3, are known. AC is a variable distance depending on the angular position θ_2 of the crank during the motion cycle. Given that the crank rotates with an angular velocity ω_2 and an angular acceleration α_2 (both in the counterclockwise direction), we will now derive general expressions to compute linear velocities and accelerations of points A, B, C, and point B relative to point C for any angular position of the crank.

In this analysis, all angular displacements, velocities, and accelerations are considered positive for counterclockwise rotation and negative for clockwise rotation. All velocity vectors are positive with respect to the positions of their respective links.

15.3   GEOMETRIC RELATIONSHIPS

To begin the analysis, we first seek to determine the angle of the connecting rod θ_3 in terms of the crank angle θ_2. Considering the typical slider-crank mechanism represented by the triangle ABC in Figure 15.8, we note that

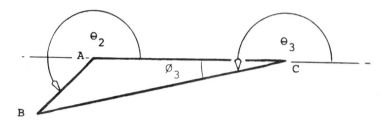

Figure 15.8   Geometric relationships.

since the link lengths AB and BC are known, we can apply geometric and trigonometric relationships to obtain

$$\theta_3 = 180° + \phi_3 \tag{15.1}$$

where

$$\phi_3 = \sin^{-1} \frac{AB}{BC} \sin(360° - \theta_2)$$

$$= -\sin^{-1} \frac{AB}{BC} \sin \theta_2$$

or

$$\theta_3 = 180° - \phi_3 \tag{15.2}$$

where

$$\phi_3 = \sin^{-1} \left(\frac{AB}{BC} \sin \theta_2 \right)$$

For the positive-offset slider crank, represented by diagram ABCD in Figure 15.2, we proceed by dropping a perpendicular from point B to meet AD at L, then extending the slider path to intersect the perpendicular BL. From this, we find the angle θ_3 as follows:

$$\theta_3 = 180° - \phi_3 \quad \text{[Equation (15.2)]}$$

$$BC \sin \phi_3 + e = AB \sin \theta_2 \tag{15.3}$$

$$\phi_3 = \sin^{-1} \left(\frac{AB}{BC} \sin \theta_2 - \frac{e}{BC} \right) \tag{15.4}$$

$$\theta_3 = 180° - \sin^{-1} \left(\frac{AB}{BC} \sin \theta_2 - \frac{e}{BC} \right) \quad (e > 0) \tag{15.5}$$

Similarly, for the negative-offset slider crank, represented by diagram ABCD in Figure 15.3, we obtain

$$\theta_3 = 180° - \sin^{-1}\left(\frac{AB}{BC} \sin \theta_2 + \frac{e}{BC}\right) \quad (e < 0) \tag{15.6}$$

Hence the general equation may be written as

$$\theta_3 = 180° - \sin^{-1}\left(\frac{AB}{BC} \sin \theta_2 - \frac{e}{BC}\right) \tag{15.7}$$

where e has a positive value for positive offset and a negative value for negative offset.

15.4   VELOCITY ANALYSIS

To determine the linear velocity relationships for $\bar{V}_B$, $\bar{V}_C$, and $\bar{V}_{B/C}$, let Bcb in Figure 15.9 represent the velocity polygon of the mechanism ABC, where:

Bb represents the magnitude of linear velocity of point B (V_B).
Bc represents the magnitude of linear velocity of point C (V_C).
bc represents the magnitude of the relative velocity of point B with respect to point C $(V_{B/C})$.

Then the internal angles of the polygon can be obtained as follows:

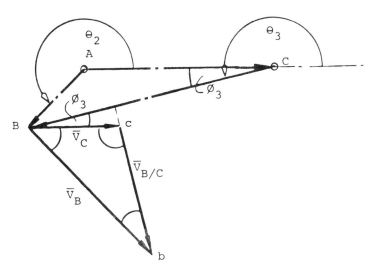

Figure 15.9   Slider-crank analysis.

$$\gamma = 360° - \theta_2 - 90°$$
$$= 270° - \theta_2$$

$$\angle b = 90° - \phi_3 - \gamma$$
$$= 90° - (\theta_3 - 180°) - (270° - \theta_2)$$
$$= 90° - \theta_3 + 180° - 270° + \theta_2$$
$$= \theta_2 - \theta_3$$

$$\angle bcB = 180° - \gamma - \angle b$$
$$= 180° - (270° - \theta_2) - (\theta_2 - \theta_3)$$
$$= 180° - 270° + \theta_2 - \theta_2 + \theta_3$$
$$= \theta_3 - 90°$$

By applying the rule of sines from trigonometry, V_B, V_C, and $V_{B/C}$ can be expressed in the form

$$\frac{V_B}{\sin(\theta_3 - 90°)} = \frac{V_C}{\sin(\theta_2 - \theta_3)} = \frac{V_{B/C}}{\sin(270° - \theta_2)}$$

which reduces to

$$\frac{V_B}{\sin(90° - \theta_3)} = \frac{V_C}{\sin(\theta_3 - \theta_2)} = \frac{V_{B/C}}{\sin(90° - \theta_2)}$$

Thus the scalar expressions for V_B, V_C, and $V_{B/C}$ are obtained as follows:

$$V_B = AB \times \omega_2 \tag{15.8}$$

$$V_C = V_B \frac{\sin(\theta_3 - \theta_2)}{\sin(90° - \theta_3)} \tag{15.9}$$

$$V_{B/C} = V_B \frac{\sin(90° - \theta_2)}{\sin(90° - \theta_3)} \tag{15.10}$$

To convert these equations to vectorial forms, we return to Figure 15.9 and note the following:

The directions of velocities $\bar{V}_B$ and $\bar{V}_{B/C}$ are assumed to be oriented at angles $(\theta_2 + 90°)$ and $(\theta_3 + 90°)$, respectively.
The direction of velocity $\bar{V}_C$ is known to act along a straight line only.

Therefore, we can write the vectorial expressions as follows:

$$\bar{V}_B = AB\omega_2 e^{i(\theta_2 + 90°)} \tag{15.11}$$

$$\bar{V}_C = AB\omega_2 \frac{\sin(\theta_3 - \theta_2)}{\sin(90° - \theta_3)} e^{i0°} \tag{15.12}$$

$$\bar{V}_{B/C} = AB\omega_2 \frac{\sin(90° - \theta_2)}{\sin(90° - \theta_3)} e^{i(\theta_3 + 90°)} \tag{15.13}$$

15.5 ACCELERATION ANALYSIS

To determine the linear acceleration relationships for point C ($\bar{A}_C$), we apply the relative motion equation, which states that

$$\bar{A}_B = \bar{A}_C + \bar{A}_{B/C} \quad \text{(vectorial sum)} \tag{15.14}$$

Expanding this equation into its normal and tangential form, we obtain

$$\bar{A}_B^N + \bar{A}_B^T = \bar{A}_C^N + \bar{A}_C^T + \bar{A}_{B/C}^N + \bar{A}_{B/C}^T \tag{15.15}$$

where

$$\bar{A}_B^N = -\omega_2^2 AB\, e^{i\theta_2}$$

$$\bar{A}_B^T = \alpha_2 AB\, e^{i(\theta_2 + 90°)}$$

$$\bar{A}_C^N = -\frac{V_C^2}{\infty} e^{i90°} = 0 \quad \text{(straight line)}$$

$$\bar{A}_C^T = A_C^T\, e^{i0°} \quad \text{(real)}$$

$$\bar{A}_{B/C}^N = -\frac{V_{B/C}^2}{BC} e^{i\theta_3}$$

$$\bar{A}_{B/C}^T = A_{B/C}^T\, e^{i(\theta_3 + 90°)}$$

Note that point C has no normal acceleration since the slider path is a straight line of infinite radius. Therefore, the absolute acceleration of C is the tangential acceleration. Thus Equation (15.15) becomes

$$-\omega_2^2 ABe^{i\theta_2} + \alpha_2 ABe^{i(\theta_2+90°)} = A_C^T - \frac{V_{B/C}^2}{BC}e^{i\theta_3} + A_{B/C}^T e^{i(\theta_3+90°)} \quad (15.16)$$

This equation contains two unknown quantities: the magnitudes A_C^T and $A_{B/C}^T$. All other quantities (magnitudes and directions) are either known or can readily be determined from problem data. Note that the directions of $\bar{A}_C^T$ and $\bar{A}_{B/C}^T$, although not known precisely, are assumed to be positive for convenience. If the actual direction of either is reversed, the numerical solution of the equation will automatically produce a negative sign for the unknown quantity. To solve the unknowns, therefore, we equate the real and imaginary parts of the equation and rearrange as follows:

Real:

$$A_C^T + A_{B/C}^T \cos(\theta_3 + 90°) = -\omega_2^2 AB \cos\theta_2 + \alpha_2 AB \cos(\theta_2 + 90°) + \frac{V_{B/C}^2}{BC}\cos\theta_3$$

$$(15.17)$$

Imaginary:

$$A_{B/C}^T \sin(\theta_3 + 90°) = -\omega_2^2 AB \sin\theta_2 + \alpha_2 AB \sin(\theta_2 + 90°) + \frac{V_{B/C}^2}{BC}\sin\theta_3$$

$$(15.18)$$

The solution of these simultaneous equations yields

$$A_C^T = \frac{C_1 B_2 - C_2 B_1}{A_1 B_2 - A_2 B_1} \quad (15.19)$$

$$A_{B/C}^T = \frac{A_1 C_2 - A_2 C_1}{A_1 B_2 - A_2 B_1} \quad (15.20)$$

where

$$A_1 = 1 \quad (15.21)$$

$$A_2 = 0 \quad (15.22)$$

$$B_1 = \cos(\theta_3 + 90°) \quad (15.23)$$

$$B_2 = \sin(\theta_3 + 90°) \quad (15.24)$$

$$C_1 = -\omega_2^2 AB \cos\theta_2 + \alpha_2 AB \cos(\theta_2 + 90°) + \frac{V_{B/C}^2}{BC}\cos\theta_3 \quad (15.25)$$

$$C_2 = -\omega_2^2 AB \sin\theta_2 + \alpha_2 AB \sin(\theta_2 + 90\ ) + \frac{V_{B/C}^2}{BC}\sin\theta_3 \quad (15.26)$$

Given that the values $V_{B/C}$, A_C^T, and $A_{B/C}^T$ have been found, the general equations for computing the linear accelerations can be summarized as follows:

$$\bar{A}_B = -\omega_2^2 AB\, e^{i\theta_2} + \alpha_2 AB\, e^{i(\theta_2 + 90°)} \tag{15.27}$$

$$\bar{A}_C = A_C^T\, e^{i0°} \quad \text{(real)} \tag{15.28}$$

$$\bar{A}_{B/C} = -\frac{V_{B/C}^2}{BC}\, e^{i\theta_3} + A_{B/C}^T\, e^{i(\theta_3 + 90°)} \tag{15.29}$$

EXAMPLE 15.1: Central Slider Crank

Considering the slider–crank mechanism in Figure 15.1, let AB = 1.5 in., BC = 3 in., $\theta_2 = 150°$, $\omega_2 = 1.0$ rad/sec (counterclockwise), and $\alpha_2 = 0$ rad/sec^2. It is required to find $\bar{V}_C$, $\bar{V}_{B/C}$, $\bar{A}_C$, and $\bar{A}_{B/C}$.

SOLUTION*

1. Determine θ_3 using Equations (15.2) and (15.1).

$$\phi_3 = \sin^{-1}\left(\frac{1.5}{3.0} \sin 150°\right)$$

$$= 14.5°$$

$$\theta_3 = 180° - 14.5°$$

$$= 165.5°$$

2. Determine V_B, V_C, and $V_{B/C}$ using Equations (15.8), (15.9), and (15.10).

$$V_B = 1.5(1.0) = 1.50 \text{ in./sec}$$

$$V_C = (1.5)\frac{\sin(165.5° - 150°)}{\sin(90° - 165.5°)} = -0.41 \text{ in./sec}$$

$$V_{B/C} = (1.5)\frac{\sin(90° - 150°)}{\sin(90° - 165.5°)} = 1.34 \text{ in./sec}$$

---

*See Figure 15.10 for velocity and acceleration profiles of complete crank cycle from $\theta_2 = 0°$ to $\theta_2 = 360°$.

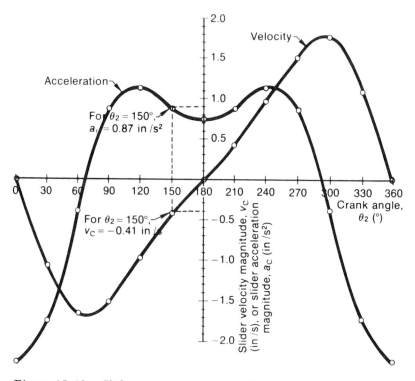

Figure 15.10   Slider motion characteristics.

3. Determine $\bar{V}_B$, $\bar{V}_C$, and $\bar{V}_{B/C}$ using Equations (15.11), (15.12), and (15.13).

$$\bar{V}_B = 1.5\,e^{i(240°)} = 1.5 \text{ in./sec } \underline{/-120°}$$

$$\bar{V}_C = -0.41\,e^{i0°} = 0.41 \text{ in./sec } \underline{/180°}$$

$$\bar{V}_{B/C} = 1.34\,e^{i(255.5°)} = 1.34 \text{ in./sec } \underline{/-104.5°}$$

4. Determine the constants A_1, A_2, B_1, B_2, C_1, and C_2 using Equations (15.21) through (15.26).

$A_1 = 1$ [from Equation (15.21)]

$A_2 = 0$ [from Equation (15.22)]

$$B_1 = \cos(165.5° + 90°)$$
$$= -0.25$$

$$B_2 = \sin(165.5° + 90°)$$
$$= -0.97$$

$$C_1 = -(1)^2(1.5)\cos 150° + 0(1.5)\cos(150° + 90°)$$
$$+ \frac{1.34^2}{3}\cos 165.5°$$
$$= -(-1.3) + 0 + (-0.58)$$
$$= 0.72$$

$$C_2 = -(1)^2(1.5)\sin 150° + 0(1.5)\sin(150° + 90°)$$
$$+ \frac{1.34^2}{3}\sin 165.5°$$
$$= -0.75 + 0 + 0.15$$
$$= -0.60$$

5.   Determine A_C^T and $A_{B/C}^T$ using Equations (15.19) and (15.20).

$$A_C^T = \frac{(0.72)(-0.97) - (-0.60)(-0.25)}{(1.00)(-0.97) - (0.00)(-0.25)}$$
$$= 0.87 \text{ in./sec}^2$$

$$A_{B/C}^T = \frac{(1.00)(-0.60) - (0.00)(0.72)}{(1.00)(-0.97) - (0.00)(-0.25)}$$
$$= 0.62 \text{ in./sec}^2$$

6.   Substitute the values found for A_C^T and $A_{B/C}^T$ into Equations (15.28) and (15.29) and solve for $\bar{A}_C$ and $\bar{A}_{B/C}$.

$$\bar{A}_C = 0.87\, e^{i0°}$$
$$= 0.87 \text{ in./sec}^2 \; \underline{/0°}$$

$$\bar{A}_{B/C} = -\frac{1.34^2}{3}\, e^{i(165.5°)} + 0.62\, e^{i(255.5°)}$$
$$= 0.60 \; \underline{/-14.5°} + 0.62 \; \underline{/-104.5°}$$
$$= 0.86 \text{ in./sec}^2 \; \underline{/-60.4°}$$

EXAMPLE 15.2: Offset Slider Crank

Considering the offset slider-crank mechanism in Figure 15.2, let
AB = 1.5 in., BC = 3 in., e = 0.5 in., $\theta_2 = 150°$, $\omega_2 = 1.0$ rad/sec
(counterclockwise), and $\alpha_2 = 0$ rad/sec^2. It is required to find $\bar{V}_C$, $\bar{V}_{B/C}$,
$\bar{A}_C$, and $\bar{A}_{B/C}$.

SOLUTION

1. Determine θ_3 using Equations (15.2) and (15.1).

$$\phi_3 = \sin^{-1}\left(\frac{1.5}{3.0}\sin 150° - \frac{0.5}{3.0}\right)$$

$$= 4.78°$$

$$\theta_3 = 180° - 4.78°$$

$$= 175.22°$$

2. Determine V_B, V_C, and $V_{B/C}$ using Equations (15.8), (15.9),
    and (15.10).

$$V_B = 1.5(1.0) = 1.50 \text{ in./sec}$$

$$V_C = (1.5)\frac{\sin(175.22° - 150°)}{\sin(90° - 175.22°)} = -0.64 \text{ in./sec}$$

$$V_{B/C} = (1.5)\frac{\sin(90° - 150°)}{\sin(90° - 175.22°)} = 1.30 \text{ in./sec}$$

3. Determine $\bar{V}_B$, $\bar{V}_C$, and $\bar{V}_{B/C}$ using Equations (15.11), (15.12),
    and (15.13).

$$\bar{V}_B = 1.5e^{i(240°)} = 1.5 \text{ in./sec } \underline{/-120°}$$

$$\bar{V}_C = -0.64\,e^{i0°} = 0.64 \text{ in./sec } \underline{/180°}$$

$$\bar{V}_{B/C} = 1.30\,e^{i(265.2°)} = 1.30 \text{ in./sec } \underline{/-94.78°}$$

4. Determine the constants A_1, A_2, B_1, B_2, C_1, and C_2 using
    Equations (15.21) through (15.26).

$$A_1 = 1 \quad \text{[from Equation (15.21)]}$$

$$A_2 = 0 \quad \text{[from Equation (15.22)]}$$

$B_1 = \cos(175.2° + 90°)$

$= -0.08$

$B_2 = \sin(175.2° + 90°)$

$= -0.99$

$C_1 = -(1^2)(1.5)\cos 150° + 0(1.5)\cos(150° + 90°)$

$+ \dfrac{1.30^2}{3}\cos(175.2°)$

$= -(-1.3) + 0 + (-0.56)$

$= 0.73$

$C_2 = -(1^2)(1.5)\sin 150° + 0(1.5)\sin(150° + 90°)$

$+ \dfrac{130^2}{3}\sin(175.2°)$

$= -(0.75) + 0 + 0.084$

$= -0.70$

5. Determine A_C^T and $A_{B/C}^T$ using Equations (15.19) and (15.20).

$A_C^T = \dfrac{(0.73)(-0.99) - (-0.70)(-0.08)}{(1.00)(-0.99) - (0.00)(-0.08)}$

$= 0.79 \text{ in./sec}^2$

$A_{B/C}^T = \dfrac{(1.00)(-0.70) - (0.00)(0.73)}{(1.00)(-0.99) - (0.00)(-0.08)}$

$= 0.70 \text{ in./sec}^2$

6. Substitute the values found for A_C^T and $A_{B/C}^T$ into Equations (15.28) and (15.29) and solve for $\bar{A}_C$ and $\bar{A}_{B/C}$.

$\bar{A}_C = 0.79\,e^{i0°}$

$= 0.79 \text{ in./sec}^2\ \underline{/0°}$

$\bar{A}_{B/C} = -\dfrac{(1.30)^2}{3.0}e^{i(175.2°)} + 0.70\,e^{i(265.2°)}$

$= 0.57\ \underline{/-4.78°} + 0.70\ \underline{/-94.8°}$

$= 0.90 \text{ in./sec}^2\ \underline{/-56.0°}$

16

Quick-Return Mechanism Analysis: Simplified Vector Method

16.1   INTRODUCTION

Quick-return mechanisms, typified by ABCD in Figure 16.1, are a special class of sliding contact linkages that provide uniform velocity motion followed by a fast return stroke to an initial position. These mechanisms, often used in equipment such as machine tools and production machinery to produce long, slow movements for cutting or feeding and fast return strokes in which no work is done, are generally found in two basic arrangements:

1. The crank-shaper (or oscillating-beam) type (Figure 16.2), where the crank arm AB is shorter than the base AD, and as a result, the follower CD oscillates as the crank arm makes a complete revolution.
2. The Whitworth type (Figure 16.3), where the crank arm AB is longer than the base AD, and as a result, both crank arm and follower make complete revolutions.

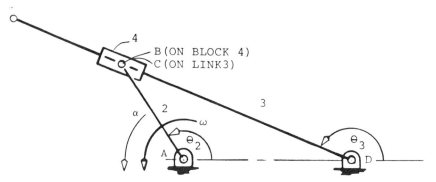

Figure 16.1   Quick-return mechanism.

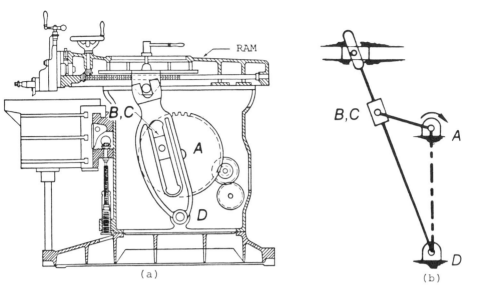

Figure 16.2   (a) Oscillating-beam shaper; (b) oscillating-beam linkage. (From Applied Kinematics by J. Harland Billings, © 1953 by D. Van Nostrand Company, Inc. Reprinted by permission of Wadsworth Publishing Company, Belmont, CA 94002.)

Quick-return mechanisms, like other basic mechanisms, can be used as models to simplify the analysis of more complex machines or they can be arranged in combination with other mechanisms to produce specific output motions.

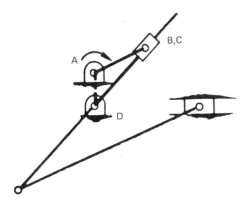

Figure 16.3   Whitworth linkage.

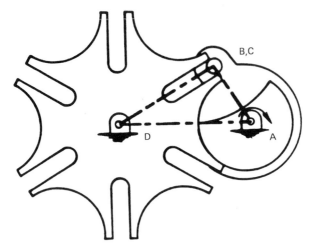

Figure 16.4   Geneva mechanism.

The Geneva mechanism shown in Figure 16.4 is a classical application of the quick-return mechanism.

Motion characteristics of quick-return mechanisms are usually difficult to analyze because of complexity of the combined linkage and slider movements which give rise to the Coriolis acceleration. The simplified mathematical method presented here quickly determines linear velocity and acceleration relationships for a quick-return mechanism for any given drive crank position in its motion cycle.

16.2   SCOPE AND ASSUMPTIONS

The mechanism ABCD in Figure 16.1 represents a typical quick-return mechanism in which the drive crank AB (or link 2) and the follower guide CD (or link 3) have known lengths. B and C are coincident points on the slider (link 4) and follower guide, respectively. AD (or link 1) is the distance between crank and follower pivot points. θ_2 is the angular position of link 2 measured from base link 1 or AD. The crank AB rotates with angular velocity ω_2 and angular acceleration α_2 (both counterclockwise).

In this analysis, equations for computing linear velocities and accelerations in terms of the variable angular position are to be determined. All angular displacements, velocities, and accelerations are considered positive for counterclockwise rotation and negative for clockwise rotation.

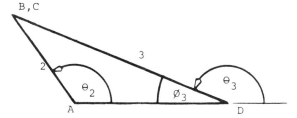

Figure 16.5   Geometric relationships.

16.3   GEOMETRIC RELATIONSHIPS

To determine θ_3, consider triangle ACD in Figure 16.5. Since AB, AD, and θ_2 are known, basic geometric and trigonometric relationships yield

$$\theta_3 = 180° - \phi_3 \tag{16.1}$$

where

$$\phi_3 = \cos^{-1} \frac{CD^2 + AD^2 - AB^2}{2(CD)(AD)} \quad (\theta_2 < 180°) \tag{16.2}$$

$$\phi_3 = -\cos^{-1} \frac{CD^2 + AD^2 - AB^2}{2(CD)(AD)} \quad (\theta_2 > 180°) \tag{16.3}$$

$$CD = [AB^2 + AD^2 - 2(AB)(AD) \cos \theta_2]^{\frac{1}{2}} \tag{16.4}$$

16.4   VELOCITY ANALYSIS

To determine the relationships for velocities $\bar{V}_B$, $\bar{V}_C$, and $\bar{V}_{B/C}$, let Bcb in Figure 16.6 represent the velocity polygon of the mechanism ABCD, where

Bb represents the magnitude of the linear velocity of point B (V_B).
Bc represents the magnitude of the linear velocity of point C (V_C).
bc represents the magnitude of the relative velocity of point B with
   respect to point C $(V_{B/C})$.

Then V_B, V_C, and $V_{B/C}$ can be determined in terms of the angular positions of the links as follows:

$$\angle Bcb = 90°$$

$$\angle bBc = \theta_3 - \theta_2$$

$$\angle b = 90° - (\theta_3 - \theta_2)$$

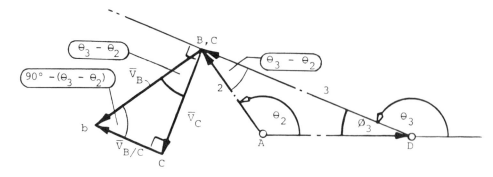

Figure 16.6   Velocity analysis.

By applying the rule of sines from trigonometry, the velocities can be
expressed in the form

$$\frac{V_B}{\sin 90°} = \frac{V_C}{\sin (90° - \theta_3 + \theta_2)} = \frac{V_{B/C}}{\sin (\theta_3 - \theta_2)}$$

Since

$$V_B = \omega_2 AB \tag{16.5}$$

then

$$V_C = \omega_2 AB \sin (90 - \theta_3 - \theta_2) = \omega_2 AB \cos (\theta_3 - \theta_2) \tag{16.6}$$

and

$$V_{B/C} = \omega_2 AB \sin (\theta_3 - \theta_2) \tag{16.7}$$

Also, since $\bar{V}_B$, $\bar{V}_C$, and $\bar{V}_{B/C}$ are assumed to be oriented at angles
$(\theta_2 + 90°)$, $(\theta_3 + 90°)$, and (θ_3), respectively, from the real or reference
axis, the required vectorial expressions can be written as follows:

$$\bar{V}_B = \omega_2 AB e^{i(\theta_2 + 90°)} \tag{16.8}$$

$$\bar{V}_C = \omega_2 AB \cos (\theta - \theta_2) e^{i(\theta_3 + 90°)} \tag{16.9}$$

$$\bar{V}_{B/C} = \omega_2 AB \sin (\theta - \theta_2) e^{i\theta_3} \tag{16.10}$$

where $e^{i(\theta_2+90°)}$, $e^{i(\theta_3+90°)}$, and $e^{i\theta_3}$ are unit vectors, used to define the directions of $\bar{V}_B$, $\bar{V}_C$, and $\bar{V}_{B/C}$, respectively. Note that $e^{i\theta} = \cos\theta +$ $i \sin\theta$, where θ is the position angle of the unit vector.

16.5 ACCELERATION ANALYSIS

To determine the linear accelerations, we apply the relative motion theory, which states that

$$\bar{A}_B = \bar{A}_C + \bar{A}_{B/C} \qquad (16.11)$$

Expanding this equation into its normal and tangential component form, we obtain

$$\bar{A}_B^N + \bar{A}_B^T = \bar{A}_C^N + \bar{A}_C^T + \bar{A}_{B/C}^N + \bar{A}_{B/C}^T + \bar{A}_{B/C}^{Cor} \qquad (16.12)$$

where

$$\bar{A}_B^N = -\omega_2^2 AB\, e^{i\theta_2}$$

$$\bar{A}_B^T = \alpha_2 AB\, e^{i(\theta_2+90°)}$$

$$\bar{A}_C^N = -\frac{V_C^2}{CD}e^{i\theta_3}$$

$$\bar{A}_C^T = A_C^T e^{i(\theta_3+90°)}$$

$$\bar{A}_{B/C}^N = -\frac{V_{B/C}^2}{\infty}e^{i(\theta_3+90°)} = 0$$

$$\bar{A}_{B/C}^T = A_{B/C}^T e^{i\theta_3}$$

$$\bar{A}_{B/C}^{Cor} = 2V_{B/C}\frac{V_C}{CD}e^{i(\theta_3+90°)}$$

Here it should be observed that an important difference between Equation (16.12) and the corresponding equation for the four-bar linkage is that since points B and C are not rigidly connected, but slide relative to each other, it is necessary to determine the Coriolis acceleration $\bar{A}_{B/C}^{Cor}$, an acceleration component which results from the sliding motion.

Note that for the Coriolis acceleration, we consider the linear velocity of the slider relative to the rotating guide and the angular velocity of the guide, namely, ω_3 or V_C/CD. Also, the Coriolis acceleration has the orientation of $\bar{V}_{B/C}$ when rotated 90° about its tail in the direction of ω_3.

Also, in Equation (16.12), note that since the sliding path of B on C is a straight line, its radius of curvature is infinite. That is, $R = \infty$. Hence $\bar{A}^N_{B/C} = 0$. Thus Equation (16.12) becomes

$$-\omega_2^2 AB\, e^{i\theta_2} + \alpha_2 AB\, e^{i(\theta_2 + 90°)} = -\frac{V^2_C}{CD} e^{i\theta_3} + A^T_C\, e^{i(\theta_3 + 90°)}$$

$$+ 2V_{B/C}\frac{V_C}{CD} e^{i(\theta_3 + 90°)} + A^T_{B/C}\, e^{i\theta_3}$$

$$(16.13)$$

This equation contains only two unknown quantities, the magnitudes A^T_C and $A^T_{B/C}$. All other quantities (magnitudes and directions) are either known or can readily be determined from the problem data. (Note that the directions of $\bar{A}^T_C$ and $\bar{A}^T_{B/C}$, although not precisely known, are assumed to be positive for convenience. If any of these directions is reversed, the numerical solution of the equation will automatically produce a negative sign for the unknown quantity.) To solve for the unknowns, therefore, we equate the real and imaginary parts of the equation and rearrange as follows:

Real:

$$A^T_C \cos(\theta_3 + 90°) + A^T_{B/C} \cos\theta_3 = -\omega_2^2 AB \cos\theta_2 + \alpha_2 AB \cos(\theta_2 + 90°)$$

$$- 2V_{B/C}\frac{V_C}{CD}\cos(\theta_3 + 90°) + \frac{V^2_C}{CD}\cos\theta_3$$

$$(16.14)$$

Imaginary:

$$A^T_C \sin(\theta_3 + 90°) + A^T_{B/C} \sin\theta_3 = -\omega_2^2 AB \sin\theta_2 + \alpha_2 AB \sin(\theta_2 + 90°)$$

$$- 2V_{B/C}\frac{V_C}{CD}\sin(\theta_3 + 90°) + \frac{V^2_C}{CD}\sin\theta_3$$

$$(16.15)$$

The solution of these simultaneous equations yields

$$A_C^T = \frac{C_1 B_2 - C_2 B_1}{A_1 B_2 - A_2 B_1} \tag{16.16}$$

$$A_{B/C}^T = \frac{A_1 C_2 - A_2 C_1}{A_1 B_2 - A_2 B_1} \tag{16.17}$$

where

$$A_1 = \cos(\theta_3 + 90°) \tag{16.18}$$

$$A_2 = \sin(\theta_3 + 90°) \tag{16.19}$$

$$B_1 = \cos\theta_3 \tag{16.20}$$

$$B_2 = \sin\theta_3 \tag{16.21}$$

$$C_1 = -\omega_2^2 AB \cos\theta_2 + \alpha_2 AB \cos(\theta_2 + 90°)$$
$$- 2V_{B/C}\frac{V_C}{CD}\cos(\theta_3 + 90°) + \frac{V_C^2}{CD}\cos\theta_3 \tag{16.22}$$

$$C_2 = -\omega_2^2 AB \sin\theta_2 + \alpha_2 AB \sin(\theta_2 + 90°)$$
$$- 2V_{B/C}\frac{V_C}{CD}\sin(\theta_3 + 90°) + \frac{V_C^2}{CD}\sin\theta_3 \tag{16.23}$$

Given that the values V_C, $V_{B/C}$, A_C^T, and $A_{B/C}^T$ are found, the general equations for computing the linear accelerations can be summarized as follows:

$$\bar{A}_B = -\omega_2^2 AB\, e^{i\theta_2} + \alpha_2 AB\, e^{i(\theta_2 + 90°)} \tag{16.24}$$

$$\bar{A}_C = -\frac{V_C^2}{CD}e^{i\theta_3} + A_C^T e^{i(\theta_3 + 90°)} \tag{16.25}$$

$$\bar{A}_{B/C} = \frac{2(V_{B/C})(V_C)}{CD}e^{i(\theta_3 + 90°)} + A_{B/C}^T e^{i\theta_3} \tag{16.26}$$

EXAMPLE 16.1: Shaper Mechanism

Considering the quick–return mechanism in Figure 16.1, let AB = 2 in., AD = 4 in., $\theta_2 = 30°$, $\omega_2 = 62.83$ rad/sec (counterclockwise), and $\alpha_2 = 0$. It is required to find $\bar{V}_C$, $\bar{V}_{B/C}$, $\bar{A}_C$, and $\bar{A}_{B/C}$.

SOLUTION

1. Determine θ_3 using Equations (16.4), (16.2), and (16.1).

$$CD = [0.167^2 + 0.333^2 - 2(0.167)(0.333) \cos 30°]^{\frac{1}{2}}$$

$$= 0.21 \text{ ft}$$

$$\phi_3 = \cos^{-1} \frac{0.21^2 + 0.33^2 - 0.167^2}{2(0.21)(0.33)}$$

$$= 23.8°$$

$$\theta_3 = 180° - 23.8°$$

$$= 156.2°$$

2. Determine V_B, V_C, and $V_{B/C}$ using Equations (16.5), (16.6), and (16.7).

$$V_B = (62.83)(0.167) = 10.47 \text{ ft/sec}$$

$$V_C = (62.83)(0.167) \cos(156.2° - 30°) = -6.18 \text{ ft/sec}$$

$$V_{B/C} = (62.83)(0.167) \sin(156.2° - 30°) = 8.45 \text{ ft/sec}$$

3. Determine $\bar{V}_B$, $\bar{V}_C$, and $\bar{V}_{B/C}$ using Equations (16.8), (16.9), and (16.10).

$$\bar{V}_B = 10.47 \, e^{i(120°)} = 10.47 \text{ ft/sec} \; \underline{/120°}$$

$$\bar{V}_C = -6.18 \, e^{i(246.2°)} = 6.18 \text{ ft/sec} \; \underline{/66.2°}$$

$$\bar{V}_{B/C} = 8.45 \, e^{i(156.2°)} = 8.45 \text{ ft/sec} \; \underline{/156.2°}$$

4. Determine constants A_1, A_2, B_1, B_2, C_1, and C_2 using Equations (16.18) through (16.23).

$$A_1 = \cos(156.2° + 90°)$$

$$= -0.4033$$

$$A_2 = \sin(156.2° + 90°)$$

$$= -0.9151$$

$$B_1 = \cos(156.2°)$$
$$= -0.9151$$

$$B_2 = \sin(156.2°)$$
$$= 0.4033$$

$$C_1 = -62.83^2(0.1666)\cos 30° + 0$$
$$- 2(8.45)\left(-\frac{6.18}{0.21}\right)\cos(156.2° + 90°)$$
$$+ \frac{(6.18)^2}{0.21}\sin 156.2°$$
$$= -942.95$$

$$C_2 = -62.83^2(0.1666)\sin 30° + 0$$
$$- 2(8.45)\left(-\frac{6.18}{0.21}\right)\sin(156.2° + 90°)$$
$$+ \frac{(6.18)^2}{0.21}\sin 156.2°$$
$$= -716.92$$

5. Determine A_C^T and $A_{B/C}^T$ using Equations (16.16) and (16.17).

$$A_C^T = \frac{(-942.95)(0.40) - (-716.92)(-0.92)}{(-0.40)(0.40) - (-0.92)(-0.92)}$$
$$= 1036.3 \text{ ft/sec}^2$$

$$A_{B/C}^T = \frac{(-0.40)(-716.92) - (-0.92)(-942.95)}{(-0.40)(0.40) - (-0.92)(-0.92)}$$
$$= 573.8 \text{ ft/sec}^2$$

6. Substitute the values found for A_C^T and $A_{B/C}^T$ into Equations (16.25) and (16.26) and solve for $\bar{A}_C$ and $\bar{A}_{B/C}$.

$$\bar{A}_C = -\frac{(-6.18)^2}{0.207}e^{i(156.2°)} + 1036.3\, e^{i(246.2°)}$$
$$= 185.2\ \underline{/-23.8°} + 1036.3\ \underline{/-113.8°}$$
$$= 1052.7 \text{ ft/sec}^2\ \underline{/-103.7°}$$

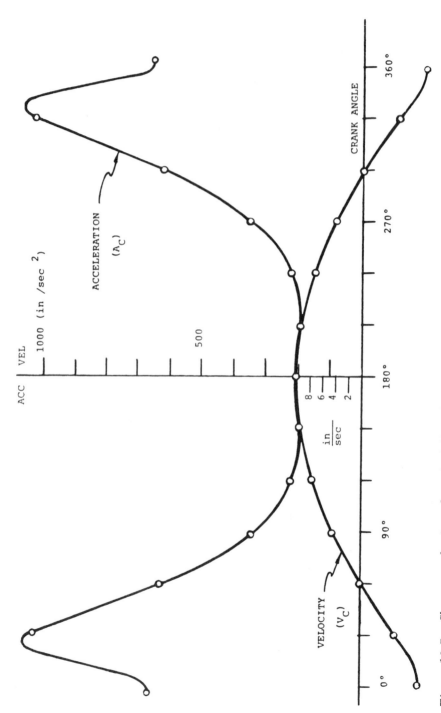

Figure 16.7   Shaper mechanism characteristics. Example 16.1.

$$\bar{A}_{B/C} = \frac{2(8.45)(-6.18)}{0.207} e^{i(246.2°)} + 573.8\,e^{i(156.2°)}$$

$$= 505.7\ \underline{/66.2°} + 573.8\ \underline{/156.2°}$$

$$= 764.8\ \text{ft/sec}^2\ \underline{/114.8°}$$

See Figure 16.7 for velocity and acceleration profiles for complete crank cycle from 0° to 360°.

EXAMPLE 16.2: Whitworth Mechanism (Figure 16.8)

Considering the quick-return mechanism in Figure 16.1, let AB = 3 in., AD = 2 in., $\theta_2 = 60°$, $\omega_2 = 30$ rad/sec (counterclockwise), and $\alpha_2 = 0$. It is required to find $\bar{V}_C$, $\bar{V}_{B/C}$, $\bar{A}_C$, and $\bar{A}_{B/C}$.

SOLUTION

1. Determine θ_3 using Equations (16.4), (16.2), and (16.1).

$$CD = [0.25^2 + 0.166^2 - 2(0.25)(0.166)\cos 60°]^{\frac{1}{2}}$$

$$= 0.22\ \text{ft}$$

$$\phi_3 = \cos\ \frac{0.22^2 + 0.166^2 - 0.25^2}{2(0.22)(0.166)}$$

$$\theta_3 = 180° - 79.1°$$

$$= 100.9°$$

2. Determine V_B, V_C, and $V_{B/C}$ using Equations (16.5), (16.6), and (16.7).

$$V_B = 30(0.25) = 7.5\ \text{ft/sec}$$

$$V_C = 30(0.25)\cos(100.9° - 60°) = 5.67\ \text{ft/sec}$$

$$V_{B/C} = 30(0.25)\sin(100.9° - 60°) = 4.9\ \text{ft/sec}$$

3. Determine $\bar{V}_B$, $\bar{V}_C$, and $\bar{V}_{B/C}$ using Equations (16.8), (16.9), and (16.10).

$$\bar{V}_B = 7.5\,e^{i150°} = 7.5\ \text{ft/sec}\ \underline{/150°}$$

$$\bar{V}_C = 5.67\,e^{i190.1°} = 5.67\ \text{ft/sec}\ \underline{/-169.1°}$$

$$\bar{V}_{B/C} = 4.9\,e^{i100.9°} = 4.9\ \text{ft/sec}\ \underline{/100.9°}$$

4. Determine constants A_1, A_2, B_1, B_2, C_1, and C_2 using Equations (16.18) through (16.23).

$$A_1 = \cos(100.9° + 90°)$$
$$= -0.98$$

$$A_2 = \sin(100.9° + 90°)$$
$$= -0.19$$

$$B_1 = \cos 100.9°$$
$$= -0.19$$

$$B_2 = \sin 100.9°$$
$$= 0.98$$

$$C_1 = -30^2(3)\cos(60°) + 0(3)\cos(60° + 90°)$$
$$- 2(4.9)\left(\frac{5.67}{0.22}\right)\cos(100.9° + 90°) + \left(\frac{5.67^2}{0.22}\right)\cos(100.9°)$$
$$= 107.9$$

$$C_2 = -30^2(3)\sin(60°) + 0(3)\sin(60° + 90°)$$
$$- 2(4.9)\left(\frac{5.67}{0.22}\right)\sin(100.9° + 90°) + \left(\frac{5.67^2}{0.22}\right)\sin(100.9°)$$
$$= -3.96$$

5. Determine A_C^T and $A_{B/C}^T$ using Equations (16.16) and (16.17).

$$A_C^T = \frac{(107.94)(0.98) - (-3.96)(-0.19)}{(-0.98)(0.98) - (-0.19)(-0.19)}$$
$$= -105.3 \text{ ft/sec}^2$$

$$A_{B/C}^T = \frac{(-0.98)(-3.96) - (-0.19)(107.94)}{(-0.98)(0.98) - (-0.19)(-0.19)}$$
$$= -24.25 \text{ ft/sec}^2$$

6. Substitute the values found for A_C^T and $A_{B/C}^T$ into Equations (16.25) and (16.26) and solve for $\bar{A}_C$ and $\bar{A}_{B/C}$.

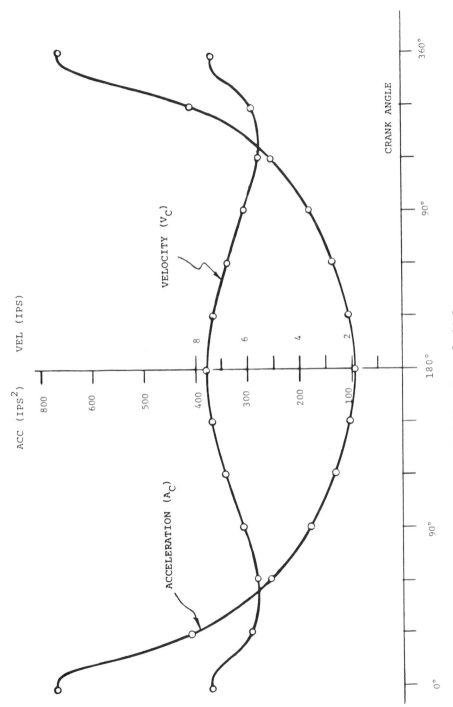

Figure 16.8   Whitworth mechanism characteristics. Example 16.2.

$$\bar{A}_C = -\frac{5.67^2}{0.22} e^{i(100.9°)} + (-105.3) e^{i(190.9°)}$$

$$= 145.9 \; \underline{/-79.1°} + 105.3 \; \underline{/10.9°}$$

$$= 179.9 \; \text{ft/sec}^2 \; \underline{/-43.3°}$$

$$\bar{A}_{B/C} = \frac{2(4.9)(5.67)}{0.22} e^{i(190.9°)} + (-24.25) e^{i(100.9°)}$$

$$= 252.5 \; \underline{/-169.1°} + 24.25 \; \underline{/-79.1°}$$

$$= 253.7 \; \text{ft/sec}^2 \; \underline{/-163.6°}$$

See Figure 16.8 for velocity and acceleration profiles for complete crank cycle from 0° to 360°.

17
Sliding Coupler Mechanism Analysis: Simplified Vector Method

17.1 INTRODUCTION

The sliding coupler mechanism, represented by ABC in Figure 17.1, is an important class of the slider crank chain where the connecting rod or coupler BC (link 3) slides through a cylinder or block which is free to rotate, via trunnions, about a fixed axis. The mechanism is generally found in two basic arrangements: an <u>oscillating cylinder</u> (or rocking block) type, where the crank arm AB is shorter than the base AC, and the <u>rotating cylinder</u> (or rotating block type), where the crank arm is longer than the base. In the first arrangement, the cylinder (or block) oscillates as the crank arm rotates, whereas in the second arrangement, the cylinder (or block) rotates with the crank.

Typical applications for the sliding coupler include steam engines, some pumps and compressors, hydraulic actuators (such as front-end loaders), variable-speed indexing drives, and various compound mechanisms (see Figures 17.2 to 17.9).

Motion characteristics of the sliding coupler mechanism are usually difficult to analyze because of complexity of the coupler and slider motions

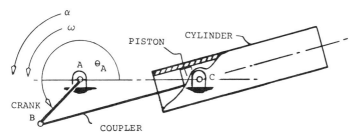

Figure 17.1 Sliding coupler mechanism.

(a)

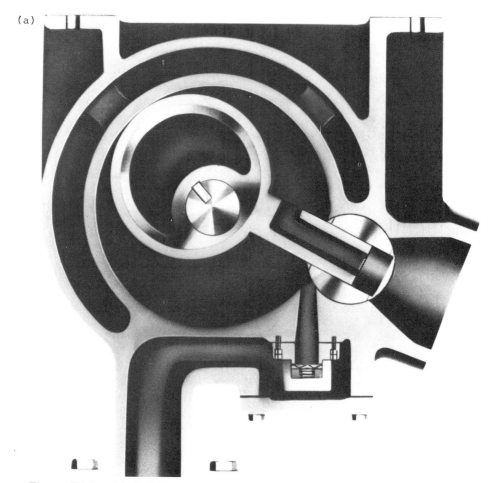

Figure 17.2   (a) Compressor. (Courtesy of Ingersoll Rand, Woodcliff
Lake, N.J.)

which give rise to the Coriolis acceleration. The simplified method presented here quickly determines linear velocity and acceleration relationships for the sliding coupler mechanism for any angular position of the
crank cycle.

17.2   SCOPE AND ASSUMPTIONS

The sliding coupler mechanism ABC in Figure 17.1 consists of a crank AB
of fixed length, a coupler BC of variable length, and a slider C which is

(b)

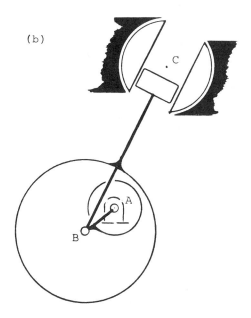

Figure 17.2 (Continued) (b) mechanism.

constrained to move within a cylinder pivoted at a fixed point C. Given that the crank rotates with an angular velocity ω_A and an angular acceleration α_A (both in a counterclockwise direction), we will now derive general expressions to compute linear velocities and accelerations of points A, B, and C relative to B for any angular position θ_A of the crank arm. All angular displacements, velocities, and accelerations are considered positive for counterclockwise rotation and negative for positive rotation.

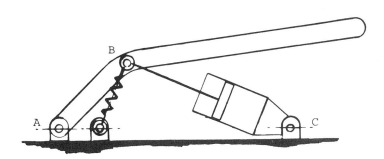

Figure 17.3 Application: foot pump.

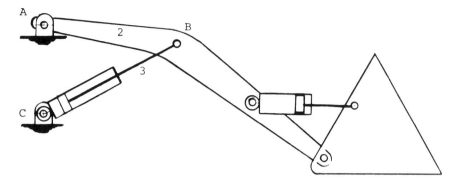

Figure 17.4   Loader actuator.

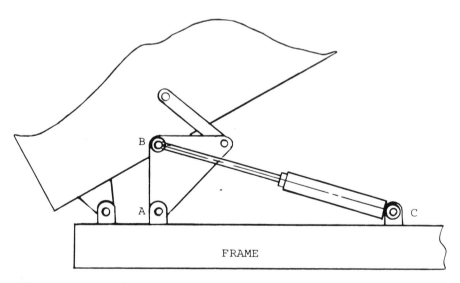

Figure 17.5   Application: dump truck.

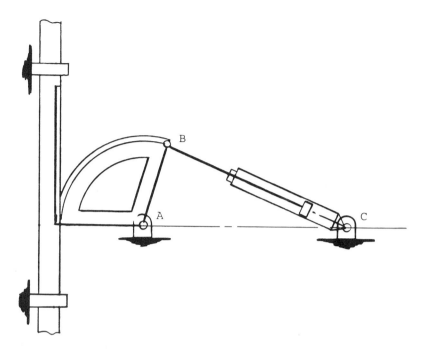

Figure 17.6   Rack and pinion mechanism.

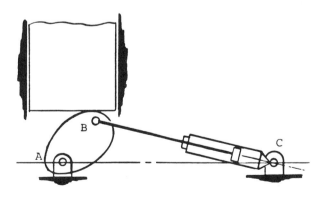

Figure 17.7   Cam-follower mechanism.

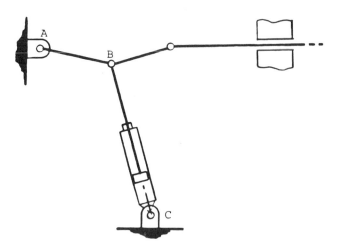

Figure 17.8   Toggle mechanism.

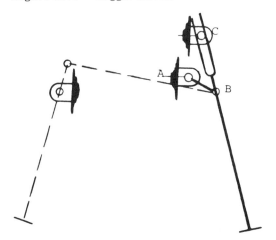

Figure 17.9   Walking mechanism. (From A. S. Hall, 1961.)

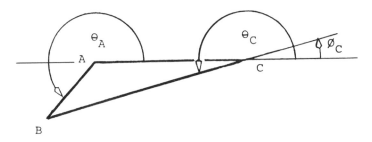

Figure 17.10   Geometric relationships.

17.3   GEOMETRIC RELATIONSHIPS

First, we determine the angle θ_C of the coupler BC in terms of crank angle θ_A. Considering triangle ABC in Figure 17.10, we note that

$$BC = [AB^2 + AC^2 - 2(AB)(AC) \cos \theta_A]^{\frac{1}{2}} \tag{17.1}$$

$$\theta_C = 180° + \phi_C \tag{17.2}$$

where it can be shown that

$$\phi_C = -\cos^{-1} \frac{BC^2 + AC^2 - AB^2}{2(BC)(AC)} \qquad (\theta_A \leq 180°)$$

$$\phi_C = \cos^{-1} \frac{BC^2 + AC^2 - AB^2}{2(BC)(AC)} \qquad (\theta_A > 180°) \tag{17.3}$$

17.4   VELOCITY ANALYSIS

To determine the relationships for velocities $\bar{V}_B$, $\bar{V}_C$, and $\bar{V}_{C/B}$, let Bcb in Figure 17.11 represent the velocity polygon of the mechanism ABC, where

Bb represents the magnitude of $\bar{V}_B$ (V_B)
Bc represents the magnitude of $\bar{V}_C$ (V_C)
bc represents the magnitude of $\bar{V}_{C/B}$ ($V_{C/B}$)

Applying the rule of sines from trigonometry, the velocity magnitudes V_B, V_C, and $V_{C/B}$ can be determined in terms of angular positions of the links as follows:

$$\frac{V_B}{\sin 90°} = \frac{V_C}{\sin(\theta_A - \theta_C)} = \frac{V_{C/B}}{\sin(90° - \theta_A + \theta_C)} \tag{17.4}$$

where

$$\sin(\theta_A - \theta_C) = \sin(\theta_A - \phi_C - 180°)$$

$$= -\sin(180° - \theta_A + \phi_C)$$

$$= -\sin(\theta_A - \phi_C) \tag{17.5}$$

$$\sin(90° - \theta_A + \theta_C) = \sin(90° - \theta_A + \phi_C + 180°)$$

$$= \sin(270° - \theta_A + \phi_C)$$

$$= -\sin(90° - \theta_A + \phi_C) \qquad (17.6)$$

Applying Equations (17.5) and (17.6) to (17.4) yields

$$\frac{V_B}{\sin 90°} = \frac{V_C}{-\sin(\theta_A - \phi_C)} = \frac{V_{C/B}}{-\sin(90° - \theta_A + \phi_C)} \qquad (17.7)$$

where

$$V_B = AB\omega \qquad (17.8)$$

$$V_C = -\frac{V_B \sin(\theta_A - \phi_C)}{\sin 90°} = -AB\omega \sin(\theta_A - \phi_C) \qquad (17.9)$$

$$V_{C/B} = -\frac{V_B \sin(90° - \theta_A + \phi_C)}{\sin 90°} = -AB\omega \cos(\theta_A - \phi_C) \qquad (17.10)$$

An inspection of the velocity polygon shows that $\bar{V}_B$ and $\bar{V}_{C/B}$ are oriented, respectively at angle $(\theta_A + 90°)$ and $(\phi_C + 90°)$ while the direction of $\bar{V}_C$ is constrained to that along the connecting rod BC, at angle ϕ_C. Therefore, the velocity vectors can be expressed in exponential form as follows:

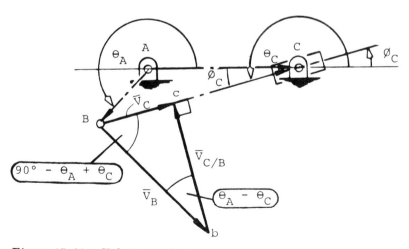

Figure 17.11   Velocity analysis.

$$\bar{V}_B = AB\omega e^{i(\theta_A + 90°)} \tag{17.11}$$

$$\bar{V}_C = -AB\omega \sin(\theta_A - \phi_C)e^{i\phi_C} \tag{17.12}$$

$$\bar{V}_{C/B} = -AB\omega \cos(\theta_A - \phi_C)e^{i(\phi_C+90°)} \tag{17.13}$$

17.5   ACCELERATION ANALYSIS

To determine the acceleration relationships, we apply the relative motion equation, which states that

$$\bar{A}_C = \bar{A}_B + \bar{A}_{C/B} \tag{17.14}$$

Expanding this equation into its normal and tangential form, we obtain

$$\bar{A}_C^N + \bar{A}_C^T + \bar{A}_C^{Cor} = \bar{A}_B^N + \bar{A}_B^T + \bar{A}_{C/B}^{N\,\cdot} + \bar{A}_{C/B}^T \tag{17.15}$$

where

$$\bar{A}_C^N = -\frac{V_C^2}{\infty}e^{i(\phi_C+90°)} = 0$$

$$\bar{A}_C^T = A_C^T e^{i\phi_C}$$

$$\bar{A}_C^{Cor} = 2V_C \frac{V_{C/B}}{BC}e^{i(\phi_C+90°)}$$

$$\bar{A}_B^N = -\omega^2 ABe^{i\theta_A}$$

$$\bar{A}_B^T = \alpha ABe^{i(\theta_A+90°)}$$

$$\bar{A}_{C/B}^N = -\frac{V_{C/B}^2}{BC}e^{i\phi_C}$$

$$\bar{A}_{C/B}^T = A_{C/B}^T e^{i(\phi_C+90°)}$$

As in the case of the quick-return mechanism, since the slider C is constrained to move relative to a rotating guide, the cylinder, it is necessary to determine Coriolis acceleration $\bar{A}_C^{Cor}$ resulting from this motion.

In this case, for Coriolis acceleration, we consider the linear velocity of the slider or V_C and the angular velocity of the guide (or connecting rod), namely ω_{CB} or $V_{C/B}/CB$. Also, Coriolis acceleration has the orientation of the $\bar{V}_{C/B}$ vector when rotated $90°$ about its tail in the direction of ω_{CB}.

Expansion of Equation (17.15) yields

$$
A_C^T e^{i\phi_C} = -\omega^2 AB\, e^{i\theta_A} + \alpha AB\, e^{i(\theta_A + 90°)} - \frac{V_{C/B}^2}{BC} e^{i\phi_C}
$$

$$
+ A_{C/B}^T e^{i(\phi_C + 90°)} - 2V_C \frac{V_{C/B}}{BC} e^{i(\phi_C + 90°)} \tag{17.16}
$$

Equation (17.15) contains two unknown quantities: the magnitudes A_C^T and $A_{C/B}^T$. All other magnitudes and unit vectors are either known or can be determined from the geometry and operating characteristics of the mechanism. Although the directions for the vectors $\bar{A}_C^T$ and $\bar{A}_{C/B}^T$ are not precisely known, they can be assumed for convenience to be positive. If any of of these directions are reversed, the numerical solution of the equation will automatically produce a negative sign, indicating the correct direction. To solve for the unknowns, therefore, we equate the real and imaginary parts of the equation and rearrange as follows:

Real:

$$
A_C^T \cos\phi_C - A_{C/B}^T \cos(\phi_C + 90°) = -\omega^2 AB \cos\theta_A + \alpha AB \cos(\theta_A + 90°)
$$

$$
- \frac{V_{C/B}^2}{BC} \cos\phi_C - 2V_C \frac{V_{C/B}}{BC} \cos(\phi_C + 90°) \tag{17.17}
$$

Imaginary:

$$
A_C^T \sin\phi_C - A_{C/B}^T \sin(\phi_C + 90°) = -\omega^2 AB \sin\theta_A + \alpha AB \sin(\theta_A + 90°)
$$

$$
- \frac{V_{C/B}^2}{BC} \sin\phi_C - 2V_C \frac{V_{C/B}}{BC} \sin(\phi_C + 90°) \tag{17.18}
$$

Solution of the simultaneous equations (17.17) and (17.18) is obtained by

$$A^T_C = \frac{C_1 B_2 - C_2 B_1}{A_1 B_2 - A_2 B_1} \tag{17.19}$$

$$A^T_{C/B} = \frac{A_1 C_2 - A_2 C_1}{A_1 B_2 - A_2 B_1} \tag{17.20}$$

where

$$A_1 = \cos \phi_C \tag{17.21}$$

$$A_2 = \sin \phi_C \tag{17.22}$$

$$B_1 = -\cos(\phi_C + 90°) \tag{17.23}$$

$$B_2 = -\sin(\phi_C + 90°) \tag{17.24}$$

$$C_1 = -\omega^2 AB \cos \theta_A + \alpha AB \cos(\theta_A + 90°)$$

$$-\frac{V^2_{C/B}}{BC} \cos \phi_C - 2V_C \frac{V_{C/B}}{BC} \cos(\phi_C + 90°) \tag{17.25}$$

$$C_2 = -\omega^2 AB \sin \theta_A + \alpha AB \sin(\theta_A + 90°)$$

$$-\frac{V^2_{C/B}}{BC} \sin \phi_C - 2V_C \frac{V_{C/B}}{BC} \sin(\phi_C + 90°) \tag{17.26}$$

Given that the values V_C, $V_{C/B}$, A^T_C, and $A^T_{C/B}$ have been determined, the general equations for computing the linear accelerations can be summarized as follows:

$$\bar{A}_B = -\omega^2 AB e^{i\theta_A} + \alpha AB e^{i(\theta_A + 90°)} \tag{17.27}$$

$$\bar{A}_C = 2V_C \frac{V_{C/B}}{BC} e^{i(\phi_C + 90°)} + A^T_C e^{i\phi_C} \tag{17.28}$$

$$\bar{A}_{C/B} = -\frac{V^2_{C/B}}{BC} e^{i\phi_C} + A^T_{C/B} e^{i(\phi_C + 90°)} \tag{17.29}$$

EXAMPLE 17.1: Oscillating Cylinder

Considering the oscillating cylinder mechanism in Figure 17.1, let AB = 8 in., AC = 10 in., $\theta_A = 120°$, $\omega_A = 18$ rad/sec (counterclockwise), and $\alpha_A = 0$. It is required to find $\bar{V}_C$, $\bar{V}_{C/B}$, $\bar{A}_C$, and $\bar{A}_{C/B}$.

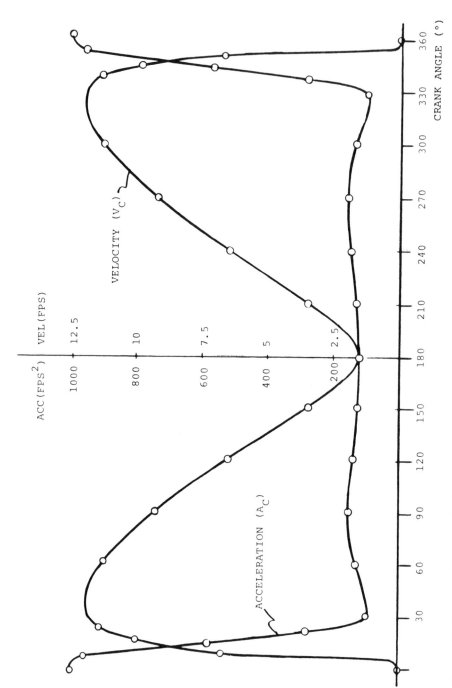

Figure 17.12   Oscillating cylinder characteristics. Example 17.1.

SOLUTION*

1. Determine ϕ_C using Equations (17.1) and (17.2).

$$BC = [0.67^2 - 0.83^2 - (0.67)(0.83)\cos 120°]^{\frac{1}{2}}$$

$$= 1.30 \text{ ft}$$

$$\phi_C = -\cos^{-1}\frac{1.3^2 + 0.83^2 - 0.67^2}{2(1.3)(0.83)}$$

$$= -26.33°$$

2. Determine V_B, V_C, and $V_{C/B}$ using Equations (17.8), (17.9), and (17.10).

$$V_B = 0.67(18) = 12 \text{ ft/sec}$$

$$V_C = -0.67(18)\sin[120° - (-26.33°)] = -6.65 \text{ ft/sec}$$

$$V_{C/B} = -(0.67)(18)\cos[120° - (-26.33°)] = 9.98 \text{ ft/sec}$$

3. Determine $\bar{V}_B$, $\bar{V}_C$, and $\bar{V}_{C/B}$ using Equations (17.11), (17.12), and (17.13).

$$\bar{V}_B = 12.0\,e^{i(120°+90°)} = 12.0 \text{ ft/sec } \underline{/-150°}$$

$$\bar{V}_C = -6.65\,e^{i(-26.3°)} = 6.65 \text{ ft/sec } \underline{/153.6°}$$

$$\bar{V}_{C/B} = 9.98\,e^{i(-26.3°+90°)} = 9.98 \text{ ft/sec } \underline{/63.7°}$$

4. Determine constants A_1, A_2, B_1, B_2, C_1, and C_2 using Equations (17.21) through (17.26).

$$A_1 = \cos(-26.3°) = 0.89$$

$$A_2 = \sin(-26.3°) = -0.44$$

$$B_1 = -\cos(-26.3° + 90°) = -0.44$$

$$B_2 = -\sin(-26.3° + 90°) = -0.89$$

---

*See Figure 17.12 for velocity and acceleration profiles for complete crank cycle from 0° to 360°.

$$C_1 = -(18^2)(0.67)\cos 120° + 0(0.67)\cos(120° + 90°)$$

$$- \frac{(9.98)^2}{1.3}\cos(-26.3°) - 2(-6.65)\left(\frac{9.98}{1.3}\right)\cos(-26.3° + 90°)$$

$$= 84.61$$

$$C_2 = -(18^2)(0.67)\sin 120° + 0(0.67)\sin(120° + 90°)$$

$$- \frac{(9.98)^2}{1.3}\sin(-26.3°) - 2(-6.65)\left(\frac{9.98}{1.3}\right)\sin(-26.3° + 90°)$$

$$= -61.58$$

5. Determine A_C^T and $A_{C/B}^T$ using Equations (17.19) and (17.20).

$$A_C^T = \frac{(84.61)(-0.89) - (-61.58)(0.44)}{(0.89)(-0.89) - (-0.44)(-0.44)}$$

$$= 103.14 \text{ ft/sec}^2$$

$$A_{C/B}^T = \frac{(0.89)(-61.58) - (-0.44)(84.61)}{(0.89)(-0.89) - (-0.44)(-0.44)}$$

$$= 17.66 \text{ ft/sec}^2$$

6. Substitute the values found for A_C^T and $A_{C/B}^T$ into Equations (17.28) and (17.29) and solve for $\bar{A}_C$ and $\bar{A}_{C/B}$.

$$\bar{A}_C = \frac{2(-6.65)(9.98)}{1.3}e^{i(-26.3° + 90°)} + 103.14\,e^{i(-26.3°)}$$

$$= 102.1 \;\underline{/-116.3°} + 103.14 \;\underline{/-26.3°}$$

$$= 145.12 \text{ ft/sec}^2 \;\underline{/-71.04°}$$

$$\bar{A}_{C/B} = -\frac{(9.98)^2}{1.3}e^{i(-26.3°)} + 17.66\,e^{i(-26.3° + 90°)}$$

$$= 76.6 \;\underline{/153.6°} + 17.66 \;\underline{/63.7°}$$

$$= 78.64 \text{ ft/sec}^2 \;\underline{/140.7°}$$

EXAMPLE 17.2: Rotating Cylinder

Considering the rotating cylinder version of mechanism in Figure 17.1, let AB = 3 in., AC = 2 in., $\theta_A = 210°$, $\omega_A = 30$ rad/sec (counterclockwise), and $\alpha_A = 0$. It is required to find $\bar{V}_C$, $\bar{V}_{C/B}$, $\bar{A}_C$, and $\bar{A}_{C/B}$.

SOLUTION

1. Determine ϕ_C using Equations (17.1) and (17.3).

$$BC = [3.0^2 - 2.0^2 - (3.0)(2.0) \cos 210°]^{\frac{1}{2}}$$

$$= 4.8 \text{ in.}$$

$$\phi_C = \cos^{-1} \frac{4.8^2 + 2.0^2 - 3.0^2}{2(4.8)(2.0)}$$

$$= 18.06°$$

2. Determine magnitudes V_B, V_C, and $V_{C/B}$ using Equations (17.8), (17.9), and (17.10).

$$V_B = 3.0(30) = 90.0 \text{ in./sec}$$

$$V_C = -3.0(30) \sin(210° - 18.06°) = 18.6 \text{ in./sec}$$

$$V_{C/B} = -3.0(30) \cos(210° - 18.06°) = 88.06 \text{ in./sec}$$

3. Determine complete vectors $\bar{V}_B$, $\bar{V}_C$, and $\bar{V}_{C/B}$ using Equations (17.11), (17.12), and (17.13).

$$\bar{V}_B = 90\,e^{i(210°+90°)} = 90 \text{ in./sec} \underline{/-60°}$$

$$\bar{V}_C = 18.6\,e^{i(18.06°)} = 18.6 \text{ in./sec} \underline{/18.06°}$$

$$\bar{V}_{C/B} = 88.06\,e^{i(18.06°+90°)} = 88.06 \text{ in./sec} \underline{/108.6°}$$

4. Determine constants A_1, A_2, B_1, B_2, C_1, and C_2 using Equations (17.21) through (17.26).

$$A_1 = \cos(18.06°) = 0.95$$

$$A_2 = \sin(18.06°) = 0.31$$

$$B_1 = -\cos(18.06° + 90°) = 0.31$$

$$B_2 = -\sin(18.06° + 90°) = -0.95$$

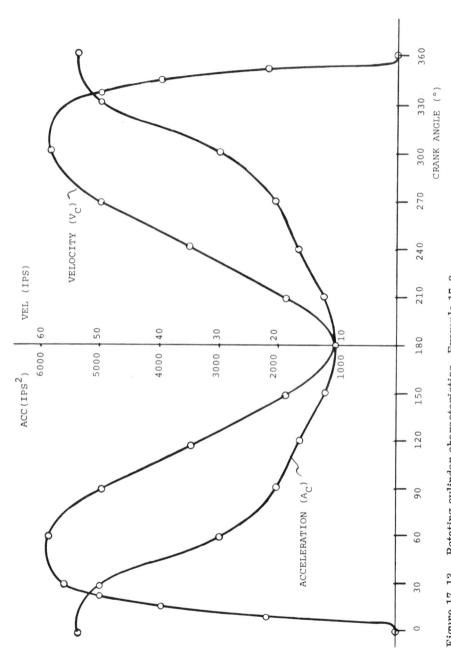

Figure 17.13   Rotating cylinder characteristics. Example 17.2.

$$C_1 = -(30.0)^2 (3.0) \cos 210° + 0(3.0) \cos (210° + 90°)$$

$$- \frac{(88.06)^2}{4.8} \cos (18.06°) - 2(18.6)\left(\frac{88.06}{4.8}\right) \cos (18.06° + 90°)$$

$$= 1024.3$$

$$C_2 = -(30.0)^2 (3.0) \sin 210° + 0(3.0) \sin (210° + 90°)$$

$$- \frac{(88.06)^2}{4.8} \sin (18.06°) - 2(18.6)\left(\frac{88.06}{4.8}\right) \sin (18.06° + 90°)$$

$$= 208.64$$

5. Determine A_C^T and $A_{C/B}^T$ using Equations (17.19) and (17.20).

$$A_C^T = \frac{(1024.3)(-0.95) - (208.6)(0.31)}{(0.95)(-0.95) - (0.31)(0.31)}$$

$$= 1038.5 \text{ in./sec}^2$$

$$A_{C/B}^T = \frac{(0.95)(208.6) - (0.31)(1024.3)}{(0.95)(-0.95) - (0.31)(0.31)}$$

$$= 119.32 \text{ in./sec}^2$$

6. Substitute the values found for A_C^T and $A_{C/B}^T$ into Equations (17.28) and (17.29) and solve for $\bar{A}_C$ and $\bar{A}_{C/B}$.

$$\bar{A}_C = \frac{2(18.6)(88.06)}{4.8} e^{i(18.06° + 90°)} + 1038.5 e^{i(18.06°)}$$

$$= 677.57 \ \underline{/108.07°} + 1038.5 \ \underline{/18.06°}$$

$$= 1240 \text{ in./sec}^2 \ \underline{/51.18°}$$

$$\bar{A}_{C/B} = -\frac{88.06^2}{4.8} e^{i(18.06°)} + 119.32 e^{i(18.06° + 90°)}$$

$$= 1603.15 \ \underline{/-161.9°} + 119.32 \ \underline{/108.06°}$$

$$= 1607.58 \text{ in./sec}^2 \ \underline{/166.18°}$$

See Figure 17.3 for velocity and acceleration profiles for complete crank cycle from 0° to 360°.

18

Slider-Crank Mechanism Analysis: Modified Vector Method

18.1   INTRODUCTION

In the slider-crank analysis by the simplified vector method (Chapter 15), the motion relationships were obtained completely by applying relative motion principles. There the velocity relationships were derived with the aid of the velocity polygon and the acceleration relationships were derived by expressing the relative acceleration equation in terms of the velocity relationships. In this analysis, the procedure is basically the same except that the acceleration equations are not obtained by relative motion but instead by differentiation of the velocity relationships.

18.2   SCOPE AND ASSUMPTIONS

Let ABC in Figure 18.1 represent a typical slider-crank mechanism where the crank AB rotates counterclockwise at an angular velocity ω about joint A, and θ_A is its instantaneous angular position away from top dead center. ϕ_C is the angle between the slider arm BC and the line AC.

It is required to develop linear velocity and acceleration relationships for points B and C and also for point B relative to point C at any position of the crank. It is assumed that the angular velocity ω for counterclockwise rotation is positive and for clockwise rotation negative.

18.3   GEOMETRIC CONSIDERATIONS

As in previous analysis, we first seek to determine the angle of the connecting rod θ_A. There is was shown that

$$\theta_C = 180° + \phi_C$$

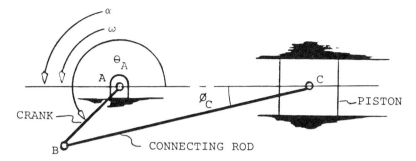

Figure 18.1   Slider-crank analysis: mechanism.

where

$$\phi_C = -\sin^{-1} \frac{AB}{BC} \sin \theta_A$$

or

$$\theta_C = 180 - \phi_C$$

where

$$\phi_C = \sin^{-1} \frac{AB}{BC} \sin \theta_A$$

18.4   VELOCITY ANALYSIS

Let Bcb be the velocity polygon for the mechanism ABC (Figure 18.2). In Section 15.4 it was shown that the scalar velocity equation for a typical slider slider crank may be expressed as

$$\frac{V_B}{\sin(90° - \theta_C)} = \frac{V_C}{\sin(\theta_C - \theta_A)} = \frac{V_{B/C}}{\sin(90° - \theta_A)} \tag{18.1}$$

where

V_B = magnitude of linear velocity of point B

V_C = magnitude of linear velocity of point C

$V_{B/C}$ = magnitude of linear velocity of point B relative to point C

Then we would obtain

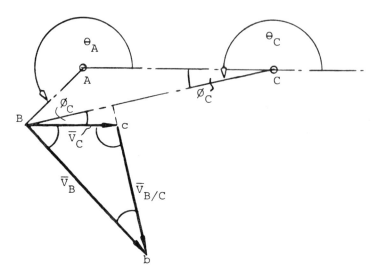

Figure 18.2   Slider-crank analysis: velocity polygon.

$$\sin(90° - \theta_C) = \sin(90° - 180° + \phi_C)$$

$$= -\sin(90° - \phi_C)$$

$$\sin(\theta_C - \theta_A) = \sin(180° - \phi_C - \theta_A)$$

$$= \sin(\phi_C + \theta_A)$$

Hence Equation (18.1) can be rewritten as

$$\frac{V_B}{-\sin(90 - \phi_C)} = \frac{V_C}{\sin(\phi_C + \theta_A)} = \frac{V_{B/C}}{\sin(90 - \theta_A)}$$

from which

$$V_B = AB\omega$$

$$V_C = -V_B \frac{\sin(\phi_C + \theta_A)}{\sin(90° - \phi_C)}$$

$$V_{B/C} = -V_B \frac{\sin(90° - \theta_A)}{\sin(90° - \phi_C)}$$

In vector form (magnitudes and directions considered), these equations can be expressed in polar form as follows:

$$\bar{V}_B = -V_B(\sin \theta_A - i \cos \theta_A)$$

$$\bar{V}_C = V_C \cos 0°$$

$$\bar{V}_{B/C} = \bar{V}_B - \bar{V}_C$$

where $-(\sin \theta_A - i \cos \theta_A)$ and $\cos 0°$ are unit vectors used to orient the vectors $\bar{V}_B$ and $\bar{V}_C$.

Finally, we obtain the following:

$$\bar{V}_B = -AB\omega (\sin \theta_A - i \cos \theta_A) \qquad (18.2)$$

$$\bar{V}_C = -AB\omega \frac{\sin (\phi_C + \theta_A)}{\sin (90° - \phi_C)} \qquad (18.3)$$

$$\bar{V}_{B/C} = -AB\omega \left[\sin \theta_A - \frac{\sin (\phi_C + \theta_A)}{\sin (90° - \phi_C)} - i \cos \theta_A \right] \qquad (18.4)$$

18.5   ACCELERATION ANALYSIS

With the velocity equations for the mechanism determined, the corresponding acceleration expressions are obtained by differentiating these equations with respect to time, noting that $\theta_A = \omega \times t$.

Acceleration of B $(\bar{A}_B)$

$$\bar{A}_B = \frac{d\bar{V}_B}{dt}$$

Using Equation (18.2), we have

$$\bar{V}_B = -AB \times \omega (\sin \theta_A - i \cos \theta_A)$$

$$\bar{A}_B = -AB \times \omega^2 (\cos \theta_A + i \sin \theta_A) - AB \times \overset{\circ}{\omega}(\sin \theta_A - i \cos \theta_A) \qquad (18.5)$$

where the first and second terms on the right-hand side represent the normal and tangential accelerations, respectively, of point B, and $\overset{\circ}{\omega}$ (or α) is the angular acceleration of the same point.

Acceleration of C ($\bar{A}_C$)

$$\bar{A}_C = \frac{d\bar{V}_C}{dt}$$

Using Equation (18.3), we have

$$\bar{V}_C = -AB \times \omega \frac{\sin(\theta_A + \phi_C)}{\sin(90° - \phi_C)}$$

and substituting for ϕ_C, it follows that

$$\bar{V}_C = -AB \times \omega \left[\sin\theta_A + \frac{AB\cos\theta_A \sin\theta_A}{(BC^2 - AB^2 \sin^2\theta_A)^{\frac{1}{2}}}\right]$$

from which

$$\bar{A}_C = -AB \times \omega^2 \left[\cos\theta_A + \frac{AB(\cos^2\theta_A - \sin^2\theta_A)}{(BC^2 - AB^2 \sin^2\theta_A)^{1/2}} + \frac{AB^3 \cos^2\theta_A \sin^2\theta_A}{(BC^2 - AB^2 \sin^2\theta_A)^{3/2}}\right]$$

$$- AB \times \overset{\circ}{\omega} \left[\sin\theta_A + \frac{AB\cos\theta_A \sin\theta_A}{(BC^2 - AB^2 \sin^2\theta_A)^{1/2}}\right] \qquad (18.6)$$

Acceleration of B Relative to C ($\bar{A}_{B/C}$)

$$\bar{A}_{B/C} = \bar{A}_B - \bar{A}_C$$

Therefore, using Equations (18.5) and (18.6), we have

$$\bar{A}_{B/C} = AB \times \omega^2 \left[\frac{AB(\cos^2\theta_A - \sin^2\theta_A)}{(BC^2 - AB^2 \sin^2\theta_A)^{1/2}} + \frac{AB^3 \cos^2\theta_A \sin^2\theta_A}{(BC^2 - AB^2 \sin^2\theta_A)^{3/2}} - i\sin\theta_A\right]$$

$$+ AB \times \overset{\circ}{\omega} \left[\frac{AB\cos\theta_A \sin\theta_A}{(BC^2 - AB^2 \sin^2\theta_A)^{1/2}} + i\cos\theta_A\right] \qquad (18.7)$$

Note that when there is no angular acceleration (or $\overset{\circ}{\omega} = 0$), the second terms on the right-hand side of both Equations (18.6) and (18.7) vanish.

EXAMPLE 18.1

Let

$AB = 1.5$ in.

$BC = 3.0$ in.

$\omega_{AB} = 1$ rad/sec (clockwise)

$\theta_A = 30°$

Find $\bar{V}_C$, $\bar{V}_{B/C}$, $\bar{A}_C$, and $\bar{A}_{B/C}$.

SOLUTION

Using Equation (18.3), we have

$$\bar{V}_C = -AB \times \omega \, \frac{\sin(\theta_A + \phi_C)}{\sin(90° - \phi_C)}$$

where

$$\omega = -\omega_{AB} = -1 \text{ rad/sec}$$

$$\phi_C = \sin^{-1}\left(\frac{AB}{BC} \sin \theta_A\right)$$

$$= \sin^{-1}(0.5 \sin 30°) = 14.5°$$

$$\bar{V}_C = -1.5(-1) \, \frac{\sin(30° + 14.5°)}{\sin(90° - 14.5°)}$$

$$= 1.5\left(\frac{0.70}{0.96}\right)$$

$$= 1.09 \text{ in./sec } \underline{/0°}$$

Using Equation (18.4), we have

$$\bar{V}_{B/C} = -AB \times \omega \left[(\sin \theta_A - i \cos \theta_A) - \frac{\sin(\theta_A + \phi_C)}{\sin(90° - \phi_C)} \right]$$

$$= -AB \times \omega \left[(\sin 30° - i \cos 30°) - \frac{\sin 44.5°}{\sin 75.5°} \right]$$

$$= -1.5(-1)(0.5 - 0.866i - 0.729)$$

$$= 1.5(-0.229 - 0.866i)$$

$$= 1.34 \tan^{-1} \frac{866}{229} \text{ in./sec } \underline{/-104.5°}$$

Using Equation (18.6), we have

$$\bar{A}_C = -1.5(1)\left[\cos 30° + \frac{1.5(\cos^2 30° - \sin^2 30°)}{(3^2 - 1.5^2 \sin^2 30°)^{1/2}} + \frac{1.5^3 \cos^2 30° \sin^2 30°}{(3^2 - 1.5^2 \sin^2 30°)^{3/2}}\right]$$

$$= -1.5(0.866 + 0.258 + 0.0258)$$

$$= 1.72 \text{ in./sec}^2 \ \underline{/180°}$$

Using Equation (18.7), we have

$$\bar{A}_{B/C} = 1.5(1)\left[\frac{1.5(\cos^2 30° - \sin^2 30°)}{(3^2 - 1.5^2 \sin^2 30°)^{1/2}} + \frac{1.5^3 \cos^2 30° \sin^2 30°}{(3^2 - 1.5^2 \sin^2 30°)^{3/2}} - i \sin 30°\right]$$

$$= 1.5(0.289 - 0.5i)$$

$$= 0.86 \tan^{-1} - \frac{500}{239} \text{ in./sec}^2 \ \underline{/-60.4°}$$

EXAMPLE 18.2

Let

$$AB = 1.5 \text{ in.}$$

$$BC = 3.0 \text{ in.}$$

$$\omega_{AB} = 1 \text{ rad/sec (counterclockwise)}$$

$$\theta_A = 120°$$

Find $\bar{V}_C$, $\bar{V}_{B/C}$, $\bar{A}_C$, and $\bar{A}_{B/C}$.

SOLUTION

Using Equation (18.3), we have

$$\bar{V}_C = -AB \times \omega \frac{\sin(\theta_A + \phi_C)}{\sin(90° - \phi_C)}$$

where

$$\omega = +\omega_{AB} = 1 \text{ rad/sec}$$

$$\phi_C = \sin^{-1}\left(\frac{AB}{BC} \sin \theta_A\right)$$

$$= \sin^{-1}(0.5 \sin 120°) = 25.6°$$

$$\bar{V}_C = -AB \times \omega \frac{\sin(120° + 25.6°)}{\sin(90° - 25.6°)}$$

$$= -1.5(1)\left(\frac{0.565}{0.90}\right)$$

$$= 0.94 \text{ in./sec } \underline{/180°}$$

Using Equation (18.4), we have

$$\bar{V}_{B/C} = -AB \times \omega \left[(\sin \theta_A - i \cos \theta_A) - \frac{\sin(\theta_A + \phi_C)}{\sin(90° - \phi_C)}\right]$$

$$= -AB \times \omega \left[(\sin 120° - i \cos 120° - \frac{\sin 145.6°}{\sin 64.4°}\right]$$

$$= -1.5(1)\left(0.866 + 0.51 - \frac{0.565}{0.90}\right)$$

$$= 1.5(-0.239 - 0.5i)$$

$$= 0.830 \tan^{-1}\frac{500}{239} \text{ in./sec } \underline{/-115.6°}$$

Using Equation (18.5), we have

$$\bar{A}_C = -1.5(1)\left[\cos 120° + \frac{1.5(\cos^2 120° - \sin^2 120°)}{(3^2 - 1.5^2 \sin^2 120°)^{1/2}} + \frac{1.5^3 \cos^2 120° \sin^2 120°}{(3^2 - 1.5^2 \sin^2 120°)^{3/2}}\right]$$

$$= -1.5(-0.5 - 0.277 + 0.032)$$

$$= 1.12 \text{ in./sec}^2 \underline{/0°}$$

Using Equation (18.7), we have

$$\bar{A}_{B/C} = 1.5(1)\left[\frac{1.5(\cos^2 120° - \sin^2 120°)}{(3^2 - 1.5^2 \sin^2 120°)^{1/2}} + \frac{1.5^2 \cos^2 120° \sin^2 120°}{(3^2 - 1.5^2 \sin^2 120°)^{3/2}} - i \sin 120°\right]$$

$$= 1.5(-0.245 - 0.866i)$$

$$= 1.35 \tan^{-1}\frac{866}{245} \text{ in./sec}^2 \underline{/-105.8°}$$

19

Slider-Crank Mechanism Analysis: Calculus Method

19.1   INTRODUCTION

An analytical approach, commonly used in analyzing the motion characteristics of a mechanism consists of:

1.   Writing a mathematical expression to describe the displacement or position of the mechanism
2.   Differentiating the displacement expression with respect to time to obtain the velocity expression
3.   Differentiating the velocity expression with respect to time to obtain the acceleration expression

This method is illustrated with the familiar slider–crank mechanism.

19.2   SCOPE AND ASSUMPTIONS

Let the mechanism ABC in Figure 19.1 represent a typical slider–crank where AB in turn represents the crank, BC the connecting rod, and C the slider. Given that the angular displacement of the crank AB at any instant is θ, we will now develop general expressions to compute the linear displacement, velocity, and acceleration of the slider C in terms of θ.

19.3   DISPLACEMENT, VELOCITY, AND ACCELERATION ANALYSIS

Let AB = R   and   BC = L.

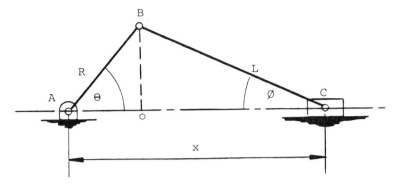

Figure 19.1   Slider-crank model.

Then

$$x = R \cos \theta + L \cos \phi \tag{19.1}$$

$$\frac{dx}{dt} = -R \sin \theta \frac{d\theta}{dt} - L \sin \phi \frac{d\phi}{dt} \tag{19.2}$$

where

$$L \sin \phi = R \sin \theta \tag{19.3}$$

and

$$L \cos \phi \frac{d\phi}{dt} = R \cos \theta \frac{d\theta}{dt} \tag{19.4}$$

or

$$\frac{d\phi}{dt} = \frac{R \cos \theta}{L \cos \phi} \frac{d\theta}{dt} \tag{19.5}$$

Therefore, by substituting Equations (19.3) and (19.5) in (19.2), we obtain

$$\frac{dx}{dt} = -R \sin \theta \frac{d\theta}{dt} - R \sin \theta \frac{R \cos \theta}{L \cos \phi} \frac{d\theta}{dt} \tag{19.6}$$

$$= -R \frac{d\theta}{dt}\left(\sin \theta + \frac{R \sin \theta \cos \theta}{L \cos \phi} \right) \tag{19.7}$$

or

$$v = -R \frac{d\theta}{dt}\left(\sin \theta + \frac{R \sin 2\theta}{2L \cos \phi} \right) \tag{19.8}$$

Let

$$A = -R \frac{d\theta}{dt} \quad \text{and} \quad B = \sin\theta + \frac{R \sin 2\theta}{2L \cos\phi} \tag{19.9}$$

Then

$$\frac{d^2 x}{dt^2} = B \frac{d}{dt} (A) + A \frac{d}{dt} (B)$$

$$= \left(\sin\theta + \frac{R \sin 2\theta}{2L \cos\phi}\right)\frac{d}{dt}\left(-R \frac{d\theta}{dt}\right) + \left(-R \frac{d\theta}{dt}\right)\frac{d}{dt}\left(\sin\theta + \frac{R \sin 2\theta}{2L \cos\phi}\right)$$

$$= -R \frac{d^2\theta}{dt^2}\left(\sin\theta + \frac{R \sin 2\theta}{2L \cos\phi}\right) - R \frac{d\theta}{dt}\frac{d}{dt}\left(\sin\theta + \frac{R \sin 2\theta}{2L \cos\phi}\right) \tag{19.10}$$

Let

$$u = R \sin 2\theta \quad \text{and} \quad v = 2L \cos\phi \tag{19.11}$$

Then

$$\frac{d}{dt}\frac{R \sin 2\theta}{2L \cos\phi} = \frac{v \, du - u \, dv}{v^2}$$

$$= \frac{2L \cos\phi(2R \cos 2\theta) \, d\theta/dt - R \sin 2\theta(-2L \sin\phi) \, d\theta/dt}{4L^2 \cos^2\phi}$$

$$= \frac{(4L \cos\phi R \cos 2\theta) \, d\theta/dt}{4L^2 \cos^2\phi}$$

$$+ \frac{2R \sin 2\theta(R \sin\theta)}{4L^2 \cos^2\phi}\frac{R \cos\theta}{L \cos\phi}\frac{d\theta}{dt}$$

$$= \left(\frac{R \cos 2\theta}{L \cos\phi} + \frac{R^3 \sin^2 2\theta}{4L^3 \cos^3\phi}\right)\frac{d\theta}{dt} \tag{19.12}$$

$$\frac{d}{dt}\left(\sin\theta + \frac{R \sin 2\theta}{2L \cos\phi}\right) = \left(\cos\theta + \frac{R \cos 2\theta}{L \cos\phi} + \frac{R^3 \sin^2 2\theta}{4L^3 \cos^3\phi}\right)\frac{d\theta}{dt} \tag{19.13}$$

$$\frac{d^2 x}{dt^2} = -R \frac{d^2\theta}{dt^2}\left(\sin\theta + \frac{R \sin 2\theta}{2L \cos\phi}\right)$$

$$- R\left(\frac{d\theta}{dt}\right)^2\left(\cos\theta + \frac{R \cos 2\theta}{L \cos\phi} + \frac{R \sin^2 2\theta}{4L^3 \cos^3\phi}\right) \tag{19.14}$$

If $L \gg R$, we can set $\phi = 0$. This yields

$$a = \frac{d^2x}{dt^2} = -R\frac{d^2\theta}{dt^2}\left(\sin\,\theta + \frac{R}{2L}\,\sin 2\theta\right)$$

$$- R\left(\frac{d\theta}{dt}\right)^2\,\cos\,\theta + \frac{R}{L}\,\cos 2\theta + \frac{R^3\,\sin^2 2\theta}{4L^3} \qquad (19.15)$$

Note that the first term of this equation becomes zero for constant angular velocity of the crank (i.e., no acceleration). Also, by making use of the relationship

$$L\,\cos\,\phi = (L^2 - R^2\,\sin^2\,\theta)^{\frac{1}{2}} \qquad (19.16)$$

we can rewrite the general expressions for velocity and acceleration Equations (19.8) and (19.15) as follows:

$$v = -R\omega\left[\sin\,\theta + \frac{R\,\sin 2\theta}{2(L^2 - R^2\,\sin^2\,\theta)^{\frac{1}{2}}}\right] \qquad (19.17)$$

$$a = -R\alpha\left[\sin\,\theta + \frac{R\,\sin 2\theta}{2(L^2 - R^2\,\sin^2\,\theta)^{\frac{1}{2}}}\right]$$

$$- R\omega^2\left[\cos\,\theta + \frac{R\,\cos 2\theta}{(L^2 - R^2\,\sin^2\,\theta)^{1/2}} + \frac{R^3\,\sin^2 2\theta}{4(L^2 - R^2\,\sin^2\,\theta)^{3/2}}\right]$$

$$(19.18)$$

Problems

KINEMATIC TERMINOLOGY

1. Define (a) kinematics, and (b) kinetics.

2. Distinguish between the terms "mechanism" and "machine."

3. How does a mechanism differ from a structure?

4. Name and give examples of the three types of plane motion.

5. Define (a) displacement, (b) velocity, and (c) acceleration.

6. Distinguish (a) between the terms "speed" and "velocity," and (b) between the terms "distance" and "displacement."

7. Explain mechanism inversion. How does it affect (a) relative motion, and (b) absolute motion of the components?

8. How does rotation differ from curvilinear motion?

9. Distinguish between absolute motion and relative motion.

10. Define radian. Determine the number of radians in 30°, 45°, 150°, and 330°.

11. What are the three basic modes of transmitting motion? Indicate using sketches, one example in each case.

12. Distinguish between the terms "reciprocal motion" and "oscillatory motion."

13. Define (a) higher pair, and (b) lower pair. Give an example of each, using sketches.

14. What is a kinematic chain? Describe the three types.

UNIFORMLY ACCELERATED MOTION

1. A train traveling at 50 mph speeds up to 70 mph in 1 min and 30 sec. Determine its acceleration and the distance traveled.

2. A flywheel turning at 200 rpm attains a speed of 300 rpm in 1 min with constant acceleration. Determine the acceleration and number of revolutions taken to attain the higher speed.

3. An engine crank pin has a linear velocity of 2400 ft/min while rotating at 150 rpm. What is the length of the crank?

4. Starting from rest, a body A held by a string, wrapped around a 12-in.-diameter pulley, falls 60 ft in 4 sec. Determine for the pulley the following:

   a. Number of revolutions
   b. Angular velocity after 4 sec
   c. Angular acceleration after 4 sec

5. An automobile accelerates from a speed of 20 mph to 55 mph in a distance of 300 ft. If the acceleration is constant, find the time taken.

6. The speed of an automobile is 55 mph. If the outside diameter of the tires is 27 in., determine the rpm of the wheels and the angular speed in rad/sec.

7. A train traveling between two stations 5 miles apart takes 10 min. It uniformly accelerates to a maximum speed at 2 ft/sec^2 and uniformly decelerates at 6 ft/sec^2. What is the maximum speed of the train and the distance traveled at this speed? What are the distances covered during the first and last minutes of the train's motion?

8. Determine the minimum time for a car to travel between two stop signs, 1 mile apart if its acceleration is 3 ft/sec^2, its deceleration is 4 ft/sec^2, and its maximum speed is 50 mph.

9. Develop a velocity-time curve to depict the motion of a body between two points, A and D, as follows:

      (1)   passing point A, its velocity is 15 mph.

      (2)   During the next 30 sec it accelerates uniformly to 50 mph to reach a point B.

      (3)   It then continues at 50 mph for 4 min to another point C.

      (4)   Finally, it comes to rest at D 6 min after passing point A.

Determine also the total distance traveled between points A and D.

10. A rotating fan, 6 ft in diameter, comes to rest with uniform accelera-tion from a speed of 600 rpm. If it turns 15 revolutions while stopping, determine the time it takes to stop.

11. Two points, A and B, lie on a radial line of a rotating disk 2 in. apart. Determine the radius of rotation of each of these points if the velocity at A is 700 ft/min and at B 800 ft/min.

12. A 20-in.-diameter wheel turns at 200 rpm. Determine:

      a.   The angular velocity in rad/sec
      b.   The linear velocity of a point on the rim
      c.   The linear velocity of a point 12 in. from the center

If the wheel speeds up to 300 rpm with uniform acceleration in 2 min, determine:

      d.   The angular acceleration
      e.   The linear acceleration of a point on the rim

VECTORS

1. Using data in Figure P.1, determine graphically the following:

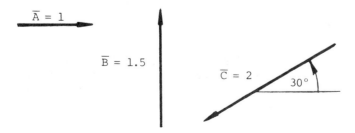

Figure P.1

a. $\bar{A} + \bar{B}$

b. $\bar{A} - \bar{B}$

c. $\bar{A} + \bar{B} - \bar{C}$

d. $\bar{A} + \bar{B} + \bar{C}$

2. Using the vector polygons given in Figure P.2, complete the following vector equations:

a. $\bar{V} =$ $\bar{A} =$

b. $\bar{B} =$ $\bar{D} =$

c. $\bar{E} =$ $\bar{H} =$

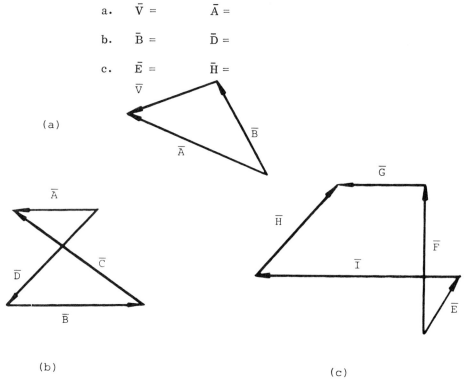

(a)

(b)   (c)

Figure P.2

3. A hiker desiring to go to a point northeast, because of various obstacles, goes $1\frac{1}{2}$ miles due east, then turns left 120°, and goes straight to the point. How far was he originally from the point, and how far did he travel to arrive at the point?

4. Graphically determine the resultant of vectors $\bar{A} + \bar{B} + \bar{C}$ in Figure P.3. Then find the effective component of this vector along axes c-c, x-x, and y-y.

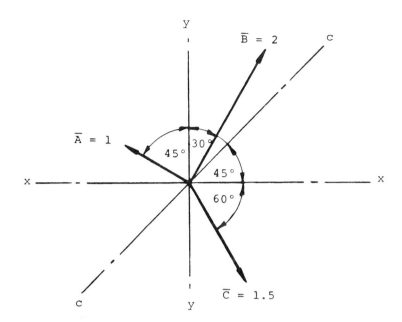

Figure P.3

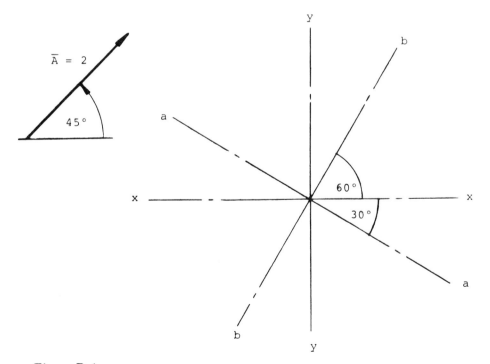

Figure P.4

5. Find the effective components of vector A in Figure P.4, along axes x-x, y-y, a-a, and b-b.

6. The effective component of a velocity vector $\bar{V}$ along the axis S-S (Figure P.5) is known. Locate this vector and also its effective components along axes T-T and R-R.

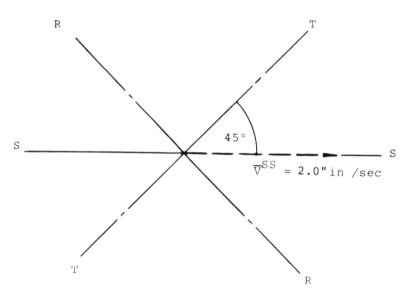

Figure P.5

7. Determine:

    a. The sum of two vectors: one 10 units due north and the other 20 units northeast
    b. The resultant of 5 (at 90°) - 4 (at 180°)
    c. The value of a vector quantity which when added to 100 units due south gives 100 units northwest

8. In Figure P.6, $\bar{N}$ and $\bar{M}$ are effective components of a vector $\bar{R}$, along axes n-n and m-m. Determine this vector.

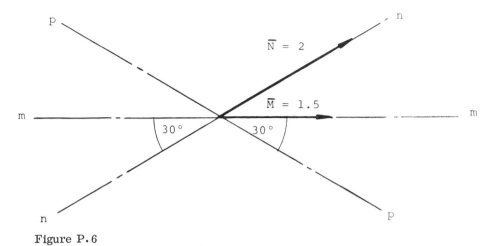

Figure P.6

9. Resolve the vector $\bar{A}$ in Figure P.7 into its components along axes b-b and c-c.

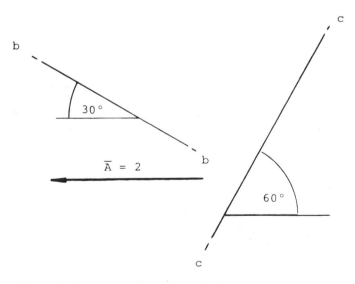

Figure P.7

10.   Indicate the directions for all vectors shown in Figure P.8, based on
      the equations

      a. $\bar{R} = \bar{T} - \bar{S} + \bar{V} - \bar{U}$
      b. $\bar{A} = \bar{B} - \bar{C} + \bar{D} - \bar{E}$

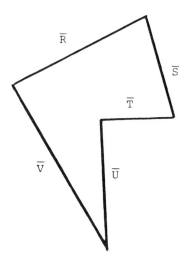
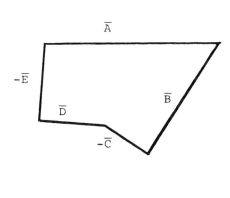

Figure P.8

11.   The velocity of point B on the link BC in Figure P.9 is shown to act
      60° with respect to the link centerline. Determine (a) the rotational
      and translational effects of this vector, and (b) the effective component
      of the same vector along link BD oriented 15° with respect to link BC.

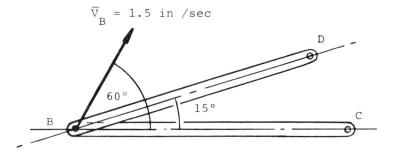

Figure P.9

12. From data given on vectors $\bar{A}$, $\bar{B}$, and $\bar{C}$ in Figure P.10, determine the magnitude and sense of all vectors when

    a. $\bar{C}$ is the resultant.
    b. $\bar{A}$ is the resultant.
    c. $\bar{B}$ is the resultant.

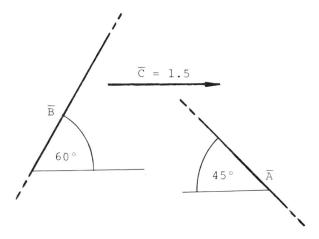

Figure P.10

13. An airplane is flying at an airspeed of 300 mph heading N45°. There is a tail wind from the west at 40 mph. What is the ground speed of the plane, and what is its actual flight direction?

14. In order to cross a stream flowing at 5 mph in a boat that travels at 12 mph, at what angle upstream should the boat be headed in order to reach a point directly opposite? What is the resultant speed of the boat?

GRAPHICAL TECHNIQUES: VELOCITY ANALYSIS

Effective Components

1. Using the effective component method, determine the linear and angular velocities of point C in Figure P.11 for any of the following conditions:

a. $\theta = 30°$, $\omega = 1$ rad/sec (clockwise)
b. $\theta = 30°$, $\omega = 1$ rad/sec (counterclockwise)
c. $\theta = 60°$, $\omega = 1$ rad/sec (clockwise)
d. $\theta = 60°$, $\omega = 1$ rad/sec (counterclockwise)

AB = 1.5"
BC = 3.0"

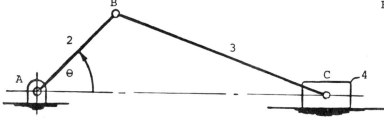

Figure P.11

2.   Repeat Problem 1 but use Figure P.12.

AB = 1.5"
BC = 3.0"

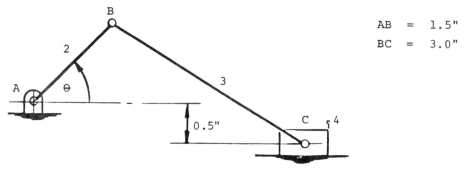

Figure P.12

3.   Repeat Problem 1 but use Figure P.13.

AB = 1.5"
BC = 3.0"

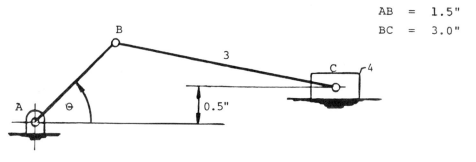

Figure P.13

4.   Repeat Problem 1 but use Figure P.14.

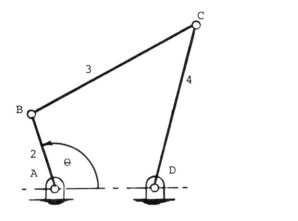

AB = 1.0"
BC = 2.5"
CD = 2.5"
AD = 1.5"

Figure P.14

5.   Repeat Problem 1 but use Figure P.15.

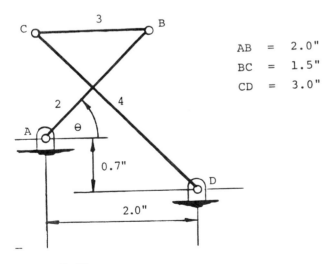

AB = 2.0"
BC = 1.5"
CD = 3.0"

Figure P.15

6.   Repeat Problem 1 but use Figure P.16.

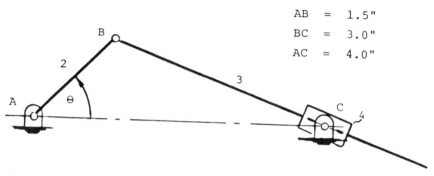

$$AB \ = \ 1.5"$$
$$BC \ = \ 3.0"$$
$$AC \ = \ 4.0"$$

Figure P.16

7.   Repeat Problem 1 but use Figure P.17.

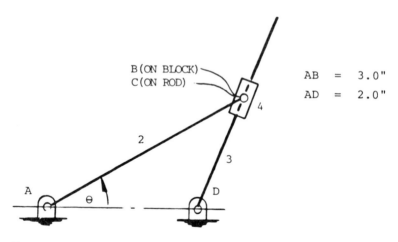

$$AB \ = \ 3.0"$$
$$AD \ = \ 2.0"$$

Figure P.17

8.   Repeat Problem 1 but use Figure P.18.

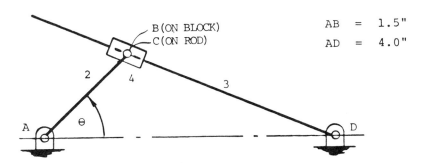

$$AB = 1.5"$$
$$AD = 4.0"$$

Figure P.18

9.   Repeat Problem 1 but use Figure P.19.

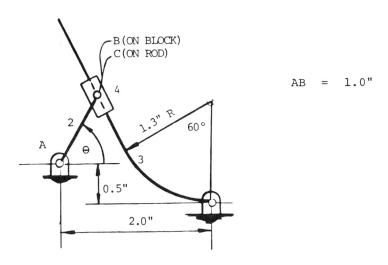

$$AB = 1.0"$$

Figure P.19

10.   Repeat Problem 1 but use Figure P.20.

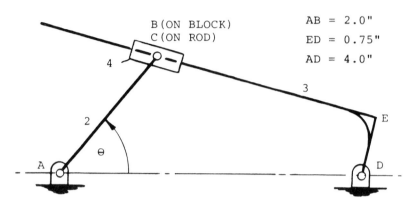

Figure P.20

11.   Determine the linear velocities of points C and E in Figure P.21.

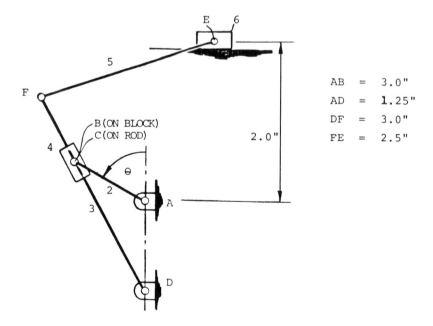

Figure P.21

12.   Repeat Problem 11 but use Figure P.22.

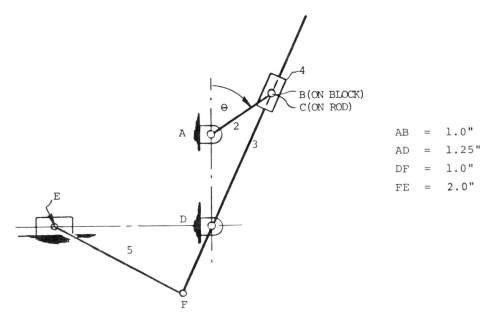

AB = 1.0"
AD = 1.25"
DF = 1.0"
FE = 2.0"

Figure P.22

13.   Repeat Problem 11 but use Figure P.23.

AB = 1.5"
BC = 3.0"
CF = 1.5"
FE = 2.0"
DE = 1.5"

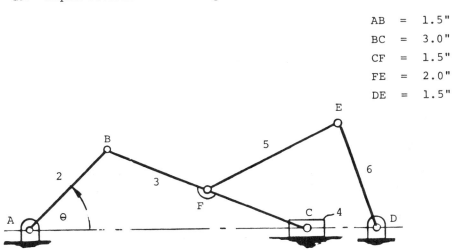

Figure P.23

14.   Repeat Problem 11 but use Figure P.24.

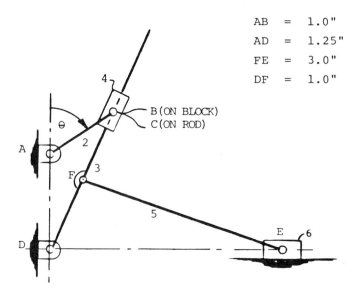

AB	=	1.0"
AD	=	1.25"
FE	=	3.0"
DF	=	1.0"

Figure P.24

15.   Repeat Problem 11 but use Figure P.25.

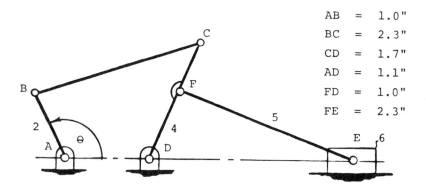

AB	=	1.0"
BC	=	2.3"
CD	=	1.7"
AD	=	1.1"
FD	=	1.0"
FE	=	2.3"

Figure P.25

16.   Repeat Problem 11 but use Figure P.26.

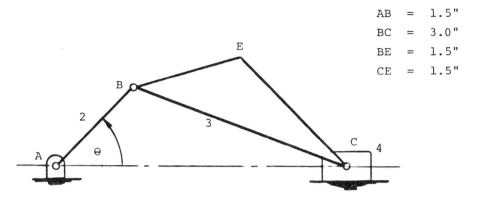

AB = 1.5"
BC = 3.0"
BE = 1.5"
CE = 1.5"

Figure P.26

17.   Repeat Problem 11 but use Figure P.27.

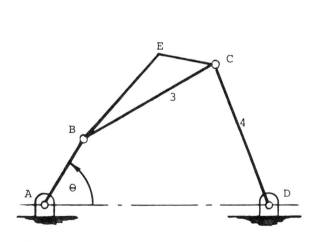

AB = 1.0"
BC = 2.0"
CD = 2.0"
AD = 3.0"
BE = 1.5"
CE = 0.75"

Figure P.27

18.   Repeat Problem 11 but use Figure P.28.

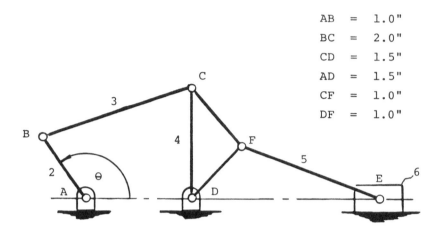

AB	=	1.0"
BC	=	2.0"
CD	=	1.5"
AD	=	1.5"
CF	=	1.0"
DF	=	1.0"

Figure P.28

19.   Repeat Problem 11 but use Figure P.29.

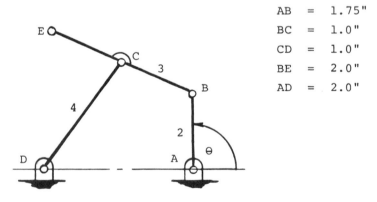

AB	=	1.75"
BC	=	1.0"
CD	=	1.0"
BE	=	2.0"
AD	=	2.0"

Figure P.29

20.   Repeat Problem 11 but use Figure P.30.

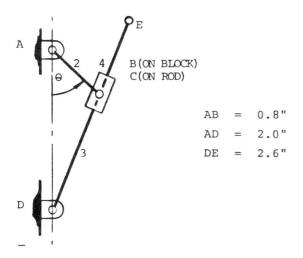

AB = 0.8"

AD = 2.0"

DE = 2.6"

Figure P.30

Instant Center

1.   First, locate all the instant centers; then, using the instant center
     method, determine the linear and angular velocities of point C in
     Figure P.11 for any of the following conditions:

     a. $\theta = 30°$, $\omega = 1$ rad/sec (clockwise)
     b. $\theta = 30°$, $\omega = 1$ rad/sec (counterclockwise)
     c. $\theta = 60°$, $\omega = 1$ rad/sec (clockwise)
     d. $\theta = 60°$, $\omega = 1$ rad/sec (counterclockwise)

2.   Repeat Problem 1 but use Figure P.12.

3.   Repeat Problem 1 but use Figure P.13.

4.   Repeat Problem 1 but use Figure P.14.

5.   Repeat Problem 1 but use Figure P.15.

6.   Repeat Problem 1 but use Figure P.16.

7.   Repeat Problem 1 but use Figure P.17.

8.   Repeat Problem 1 but use Figure P.18.

9.   Repeat Problem 1 but use Figure P.19.

10. Repeat Problem 1 but use Figure P.20.

11. Locate all instant centers in Figure P.21.

12. Locate all instant centers in Figure P.22.

13. Locate all instant centers in Figure P.23.

14. Locate all instant centers in Figure P.24.

15. Locate all instant centers in Figure P.25.

Relative Velocity

1.   Using the relative velocity method, determine the linear and angular
     velocities of point C in Figure P.11 for any of the following conditions:

     a. $\theta = 30°$, $\omega = 1$ rad/sec (clockwise)
     b. $\theta = 30°$, $\omega = 1$ rad/sec (counterclockwise)
     c. $\theta = 60°$, $\omega = 1$ rad/sec (clockwise)
     d. $\theta = 60°$, $\omega = 1$ rad/sec (counterclockwise)

2.   Repeat Problem 1 but use Figure P.12.

3.   Repeat Problem 1 but use Figure P.13.

4.   Repeat Problem 1 but use Figure P.14.

5.   Repeat Problem 1 but use Figure P.15.

6.   Repeat Problem 1 but use Figure P.16.

7.   Repeat Problem 1 but use Figure P.17.

8.   Repeat Problem 1 but use Figure P.18.

9.   Repeat Problem 1 but use Figure P.19.

10. Repeat Problem 1 but use Figure P.20.

11. Determine the linear velocities of points C and E in Figure P.21.

12. Repeat Problem 11 but use Figure P.22.

13.   Repeat Problem 11 but use Figure P.23.

14.   Repeat Problem 11 but use Figure P.24.

15.   Repeat Problem 11 but use Figure P.25.

16.   Repeat Problem 11 but use Figure P.26.

17.   Repeat Problem 11 but use Figure P.27.

18.   Repeat Problem 11 but use Figure P.28.

19.   Repeat Problem 11 but use Figure P.29.

20.   Repeat Problem 11 but use Figure P.30.

GRAPHICAL TECHNIQUES: ACCELERATION ANALYSIS

Effective Components

1.   Determine the linear acceleration of point C in Figure P.11 for any
     of the following conditions:

   a. $\theta = 45°$ rad, $\omega = 0.5$ rad/sec (clockwise),
        $\alpha = 1$ rad/sec^2 (clockwise)
   b. $\theta = 45°$ rad, $\omega = 1$ rad/sec (counterclockwise),
        $\alpha = 0.5$ rad/sec^2 (clockwise)
   c. $\theta = 135°$ rad, $\omega = 0.5$ rad/sec (clockwise),
        $\alpha = 1$ rad/sec^2 (counterclockwise)
   d. $\theta = 135°$ rad, $\omega = 1$ rad/sec (counterclockwise),
        $\alpha = 0.5$ rad/sec^2 (counterclockwise)

2.   Repeat Problem 1 but use Figure P.12.

3.   Repeat Problem 1 but use Figure P.13.

4.   Repeat Problem 1 but use Figure P.14.

5.   Repeat Problem 1 but use Figure P.15.

6.   Repeat Problem 1 but use Figure P.16.

7.   Repeat Problem 1 but use Figure P.17.

8.   Repeat Problem 1 but use Figure P.18.

9. Repeat Problem 1 but use Figure P.19.

10. Repeat Problem 1 but use Figure P.20.

11. Determine the linear acceleration of points C and E in Figure P.21.

12. Repeat Problem 11 but use Figure P.22.

13. Repeat Problem 11 but use Figure P.23.

14. Repeat Problem 11 but use Figure P.24.

15. Repeat Problem 11 but use Figure P.25.

16. Repeat Problem 11 but use Figure P.26.

17. Repeat Problem 11 but use Figure P.27.

18. Repeat Problem 11 but use Figure P.28.

19. Repeat Problem 11 but use Figure P.29.

20. Repeat Problem 11 but use Figure P. 30.

Relative Acceleration

1. Determine the linear acceleration of point C in Figure P.11 for any of the following conditions:

   a. $\theta = 45°$ rad, $\omega = 0.5$ rad/sec (clockwise),
      $\alpha = 1$ rad/sec^2 (clockwise)
   b. $\theta = 45°$ rad, $\omega = 1$ rad/sec (counterclockwise),
      $\alpha = 0.5$ rad/sec^2 (clockwise)
   c. $\theta = 135°$ rad, $\omega = 0.5$ rad/sec (clockwise),
      $\alpha = 1$ rad/sec^2 (counterclockwise)
   d. $\theta = 135°$ rad, $\omega = 1$ rad/sec (counterclockwise),
      $\alpha = 0.5$ rad/sec^2 (counterclockwise)

2. Repeat Problem 1 but use Figure P.12.

3. Repeat Problem 1 but use Figure P.13.

4. Repeat Problem 1 but use Figure P.14.

5. Repeat Problem 1 but use Figure P.15.

6.   Repeat Problem 1 but use Figure P.16.

7.   Repeat Problem 1 but use Figure P.17.

8.   Repeat Problem 1 but use Figure P.18.

9.   Repeat Problem 1 but use Figure P.19.

10.   Repeat Problem 1 but use Figure P.20.

11.   Determine the linear acceleration of points C and E in Figure P.21.

12.   Repeat Problem 11 but use Figure P.22.

13.   Repeat Problem 11 but use Figure P.23.

14.   Repeat Problem 11 but use Figure P.24.

15.   Repeat Problem 11 but use Figure P.25.

16.   Repeat Problem 11 but use Figure P.26.

17.   Repeat Problem 11 but use Figure P.27.

18.   Repeat Problem 11 but use Figure P.28.

19.   Repeat Problem 11 but use Figure P.29.

20.   Repeat Problem 11 but use Figure P.30.

Velocity Difference

1.   Determine the linear acceleration of point C in Figure P.11 for any of the following conditions:

    a. $\theta = 45°$ rad, $\omega = 0.5$ rad/sec (clockwise),
        $\alpha = 0$ rad/sec   (clockwise)
    b. $\theta = 45°$ rad, $\omega = 1$ rad/sec (counterclockwise),
        $\alpha = 0$ rad/sec   (clockwise)
    c. $\theta = 135°$ rad, $\omega = 0.5$ rad/sec (clockwise),
        $\alpha = 1$ rad/sec   (counterclockwise)
    d. $\theta = 135°$ rad, $\omega = 1$ rad/sec (counterclockwise),
        $\alpha = 0.5$ rad/sec   (counterclockwise)

2.   Repeat Problem 1 but use Figure P.12.

3.   Repeat Problem 1 but use Figure P.13.

4.   Repeat Problem 1 but use Figure P.14.

5.   Repeat Problem 1 but use Figure P.15.

6.   Repeat Problem 1 but use Figure P.16.

7.   Repeat Problem 1 but use Figure P.17.

8.   Repeat Problem 1 but use Figure P.18.

9.   Repeat Problem 1 but use Figure P.19.

10.   Repeat Problem 1 but use Figure P.20.

GRAPHICAL TECHNIQUES: MISCELLANEOUS

1.   Determine the linear velocity of point C (Figure P.31).

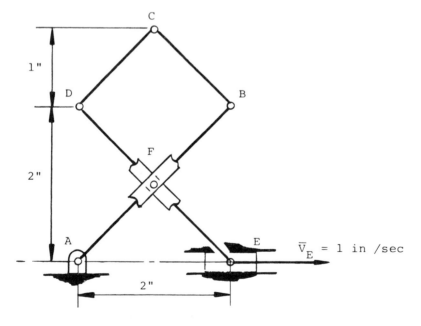

Figure P.31

2.   Determine the linear velocity of point B (Figure P.32).

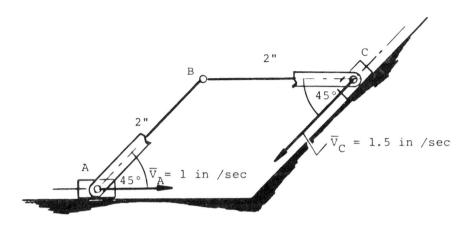

Figure P.32

3.   Determine the linear velocities of points C and A (Figure P.33).

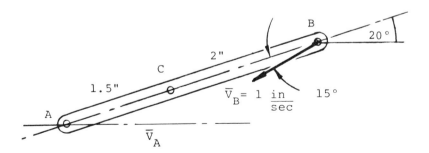

Figure P.33

4. Determine the linear velocities of points E and F (Figure P.34).

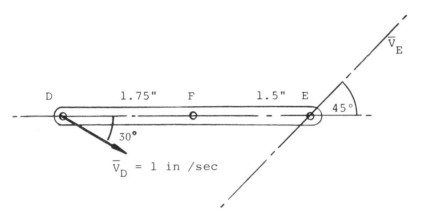

Figure P.34

5. Determine the linear acceleration of point B (Figure P.35).

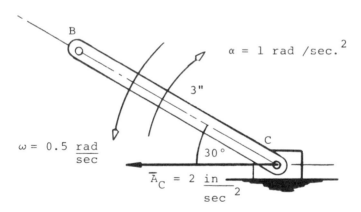

Figure P.35

6.   Determine the following (Figure P.36):

   a.   The linear acceleration of point B
   b.   The linear acceleration of point C
   c.   The linear acceleration of point B relative to point C

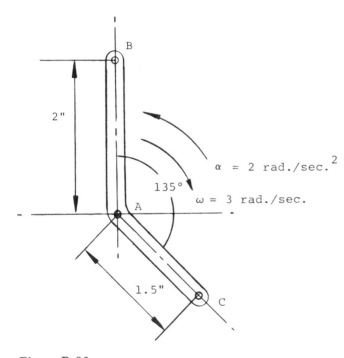

$\alpha = 2$ rad./sec.^{2}

$\omega = 3$ rad./sec.

Figure P.36

7. Using instant center 24, determine the linear velocity of the pivot 34
    (Figure P.37).

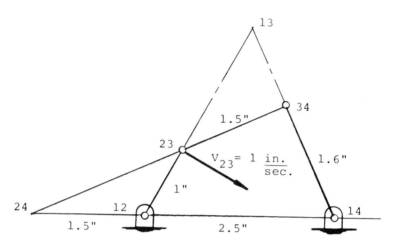

Figure P.37

8. Determine the linear velocities of points B and C and the linear veloc-
    ity C relative to B (Figure P.38).

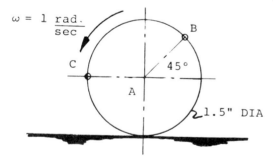

Figure P.38

9.   Determine the linear acceleration of point C (Figure P.39) if the wheel rolls without slipping.

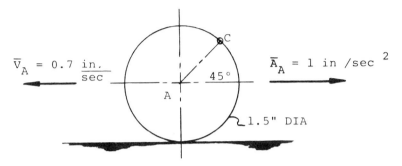

Figure P.39

10. Determine the linear acceleration of point P on the follower (Figure P.40), using

   a.   The relative acceleration method
   b.   The equivalent linkage method

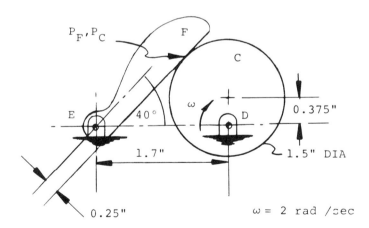

Figure P.40

11.   Determine the following (Figure P.41):

    a.   The linear acceleration of point B
    b.   The angular acceleration of point C relative to point A
    c.   The angular velocity of point C relative to point A

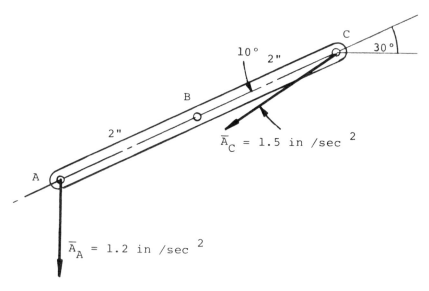

Figure P.41

12.   Determine the following (Figure P.42):

    a.   The linear acceleration of point B
    b.   The angular acceleration of point A relative to point C
    c.   The angular velocity of point A relative to point C

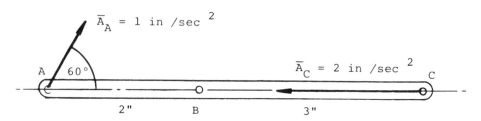

Figure P.42

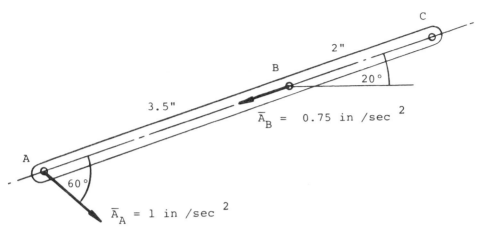

Figure P.43

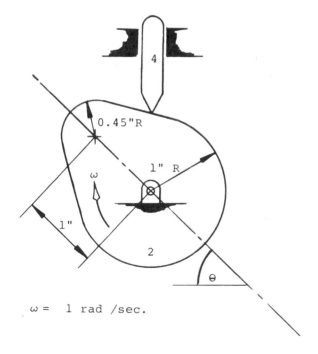

$\omega = $ 1 rad /sec.

Figure P.44

13.   Determine the following (Figure P.43):

     a.   The linear acceleration of point C
     b.   The angular acceleration of point A relative to point C
     c.   The angular velocity of point A relative to point C

14.   Determine the linear acceleration of the cam follower, link 4 of Figure P.44, for the following positions:

     a. $\theta = 45°$
     b. $\theta = 75°$

15.   Determine the linear acceleration of the cam follower, link 4 of Figure P.45, using:

     a.   The relative acceleration method
     b.   The equivalent linkage method

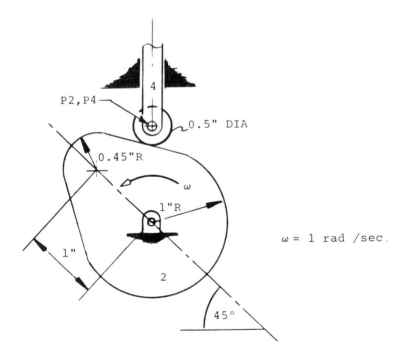

Figure P.45

16. a. Develop the linear velocity versus angular displacement curve for
        the complete cycle of the Scotch yoke mechanism (Figure P.46),
        based on a crank angular velocity of 2 rad/sec and a crank dis-
        placement of 30°.
     b. From linear velocity versus angular displacement curve in part (a),
        develop the linear acceleration-displacement curve, using graph-
        ical differentiation.

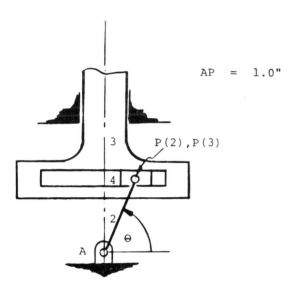

AP = 1.0"

Figure P.46

17. a. Develop the complete linear velocity-time curve for point E on the
        slider-crank mechanism (Figure P.47), using convenient time
        intervals, based on a crank angular velocity of 6.28 rad/sec.
     b. From the linear velocity-time curve in part (a), develop the linear
        displacement-time curve, using graphical integration.

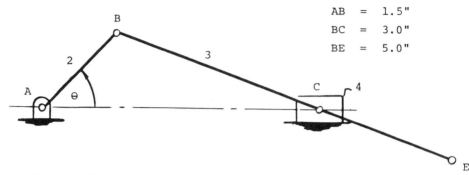

AB = 1.5"
BC = 3.0"
BE = 5.0"

Figure P.47

18. Develop the linear acceleration versus angular displacement curve
     for a complete cycle of the drag-link mechanism (Figure P.48), based
     on a crank angular velocity of 1 rad/sec and a crank angular accel-
     eration of 0.5 rad/sec.

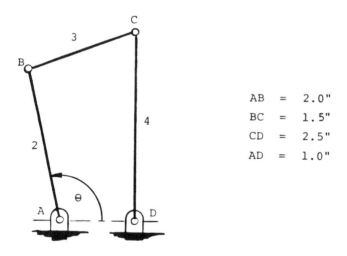

AB = 2.0"
BC = 1.5"
CD = 2.5"
AD = 1.0"

Figure P.48

19. Develop the complete linear acceleration-time curve for the sliding
     coupler mechanism (Figure P.49), based on a crank angular velocity
     of 30 rad/sec.

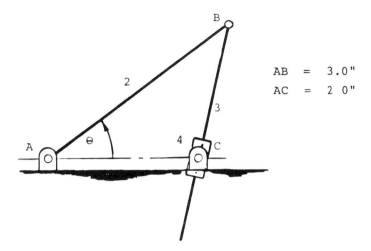

$$AB = 3.0"$$
$$AC = 2\ 0"$$

Figure P.49

20. A vehicle starting from rest is observed to have the following speeds at the times given:

Time (sec)	1	2	4	6	8	10	12
Velocity (mph)	1.5	3.0	8.5	16	21.5	25	26.5

Draw the velocity diagram and obtain from it the acceleration and displacement curves.

ANALYTICAL TECHNIQUES: VELOCITY AND ACCELERATION

1. Using the simplified vector method equations, calculate the acceleration of point C and the acceleration of point C relative to B in Figure P.11 for any of the following conditions:

   a. $\theta = 45°$ rad, $\omega = 1$ rad/sec (clockwise),
         $\alpha = 0$ rad/sec^2 (clockwise)
   b. $\theta = 135°$ rad, $\omega = 1$ rad/sec (counterclockwise),
         $\alpha = 0.5$ rad/sec^2 (clockwise)
   c. $\theta = 225°$ rad, $\omega = 1$ rad/sec (clockwise),
         $\alpha = 0$ rad/sec^2 (counterclockwise)
   d. $\theta = 315°$ rad, $\omega = 1$ rad/sec (counterclockwise),
         $\alpha = 0.5$ rad/sec^2 (counterclockwise)

Check the results using alternative equations, where possible, or a graphical method.

2.   Repeat Problem 1 but use Figure P.12.

3.   Repeat Problem 1 but use Figure P.13.

4.   Repeat Problem 1 but use Figure P.14.

5.   Repeat Problem 1 but use Figure P.15.

6.   Repeat Problem 1 but use Figure P.16.

7.   Repeat Problem 1 but use Figure P.17.

8.   Repeat Problem 1 but use Figure P.18.

Appendix A

A.1 INTRODUCTION

The computer programs listed in this section were developed to analyze the linkages covered in Part III, based on the mathematical methods discussed in that section. The programs are written in two common languages:

1. Fortran language applicable to a Univac 1108 digital computer
2. Calculator Keystroke language applicable to a Hewlett-Packard HP-41C calculator

To apply these programs, typically, the link lengths of the mechanism, the crank angle, the angular velocity, and the angular acceleration must be known. In the slider-crank case, additional information on slider offset (or eccentricity) must also be known. From these data, the required velocities and accelerations can be computed for any specified angular position of the crank in its motion cycle.

As illustrations of typical outputs obtained from these programs, printouts of example problem results are given following each program listing. The Fortran programs begin on page 358.

A.2 CALCULATOR OPERATING PROCEDURE

This procedure is based on use of the Hewlett-Packard HP-41C programmable calculator in conjunction with the Math Pack Module for computation involving complex numbers and a HP peripheral printer for printing out the results. The calculator programs begin on page 388.

PROCEDURE

1. Install the Math Pack Module and connect the printer.
2. Turn on the calculator.
3. Allocate the number of storage registers required for program.

> Press: XEQ
> alpha
> SIZE
> alpha
> 080

4. Set a flag to facilitate use of the Math Pack Module which computes complex numbers.

> Press: □ SF04

5. Prepare to load the program. It may be necessary to remove a previously stored program from program memory to create room for the new program.

> Press: □ GTO..

6. Load the program.

> Press: PRGM

Write the program steps into the program memory following the step-by-step instructions given in the printed program. Upon entering the last instruction, "END," exit the program mode.

> Press: PRGM

7. Execute and run the program.

> Press: XEQ
> alpha
> "PROGRAM NAME"
> alpha

If the program name is spelled incorrectly, the display panel will flash "NONEXISTENT." To correct this, repeat the procedure more carefully. Also, if the printer is not on-line, the signal "PRINTER OFF" will be flashed on the display panel. Be sure that the printer is on-line, switched on, and set in the manual position or mode. Check to be sure that the correct program is being executed.

8. Input the data for variables as requested in the display panel.

> Press: (numerical keys for the appropriate value of each
> variable displayed), then
> Press: R/S (after each variable input)

Note that, depending on the nature of the problem, some variable values are given, whereas others are calculated by the program.

9. Program execution and printing will begin after the data for the last requested variable are entered.

10. At the completion of program execution and printing, turn the calculator and printer to the "OFF" position.

FOUR-BAR: SIMPLIFIED VECTOR METHOD

```
LINKAGE*ANALYSIS(1).FOUR-BAR/ANALYSIS(2)
    1     C      FOUR-BAR LINKAGE ANALYSIS - SIMPLIFIED VECTOR METHOD
    2     C      AB      = CRANK
    3     C      BC      = COUPLER
    4     C      CD      = FOLLOWER
    5     C      AD      = FRAME
    6     C      C       = DEGREES/RADIAN
    7     C      THETAA  = THETA(A), POSITION ANGLE OF LINK AB
    8     C      THETAB  = THETA(B), POSITION ANGLE OF LINK BC
    9     C      THETAC  = THETA(C), POSITION ANGLE OF LINK CD
   10     C      THETAX  = THETA(A) + 90 DEGS.
   11     C      THETAY  = THETA(B) + 90 DEGS.
   12     C      THETAZ  = THETA(C) + 90 DEGS.
   13     C      PHID    = PHI(D),ANGLE BETWEEN CB AND AD
   14     C      GAMMAB  = GAMMA(B)
   15     C      GAMMAD  = GAMMA(D)
   16     C      OMEGAA  = OMEGA(A),ANGULAR VELOCITY OF CRANK
   17     C      ALPHAA  = ALPHA(A),ANGULAR ACCELERATION OF CRANK
   18     C      SMB     = THETA(B)-THETA(C)
   19     C      SMC     = THETA(B)-THETA(A)
   20     C      SMCB    = THETA(C)-THETA(A)
   21     C      VB      = VELOCITY OF B
   22     C      VC      = VELOCITY OF C
   23     C      VCB     = VELOCITY OF C RELATIVE TO B
   24     C      CVB     = VEL OF B (COMPLEX)
   25     C      CVC     = VEL OF C (COMPLEX)
   26     C      CVCB    = VEL OF C WITH RESPECT TO B (COMPLEX)
   27     C      ABSVB   = ABSOLUTE VELOCITY OF B
   28     C      ABSVC   = ABSOLUTE VELOCITY OF C
   29     C      ABSVCB  = ABSOLUTE VELOCITY OF C WITH RESPECT TO B
   30     C      ACCNB   = NORMAL ACC OF B (COMPLEX)
   31     C      ACCTB   = TANGENTIAL ACC OF B (COMPLEX)
   32     C      ACCNC   = NORMAL ACC OF C (COMPLEX)
   33     C      ACCTC   = TANGENTIAL ACC OF C (COMPLEX)
   34     C      ACCNCB  = NORMAL ACC OF C WITH RESPECT TO B(COMPLEX)
   35     C      ACCTCB  = TANGENTIAL ACC OF C WITH RESPECT TO B(COMPLEX)
   36     C      ATC     = TANGENTIAL ACC OF C
   37     C      ATCB    = TANGENTIAL ACC OF C WITH RESPECT TO B
   38     C      ACCB    = ACC OF B (COMPLEX)
   39     C      ACCC    = ACC OF C (COMPLEX)
   40     C      ACCCB   = ACC OF C WITH RESPECT TO B (COMPLEX)
   41     C      ABSNB   = ABSOLUTE NORMAL ACC OF B
   42     C      ABSTB   = ABSOLUTE TANGENTIAL ACC OF B
   43     C      ABSNC   = ABSOLUTE NORMAL ACC OF C
   44     C      ABSTB   = ABSOLUTE TANGENTIAL ACC OF C
   45     C      ABSNCB  = ABSOLUTE NORMAL ACC OF C WITH RESPECT TO B
   46     C      ABSTCB  = ABSOLUTE TANGENTIAL ACC OF C WITH RESPECT TO B
   47     C      ABSAB   = ABSOLUTE ACCELERATION OF B
   48     C      ABSAC   = ABSOLUTE ACCELERATION OF C
   49     C      ABSACB  = ABSOLUTE ACC OF C RELATIVE TO B
   50     C      PHZVB   = PHASE ANGLE OF VELOCITY OF B
   51     C      PHZVC   = PHASE ANGLE OF VELOCITY OF C
   52     C      PHZVCB  = PHASE ANGLE OF VELOCITY OF C WITH RESPECT TO B
   53     C      PHZNB   = PHASE ANGLE OF NORMAL ACC OF B
   54     C      PHZTB   = PHASE ANGLE OF TANGENTIAL ACC OF B
   55     C      PHZNC   = PHASE ANGLE OF NORMAL ACC OF C
   56     C      PHZTC   = PHASE ANGLE OF TANGENTIAL ACC OF C
   57     C      PHZNCB  = PHASE ANGLE OF NORMAL ACC OF C WITH RESPECT TO B
   58     C      PHZTCB  = PHASE ANGLE OF TANGENTIAL ACC OF C WITH RESPECT TO B
   59     C      PHZAB   = PHASE ANGLE OF ABSOLUTE ACC OF B
   60     C      PHZAC   = PHASE ANGLE OF ABSOLUTE ACC OF C
   61     C      PHZACB  = PHASE ANGLE OF ABSOLUTE ACC OF C WITH RESPECT TO B
   62     C
   63            C = 57.29578
   64            COMPLEX ACCNB,ACCTB,ACCB,ACCNC,ACCTC,ACCC,ACCNCB,ACCTCB,ACCCB
   65            COMPLEX CVB,CVC,CVCB
   66     C
   67          1 READ(5,100)AB,BC,CD,AD,THETAA,OMEGAA,ALPHAA
   68        100 FORMAT(7F10.4)
   69            IF(AB.EQ.0.0) GO TO 999
   70            WRITE(6,104)
   71        104 FORMAT(1H1,9X,'PROBLEM DATA')
```

```
72              WRITE(6,105)
73        105 FORMAT(//12X,'AB'10X'BC'10X'CD'10X'AD'4X'THETA(A)'4X'OMEGA(A)'
74           14X'ALPHA(A)')
75              WRITE(6,106)AB,BC,CD,AD,THETAA,OMEGAA,ALPHAA
76        106 FORMAT(/2X,8F12.4)
77        C
78              THETAA= THETAA/C
79              BD=SQRT(AB**2+AD**2-2.0*AB*AD*COS(THETAA))
80              PHID= ASIN(AB/BD*SIN(THETAA))
81              PHID  =PHID*C
82              S2=(BC+BD+CD)/2.0
83        C
84              GAMMAB=ACOS((BC**2+BD**2-CD**2)/(2.0*BC*BD))
85              GAMMAD=ACOS((BD**2+CD**2-BC**2)/(2.0*BD*CD))
86        C
87              GAMMAB= GAMMAB*C
88              GAMMAD= GAMMAD*C
89        C
90              THETAA = THETAA*C
91              THETAB = GAMMAB - PHID
92              THETAC = 180. - PHID - GAMMAD
93        C
94              WRITE(6,901)
95        901 FORMAT(//12X'BD'6X 'PHI(D)'4X'GAMMA(B)'4X'GAMMA(D)'
96           14X'THETA(A)'4X'THETA(B)'4X'THETA(C)')
97              WRITE(6,107)BD,PHID,GAMMAB,GAMMAD,THETAA,THETAB,THETAC
98        C
99              SMB = THETAB - THETAC
100             SMC = THETAB - THETAA
101             SMCB = THETAC - THETAA
102             SRB=SMB/C
103             SRC=SMC/C
104             SRCB=SMCB/C
105             SNB = SIN(SRB)
106             SNC = SIN(SRC)
107             SNCB = SIN(SRCB)
108             VB = OMEGAA * AB
109             VC = VB*SNC/SNB
110             VCB = VB*SNCB/SNB
111       C
112             WRITE(6,902)
113       902 FORMAT(//11X,'SMB'9X'SMC'8X'SMCB')
114             WRITE(6,107)SMB,SMC,SMCB
115             WRITE(6,903)
116       903 FORMAT(/10X,'LINEAR VELOCITIES')
117             WRITE(6,904)
118       904 FORMAT(/10X'V(B)'8X'V(C)'7X'V(CB)')
119             WRITE(6,107)VB,VC,VCB
120       C
121             THETAA = (THETAA/C)
122             THETAB = (THETAB/C)
123             THETAC = (THETAC/C)
124       C
125             THETAY = (THETAB + 90.0/C)
126             THETAX = (THETAA + 90.0/C)
127             THETAZ = (THETAC + 90.0/C)
128       C
129             X = COS(THETAX)
130             Y = SIN(THETAX)
131             CVB = VB * CMPLX(X,Y)
132             X= COS(THETAZ)
133             Y = SIN(THETAZ)
134             CVC = VC* CMPLX(X,Y)
135             X = COS(THETAY)
136             Y = SIN(THETAY)
137             CVCB = VCB * CMPLX(X,Y)
138       C
139             ABSVB = CABS(CVB)
140             PHZVB=C*ATAN2(AIMAG(CVB),REAL(CVB))
141             ABSVC = CABS(CVC)
142       C     TEST ABSVC
```

```
143                 IF(ABSVC.GT. O.0001)GO TO 80
144                 PHZVC=0.0
145                 GO TO 85
146              80 PHZVC=C*ATAN2(AIMAG(CVC),REAL(CVC))
147              85 CONTINUE
148                 ABSVCB = CABS(CVCB)
149        C        TEST ABSVCB
150                 IF(ABSVCB.GT. O.0001)GO TO 90
151                 PHZVCB=0.0
152                 GO TO 95
153              90 PHZVCB=C*ATAN2(AIMAG(CVCB),REAL(CVCB))
154              95 CONTINUE
155                 WRITE(6,905)
156             905 FORMAT(/1OX,'REAL'8X'IMAG'9X'ABS'7X'PHASE')
157                 WRITE(6,301)CVB,ABSVB,PHZVB
158                 WRITE(6,302)CVC,ABSVC,PHZVC
159                 WRITE(6,303)CVCB,ABSVCB,PHZVCB
160                 D1 = VC**2*COS(THETAC)/CD
161                 D2 = -OMEGAA**2*AB*COS(THETAA)
162                 D3 = ALPHAA*AB*COS(THETAX)
163                 D4 =-VCB**2*COS(THETAB)/BC
164                 D5 = VC**2*SIN(THETAC)/CD
165                 D6 = -OMEGAA**2*AB*SIN(THETAA)
166                 D7 = ALPHAA*AB*SIN(THETAX)
167                 D8 =-VCB**2*SIN(THETAB)/BC
168                 C1 = D1+D2+D3+D4
169                 C2 = D5+D6+D7+D8
170                 A1 = COS(THETAZ)
171                 A2 = SIN(THETAZ)
172                 B1 =-COS(THETAY)
173                 B2 =-SIN(THETAY)
174        C
175                 ATC = (C1*B2 - C2*B1)/(A1*B2 - A2*B1)
176                 ATCB = (A1*C2 - A2*C1)/(A1*B2 - A2*B1)
177        C
178                 X = COS(THETAA)
179                 Y = SIN(THETAA)
180                 ACCNB = -OMEGAA**2*AB*CMPLX(X,Y)
181                 X=COS(THETAX)
182                 Y=SIN(THETAX)
183                 ACCTB = ALPHAA*AB*CMPLX(X,Y)
184                 ACCB = ACCNB + ACCTB
185                 X = COS(THETAC)
186                 Y = SIN(THETAC)
187                 ACCNC = -VC**2*CMPLX(X,Y)/CD
188                 X = COS(THETAZ)
189                 Y = SIN(THETAZ)
190                 ACCTC = ATC*CMPLX(X,Y)
191                 ACCC = ACCNC + ACCTC
192                 X = COS(THETAB)
193                 Y = SIN(THETAB)
194                 ACCNCB =-VCB**2*CMPLX(X,Y)/BC
195                 X = COS(THETAY)
196                 Y = SIN(THETAY)
197                 ACCTCB = ATCB * CMPLX(X,Y)
198                 ACCCB = ACCNCB + ACCTCB
199        C
200        C
201                 ABSNB=CABS(ACCNB)
202                 PHZNB = C*ATAN2(AIMAG(ACCNB),REAL(ACCNB))
203                 ABSTB=CABS(ACCTB)
204        C        TEST ABSTB
205                 IF(ABSTB.GT.0.0001)GO TO 10
206                 PHZTB = O.0
207                 GO TO 15
208              10 PHZTB = C*ATAN2(AIMAG(ACCTB),REAL(ACCTB))
209              15 CONTINUE
210                 ABSNC=CABS(ACCNC)
211        C        TEST ABSNC
212                 IF(ABSNC.GT. O.0001)GO TO 20
213                 PHZNC=0.0
214                 GO TO 25
```

```
215           20 PHZNC = C*ATAN2(AIMAG(ACCNC),REAL(ACCNC))
216           25 CONTINUE
217              ABSTC=CABS(ACCTC)
218      C       TEST ABSTC
219              IF(ABSTC.GT. 0.0001)GO TO 30
220              PHZTC=0.0
221              GO TO 35
222           30 PHZTC = C*ATAN2(AIMAG(ACCTC),REAL(ACCTC))
223           35 CONTINUE
224              ABSNCB=CABS(ACCNCB)
225      C       TEST ABSNCB
226              IF(ABSNCB. GT. 0.0001)GO TO 40
227              PHZNCB=0.0
228              GO TO 45
229           40 PHZNCB = C*ATAN2(AIMAG(ACCNCB),REAL(ACCNCB))
230           45 CONTINUE
231              ABSTCB=CABS(ACCTCB)
232      C       TEST ABSTCB
233              IF(ABSTCB.GT. 0.0001)GO TO 50
234              PHZTCB=0.0
235              GO TO 55
236           50 PHZTCB = C*ATAN2(AIMAG(ACCTCB),REAL(ACCTCB))
237           55 CONTINUE
238              WRITE(6,906)
239          906 FORMAT(/10X,'NORMAL ACCELERATIONS'38X,'TANGENTIAL ACCELERATIONS')
240              WRITE(6,915)
241          915 FORMAT(/10X,'REAL'8X'IMAG'9X'ABS'7X'PHASE'18X'REAL'8X'IMAG'
242             19X'ABS'7X'PHASE')
243              WRITE(6,301)ACCNB,ABSNB,PHZNB,ACCTB,ABSTB,PHZTB
244              WRITE(6,302)ACCNC,ABSNC,PHZNC,ACCTC,ABSTC,PHZTC
245              WRITE(6,303)ACCNCB,ABSNCB,PHZNCB,ACCTCB,ABSTCB,PHZTCB
246      C
247              WRITE(6,907)
248          907 FORMAT(/10X'ABSOLUTE ACCELERATIONS')
249              ABSAB = CABS(ACCB)
250              PHZAB = C*ATAN2(AIMAG(ACCB),REAL(ACCB))
251              ABSAC = CABS(ACCC)
252      C       TEST ABSAC
253              IF(ABSAC.GT. 0.0001) GO TO 60
254              PHZAC=0.0
255              GO TO 65
256           60 PHZAC = C*ATAN2(AIMAG(ACCC),REAL(ACCC))
257           65 CONTINUE
258              ABSACB = CABS(ACCCB)
259      C       TEST ABSACB
260              IF(ABSACB.GT. 0.0001)GO TO 70
261              PHZACB=0.0
262              GO TO 75
263           70 PHZACB = C*ATAN2(AIMAG(ACCCB),REAL(ACCCB))
264           75 CONTINUE
265      C
266              WRITE(6,301)ACCB,ABSAB,PHZAB
267              WRITE(6,302)ACCC,ABSAC,PHZAC
268              WRITE(6,303)ACCCB,ABSACB,PHZACB
269              THETAX=THETAX*C
270              THETAY=THETAY*C
271              THETAZ=THETAZ*C
272              WRITE(6,908)
273          908 FORMAT(/6X,'THETA(X)'4X'THETA(Y)'4X'THETA(Z)')
274              WRITE(6,107)THETAX,THETAY,THETAZ
275          107 FORMAT(/2X,8F12.4)
276              WRITE(6,909)
277          909 FORMAT(/12X,'A1'10X'A2'10X'B1'10X'B2'10X'C1'10X'C2'9X'ATC'
278             18X'ATCB')
279              WRITE(6,107)A1,A2,B1,B2,C1,C2,ATC,ATCB
280          301 FORMAT(/2X,4F12.4,8X'B ',4F12.4)
281          302 FORMAT(/2X,4F12.4,8X'C ',4F12.4)
282          303 FORMAT(/2X,4F12.4,8X'CB',4F12.4)
283      C
284              GO TO 1
285          999 STOP
286              END
```

PROBLEM DATA

AB	BC	CD	AD	THETA(A)	OMEGA(A)	ALPHA(A)
1.5000	3.0000	3.0000	4.0000	30.0000	2.0000	1.0000

BD	PHI(D)	GAMMA(B)	GAMMA(D)	THETA(A)	THETA(B)	THETA(C)
2.8032	15.5188	62.1478	62.1478	30.0000	46.6289	102.3334

SMB	SMC	SMCB
-55.7045	16.6289	72.3334

LINEAR VELOCITIES

V(B)	V(C)	V(CB)
3.0000	-1.0392	-3.4601

REAL	IMAG	ABS	PHASE	
-1.5000	2.5981	3.0000	120.0000	B
1.0152	.2220	1.0392	12.3334	C
2.5152	-2.3761	3.4601	-43.3711	CB

NORMAL ACCELERATIONS   TANGENTIAL ACCELERATIONS

REAL	IMAG	ABS	PHASE		REAL	IMAG	ABS	PHASE
-5.1962	-3.0000	6.0000	-150.0000	B	-.7500	1.2990	1.5000	120.0000
.0769	-.3517	.3600	-77.6666	C	-10.7699	-2.3548	11.0244	-167.6666
-2.7405	-2.9009	3.9907	-133.3711	CB	-2.0064	1.8954	2.7601	136.6289

ABSOLUTE ACCELERATIONS

-5.9462	-1.7010	6.1847	-164.0362	B
-10.6930	-2.7065	11.0302	-165.7964	C
-4.7469	-1.0055	4.8522	-168.0402	CB

THETA(X)	THETA(Y)	THETA(Z)
120.0000	136.6289	192.3334

A1	A2	B1	B2	C1	C2	ATC	ATCB
-.9769	-.2136	.7269	-.6867	-8.7635	-4.2502	11.0244	2.7601

PROBLEM DATA

AB	BC	CD	AD	THETA(A)	OMEGA(A)	ALPHA(A)
2.0000	1.5000	2.5000	1.0000	240.0000	1.0000	.5000

BD	PHI(D)	GAMMA(B)	GAMMA(D)	THETA(A)	THETA(B)	THETA(C)
2.6458	-40.8934	67.7923	33.7446	240.0000	108.6857	187.1488

SMB	SMC	SMCB
-78.4630	-131.3143	-52.8512

LINEAR VELOCITIES

V(B)	V(C)	V(CB)
2.0000	1.5332	1.6270

REAL	IMAG	ABS	PHASE	
1.7321	-1.0000	2.0000	-30.0000	B
.1908	-1.5213	1.5332	-82.8512	C
-1.5413	-.5213	1.6270	-161.3143	CB

NORMAL ACCELERATIONS   TANGENTIAL ACCELERATIONS

REAL	IMAG	ABS	PHASE		REAL	IMAG	ABS	PHASE
1.0000	1.7321	2.0000	60.0000	B	.8660	-.5000	1.0000	-30.0000
.9329	.1170	.9403	7.1488	C	.1280	-1.0202	1.0282	-82.8512
.5654	-1.6718	1.7648	-71.3143	CB	-1.3705	-.4635	1.4468	-161.3143

ABSOLUTE ACCELERATIONS

1.8660	1.2321	2.2361	33.4349	B
1.0609	-.9032	1.3933	-40.4102	C
-.8051	-2.1353	2.2820	-110.6594	CB

THETA(X)	THETA(Y)	THETA(Z)
330.0000	198.6857	277.1488

A1	A2	B1	B2	C1	C2	ATC	ATCB
.1244	-.9922	.9473	.3204	1.4985	-.5567	1.0282	1.4468

SLIDER-CRANK: SIMPLIFIED VECTOR METHOD

```
LINKAGE*ANALYSIS(1).SL-CRANK/ANALYSIS-1(2)
     1     C         SLIDER-CRANK MECHANISM ANALYSIS-SIMPLIFIED VECTOR METHOD
     2     C    C      = DEGREES/RADIAN(CONSTANT)
     3     C    AB     = CRANK
     4     C    BC     = CONNECTING ROD
     5     C    PHI    = PHI(3),ANGLE BETWEEN CONNECTING ROD AND DEAD CENTER
     6     C    THETA2 = POSITION ANGLE OF LINK AB
     7     C    THETA3 = POSITION ANGLE OF LINK BC
     8     C    THETAX = THETA(2)+90DEGS.
     9     C    THETAY = THETA(3)+90DEGS.
    10     C    OMEGA2 = ANGULAR VELOCITY OF LINK AB
    11     C    ALPHA2 = ANGULAR ACC OF LINK AB
    12     C    SNB    = 90 - THETA(DEGS)
    13     C    SNC    = THETA(2)-THETA(3)
    14     C    SNBC   = 90 - THETA(2)
    15     C    VB     = VELOCITY OF B
    16     C    VC     = VELOCITY OF C
    17     C    VBC    = VELOCITY OF B RELATIVE TO C
    18     C    CVB    = VEL OF B (COMPLEX)
    19     C    CVC    = VEL OF C (COMPLEX)
    20     C    CVBC   = VEL OF B RELATIVE TO C (COMPLEX)
    21     C    ABSVB  = ABSOLUTE VELOCITY OF B
    22     C    ABSVC  = ABSOLUTE VELOCITY OF C
    23     C    ABSVBC = ABSOLUTE VELOCITY OF B RELATIVE TO C
    24     C    ACCNB  = NORMAL ACC OF B (COMPLEX)
    25     C    ACCTB  = TANGENTIAL ACC OF B (COMPLEX)
    26     C    ACCNC  = NORMAL ACC OF C (COMPLEX)
    27     C    ACCTC  = TANGENTIAL ACC OF C (COMPLEX)
    28     C    ACCNBC = NORMAL ACC OF B REL TO C (COMPLEX)
    29     C    ACCTBC = TANGENTIAL ACC OF B RELATIVE TO C(COMPLEX)
    30     C    ATC    = TANGENTIAL ACC OF C
    31     C    ATBC   = TANGENTIAL ACC OF B RELATIVE TO C
    32     C    ACCB   = ACC OF B (COMPLEX)
    33     C    ACCC   = ACC OF C (COMPLEX)
    34     C    ACCBC  = ACC OF B RELATIVE TO C (COMPLEX)
    35     C    ABSNB  = ABSOLUTE NORMAL ACC OF B
    36     C    ABSTB  = ABSOLUTE TANGENTIAL ACC OF B
    37     C    ABSNC  = ABSOLUTE NORMAL ACC OF C
    38     C    ABSTC  = ABSOLUTE TANGENTIAL ACC OF C
    39     C    ABSNBC = ABSOLUTE NORMAL ACC OF B REL TO C
    40     C    ABSTBC = ABSOLUTE TANGENTIAL ACC OF B RELATIVE TO C
    41     C    ABSAB  = ABSOLUTE ACC OF B
    42     C    ABSAC  = ABSOLUTE ACC OF C
    43     C    ABSABC = ABSOLUTE ACC OF B RELATIVE TO C
    44     C    PHZVB  = PHASE ANGLE OF VELOCITY OF B
    45     C    PHZVC  = PHASE ANGLE OF VELOCITY OF C
    46     C    PHZVBC = PHASE ANGLE OF VELOCITY OF B RELATIVE T C
    47     C    PHZNB  = PHASE ANGLE OF NORMAL ACC OF B
    48     C    PHZTB  = PHASE ANGLE OF TANGENTIAL ACC OF B
    49     C    PHZNC  = PHASE ANGLE OF NORMAL ACC OF C
    50     C    PHZTC  = PHASE ANGLE OF TANGENTIAL ACC OF C
    51     C    PHZNBC = PHASE ANGLE OF NORMAL ACC OF B REL TO C
    52     C    PHZTBC = PHASE ANGLE OF TANGENTIAL ACC OF B RELATIVE TO C
    53     C    PHZAB  = PHASE ANGLE OF ABSOLUTE ACC OF B
    54     C    PHZAC  = PHASE ANGLE OF ABSOLUTE ACC OF C
    55     C    PHZABC = PHASE ANGLE OF ABSOLUTE ACC OF B RELATIVE TO C
    56          COMPLEX CVB,CVC,CVBC,ACCC,ACCNC,ACCTC
    57          COMPLEX ACCNB,ACCTB,ACCB,ACCNBC,ACCTBC,ACCBC
    58          C=57.29578
```

```
59              1 READ(5,99)AB,BC,THETA2,OMEGA2,ALPHA2,E
60             99 FORMAT(6F12.4)
61                IF(AB.EQ.0.0) GO TO 999
62                WRITE(6,899)
63            899 FORMAT(1H1,9X,'PROBLEM DATA')
64                WRITE(6,900)
65            900 FORMAT(//12X,'AB'10X'BC'4X'THETA(2)'4X'OMEGA(2)'4X'ALPHA(2)'6X'ECC
66               1(E)')
67                WRITE(6,100)AB,BC,THETA2,OMEGA2,ALPHA2,E
68            100 FORMAT(/2X,6F12.4)
69                THETA2=THETA2/C
70                PHI=ASIN(AB/BC*SIN(THETA2)-E/BC)
71                PHI=PHI*C
72                THETA3=180.0-PHI
73                WRITE(6,901)
74            901 FORMAT(//8X,'PHI(3)'4X'THETA(3)')
75                WRITE(6,101)PHI,THETA3
76                THETA2=THETA2*C
77                SNB=90.0-THETA3
78                SNC=THETA3-THETA2
79                SNBC=90.0-THETA2
80                WRITE(6,902)
81            902 FORMAT(/11X,'SNB'9X'SNC'8X'SNBC')
82                WRITE(6,101)SNB,SNC,SNBC
83                SMB=SNB/C
84                SMC=SNC/C
85                SMBC=SNBC/C
86                VB= AB*OMEGA2
87                VC= VB*SIN(SMC)/SIN(SMB)
88                VBC=VB*SIN(SMBC)/SIN(SMB)
89                WRITE(6,903)
90            903 FORMAT(/10X,'LINEAR VELOCITIES')
91                WRITE(6,904)
92            904 FORMAT(/10X,'V(B)'8X'V(C)'7X'V(BC)')
93                WRITE(6,101)VB,VC,VBC
94                THETAX=(THETA2+90.0)/C
95                THETAY=(THETA3+90.0)/C
96                THETA2=THETA2/C
97                THETA3=THETA3/C
98                X=COS(THETAX)
99                Y=SIN(THETAX)
100               CVB=VB*CMPLX(X,Y)
101               ABSVB=CABS(CVB)
102               PHZVB=C*ATAN2(AIMAG(CVB),REAL(CVB))
103               WRITE(6,905)
104           905 FORMAT(/10X,'REAL'8X'IMAG'9X'ABS'7X'PHASE')
105               WRITE(6,301)CVB,ABSVB,PHZVB,
106               CVC=VC
107               ABSVC=CABS(CVC)
108    C          TEST FOR ABS VEL OF C
109               IF(ABSVC.GT.0.0001)GO TO 20
110               PHZVC=0.0
111               GO TO 25
112            20 PHZVC=C*ATAN2(AIMAG(CVC),REAL(CVC))
113            25 CONTINUE
114               WRITE(6,302)CVC,ABSVC,PHZVC
115               X=COS(THETAY)
116               Y=SIN(THETAY)
117               CVBC=VBC*CMPLX(X,Y)
```

```
118             ABSVBC=CABS(CVBC)
119       C     TEST VEL OF B REL TO C
120             IF(ABSVBC.GT.0.0001) GO TO 30
121             PHZVBC=0.0
122             GO TO 35
123          30 PHZVBC=C*ATAN2(AIMAG(CVBC),REAL(CVBC))
124          35 CONTINUE
125             WRITE(6,303)CVBC,ABSVBC,PHZVBC
126             A1=1
127             A2=0
128             B1=COS(THETAY)
129             B2=SIN(THETAY)
130             C1=-(OMEGA2**2)*AB*COS(THETA2)+(ALPHA2*AB)*COS(THETAX)+(VBC**2/BC)
131            1*COS(THETA3)
132             C2=-(OMEGA2**2)*AB*SIN(THETA2)+(ALPHA2*AB)*SIN(THETAX)+(VBC**2/BC)
133            1*SIN(THETA3)
134             ATC=(C1*B2-C2*B1)/(A1*B2-A2*B1)
135             ATBC=(A1*C2-A2*C1)/(A1*B2-A2*B1)
136             X=COS(THETA2)
137             Y=SIN(THETA2)
138             ACCNB=-(VB**2/AB)*CMPLX(X,Y)
139             ABSNB=CABS(ACCNB)
140             PHZNB=C*ATAN2(AIMAG(ACCNB),REAL(ACCNB))
141             X=COS(THETAX)
142             Y=SIN(THETAX)
143             ACCTB=AB*ALPHA2*CMPLX(X,Y)
144             ABSTB=CABS(ACCTB)
145       C     TEST ABS TAN ACC OF B
146             IF(ABSTB.GT.0.0001) GO TO 40
147             PHZTB=0.0
148             GO TO 45
149          40 PHZTB=C*ATAN2(AIMAG(ACCTB),REAL(ACCTB))
150          45 CONTINUE
151             WRITE(6,906)
152         906 FORMAT(/10X,'NORMAL ACCELERATIONS'40X,'TANGENTIAL ACCELERATIONS')
153             WRITE(6,915)
154         915 FORMAT(/10X,'REAL'8X'IMAG'9X'ABS'7X'PHASE'20X,'REAL'8X'IMAG'
155            19X'ABS'7X'PHASE')
156             WRITE(6,301)ACCNB,ABSNB,PHZNB,ACCTB,ABSTB,PHZTB
157             ACCNC=0.0
158             ABSNC=0.0
159             PHZNC=0.0
160             ACCTC=ATC
161             ABSTC=CABS(ACCTC)
162       C     TEST FOR TAN ACC OF C
163             IF(ABSTC.GT.0.0001)GO TO 90
164             PHZTC= 0.0
165             GO TO 95
166          90 PHZTC=C*ATAN2(AIMAG(ACCTC),REAL(ACCTC))
167          95 CONTINUE
168             WRITE(6,302)ACCNC,ABSNC,PHZNC,ACCTC,ABSTC,PHZTC
169             X=COS(THETA3)
170             Y=SIN(THETA3)
171             ACCNBC=-(VBC**2/BC)*CMPLX(X,Y)
172             ABSNBC=CABS(ACCNBC)
173       C     TEST FOR ABS NOR ACC OF B REL TO C
174             IF(ABSNBC.GT.0.0001) GO TO 50
175             PHZNBC=0.0
176             GO TO 55
```

```
177          50 PHZNBC=C*ATAN2(AIMAG(ACCNBC),REAL(ACCNBC))
178          55 CONTINUE
179             X=COS(THETAY)
180             Y=SIN(THETAY)
181             ACCTBC=ATBC*CMPLX(X,Y)
182             ABSTBC=CABS(ACCTBC)
183    C        TEST TAN ACC B REL TO C
184             IF(ABSTBC.GT.0.0001) GO TO 60
185             PHZTBC=0.0
186             GO TO 65
187          60 PHZTBC=C*ATAN2(AIMAG(ACCTBC),REAL(ACCTBC))
188          65 CONTINUE
189             WRITE(6,303)ACCNBC,ABSNBC,PHZNBC,ACCTBC,ABSTBC,PHZTBC
190             ACCB=ACCNB+ACCTB
191             ABSAB=CABS(ACCB)
192             PHZAB=C*ATAN2(AIMAG(ACCB),REAL(ACCB))
193             WRITE(6,907)
194         907 FORMAT(/10X,'ABSOLUTE ACCELERATIONS')
195             WRITE(6,301)ACCB,ABSAB,PHZAB
196             ACCC=ACCNC+ACCTC
197             ABSAC=CABS(ACCC)
198    C        TEST FOR ABS ACC OF C
199             IF(ABSAC.GT.0.0001)GO TO 10
200             PHZAC=0.0
201             GO TO 15
202          10 PHZAC=C*ATAN2(AIMAG(ACCC),REAL(ACCC))
203          15 CONTINUE
204             WRITE(6,302)ACCC,ABSAC,PHZAC
205             ACCBC=ACCNBC+ACCTBC
206             ABSABC=CABS(ACCBC)
207    C        TEST ABS ACC OF B REL TO C
208             IF(ABSABC.GT.0.0001) GO TO 70
209             PHZABC=0.0
210             GO TO 75
211          70 PHZABC=C*ATAN2(AIMAG(ACCBC),REAL(ACCBC))
212          75 CONTINUE
213             WRITE(6,303)ACCBC,ABSABC,PHZABC
214             THETA2=THETA2*C
215             THETA3=THETA3*C
216             THETAX=THETAX*C
217             THETAY=THETAY*C
218             WRITE(6,908)
219         908 FORMAT( /6X,'THETA(X)'4X'THETA(Y)')
220             WRITE(6,201) THETAX,THETAY
221         201 FORMAT(/2X,4F12.4,12X,4F12.4)
222             WRITE(6,909)
223         909 FORMAT(/12X,'A1'10X'A2'10X'B1'10X'B2'10X'C1'10X'C2'9X'ATC'8X
224        1'ATBC')
225             WRITE(6,101) A1,A2,B1,B2,C1,C2,ATC,ATBC
226         101 FORMAT(/2X,8F12.4)
227         301 FORMAT(/2X,4F12.4,8X'(B) ',4F12.4)
228         302 FORMAT(/2X,4F12.4,8X'(C) ',4F12.4)
229         303 FORMAT(/2X,4F12.4,8X'(BC)',4F12.4)
230             GO TO 1
231         999 STOP
232             END
```

PROBLEM DATA

AB	BC	THETA(2)	OMEGA(2)	ALPHA(2)	ECC(E)
1.5000	3.0000	150.0000	1.0000	.0000	.0000

PHI(3)	THETA(3)
14.4775	165.5225

SNB	SNC	SNBC
-75.5225	15.5225	-60.0000

LINEAR VELOCITIES

V(B)	V(C)	V(BC)
1.5000	-.4146	1.3416

REAL	IMAG	ABS	PHASE	
-.7500	-1.2990	1.5000	-120.0000	(B)
-.4146	.0000	.4146	180.0000	(C)
-.3354	-1.2990	1.3416	-104.4775	(BC)

NORMAL ACCELERATIONS					TANGENTIAL ACCELERATIONS			
REAL	IMAG	ABS	PHASE		REAL	IMAG	ABS	PHASE
1.2990	-.7500	1.5000	-30.0000	(B)	.0000	.0000	.0000	.0000
.0000	.0000	.0000	.0000	(C)	.8730	.0000	.8730	.0000
.5809	-.1500	.6000	-14.4775	(BC)	-.1549	-.6000	.6197	-104.4775

ABSOLUTE ACCELERATIONS

REAL	IMAG	ABS	PHASE	
1.2990	-.7500	1.5000	-30.0000	(B)
.8730	.0000	.8730	.0000	(C)
.4260	-.7500	.8626	-60.4018	(BC)

THETA(X)	THETA(Y)
240.0000	255.5225

A1	A2	B1	B2	C1	C2	ATC	ATBC
1.0000	.0000	-.2500	-.9682	.7181	-.6000	.8730	.6197

PROBLEM DATA

AB	BC	THETA(2)	OMEGA(2)	ALPHA(2)	ECC(E)
1.5000	3.0000	150.0000	1.0000	.0000	.5000

PHI(3)	THETA(3)
4.7802	175.2198

SNB	SNC	SNBC
-85.2198	25.2198	-60.0000

LINEAR VELOCITIES

V(B)	V(C)	V(BC)
1.5000	-.6414	1.3036

REAL	IMAG	ABS	PHASE	
-.7500	-1.2990	1.5000	-120.0000	(B)
-.6414	.0000	.6414	180.0000	(C)
-.1086	-1.2990	1.3036	-94.7802	(BC)

NORMAL ACCELERATIONS

TANGENTIAL ACCELERATIONS

REAL	IMAG	ABS	PHASE		REAL	IMAG	ABS	PHASE
1.2990	-.7500	1.5000	-30.0000	(B)	.0000	.0000	.0000	.0000
.0000	.0000	.0000	.0000	(C)	.7933	.0000	.7933	.0000
.5645	-.0472	.5664	-4.7802	(BC)	-.0588	-.7028	.7053	-94.7802

ABSOLUTE ACCELERATIONS

REAL	IMAG	ABS	PHASE	
1.2990	-.7500	1.5000	-30.0000	(B)
.7933	.0000	.7933	.0000	(C)
.5057	-.7500	.9046	-56.0099	(BC)

THETA(X)	THETA(Y)
240.0000	265.2198

A1	A2	B1	B2	C1	C2	ATC	ATBC
1.0000	.0000	-.0833	-.9965	.7346	-.7028	.7933	.7053

PROBLEM DATA

AB	BC	THETA(2)	OMEGA(2)	ALPHA(2)	ECC(E)
1.5000	3.0000	150.0000	1.0000	.0000	-.5000

PHI(3)	THETA(3)
24.6243	155.3757

SNB	SNC	SNBC
-65.3757	5.3757	-60.0000

LINEAR VELOCITIES

V(B)	V(C)	V(BC)
1.5000	-.1546	1.4290

REAL	IMAG	ABS	PHASE	
-.7500	-1.2990	1.5000	-120.0000	(B)
-.1546	.0000	.1546	180.0000	(C)
-.5954	-1.2990	1.4290	-114.6243	(BC)

NORMAL ACCELERATIONS   TANGENTIAL ACCELERATIONS

REAL	IMAG	ABS	PHASE		REAL	IMAG	ABS	PHASE
1.2990	-.7500	1.5000	-30.0000	(B)	.0000	.0000	.0000	.0000
.0000	.0000	.0000	.0000	(C)	.8940	.0000	.8940	.0000
.6188	-.2836	.6807	-24.6243	(BC)	-.2138	-.4664	.5130	-114.6243

ABSOLUTE ACCELERATIONS

REAL	IMAG	ABS	PHASE	
1.2990	-.7500	1.5000	-30.0000	(B)
.8940	.0000	.8940	.0000	(C)
.4050	-.7500	.8524	-61.6308	(BC)

THETA(X)	THETA(Y)
240.0000	245.3757

A1	A2	B1	B2	C1	C2	ATC	ATBC
1.0000	.0000	-.4167	-.9091	.6803	-.4664	.8940	.5130

QUICK-RETURN: SIMPLIFIED VECTOR METHOD

```
LINKAGE*ANALYSIS(1).QUICK-RETURN/ANALYSIS(6)
    1     C       QUICK-RETURN LINKAGE ANALYSIS   - SIMPLIFIED VECTOR METHOD
    2     C       AB,CD,AD = LINK LENGTHS
    3     C       C        = DEGREES/RADIAN(CONSTANT)
    4     C       THETAA   = THETA(2), POSITION ANGLE OF LINK AB
    5     C       PHID     = PHI(3)
    6     C       THETAD   = THETA(3), POSITION ANGLE OF LINK CD
    7     C       OMEGAA   = OMEGA(2), ANGULAR VELOCITY OF LINK AB
    8     C       ALPHAA   = ALPHA(2), ANGULAR VELOCITY OF AB
    9     C       SNB      = 90(DEGS)
   10     C       SNC      = 90.0+THETA(D)+THETA(A)
   11     C       SNBC     = THETA(D)-THETA(A)
   12     C       VB       = VELOCITY OF B
   13     C       VC       = VELOCITY OF C
   14     C       VBC      = VELOCITY OF B RELATIVE TO C
   15     C       CVB      = VEL OF B (COMPLEX)
   16     C       CVC      = VEL OF C (COMPLEX)
   17     C       CVBC     = VEL OF B RELATIVE TO C (COMPLEX)
   18     C       ABSVB    = ABSOLUTE VELOCITY OF B
   19     C       ABSVC    = ABSOLUTE VELOCITY OF C
   20     C       ABSVBC   = ABSOLUTE VELOCITY OF B RELATIVE TO C
   21     C       ACCNB    = NORMAL ACC OF B (COMPLEX)
   22     C       ACCTB    = TANGENTIAL ACC OF B (COMPLEX)
   23     C       ACCNC    = NORMAL ACC OF C (COMPLEX)
   24     C       ACCTC    = TANGENTIAL ACC OF C (COMPLEX)
   25     C       ACCNBC   = NORMAL ACC OF B RELATIVE TO C (COMPLEX)
   26     C       ACCTBC   = TANGENTIAL ACC OF B RELATIVE TO C(COMPLEX)
   27     C       ACCCOR   = CORIOLIS ACCELERATION (COMPLEX)
   28     C       ATC      = TANGENTIAL ACC OF C
   29     C       ATBC     = TANGENTIAL ACC OF B RELATIVE TO C
   30     C       ACCB     = ACC OF B (COMPLEX)
   31     C       ACCC     = ACC OF C (COMPLEX)
   32     C       ACCBC    = ACC OF B RELATIVE TO C (COMPLEX)
   33     C       ABSNB    = ABSOLUTE NORMAL ACC OF B
   34     C       ABSTB    = ABSOLUTE TANGENTIAL ACC OF B
   35     C       ABSNC    = ABSOLUTE NORMAL ACC OF C
   36     C       ABSTC    = ABSOLUTE TANGENTIAL ACC OF C
   37     C       ABSNBC   = ABSOLUTE NORMAL ACC OF B RELATIVE TO C
   38     C       ABSTBC   = ABSOLUTE TANGENTIAL ACC OF B RELATIVE TO C
   39     C       ABSCOR   = ABSOLUTE CORIOLIS ACCELERATION
   40     C       ABSAB    = ABSOLUTE ACC OF B
   41     C       ABSAC    = ABSOLUTE ACC OF C
   42     C       ABSABC   = ABSOLUTE ACC OF B RELATIVE TO C
   43     C       PHZVB    = PHASE ANGLE OF VELOCITY OF B
   44     C       PHZVC    = PHASE ANGLE OF VELOCITY OF C
   45     C       PHZVBC   = PHASE ANGLE OF VELOCITY OF B RELATIVE TO C
   46     C       PHZNB    = PHASE ANGLE OF NORMAL ACC OF B
   47     C       PHZTB    = PHASE ANGLE OF TANGENTIAL ACC OF B
   48     C       PHZNC    = PHASE ANGLE OF NORMAL ACC OF C
   49     C       PHZTC    = PHASE ANGLE OF TANGENTIAL ACC OF C
   50     C       PHZNBC   = PHASE ANGLE OF NORMAL ACC OF B RELATIVE TO C
   51     C       PHZTBC   = PHASE ANGLE OF TANGENTIAL ACC OF B RELATIVE TO C
   52     C       PHZCOR   = PHASE ANGLE OF CORIOLIS ACCELERATION
   53     C       PHZAB    = PHASE ANGLE OF ABSOLUTE ACC OF B
   54     C       PHZAC    = PHASE ANGLE OF ABSOLUTE ACC OF C
   55     C       PHZABC   = PHASE ANGLE OF ABSOLUTE ACC OF B RELATIVE TO C
   56             COMPLEX CVB,CVC,CVBC,ACCNC,ACCTC,ACCC
   57             COMPLEX ACCNB,ACCTB,ACCB,ACCCOR,ACCTBC,ACCBC
   58             COMPLEX ACCNBC
   59             C=57.29578
   60     1 READ(5,99)AB,AD,THETAA,OMEGAA,ALPHAA
   61    99 FORMAT(5F12.4)
   62             IF(AB.EQ.0.0)GO TO 999
```

```
63              WRITE(6,899)
64          899 FORMAT(1H1,9X,'PROBLEM DATA')
65              WRITE(6,900)
66          900 FORMAT  ( /12X,'AB'10X'AD'4X'THETA(2)'4X'OMEGA(2)'4X'ALPHA(2)')
67              WRITE(6,100)AB,AD,THETAA,OMEGAA,ALPHAA
68          100 FORMAT(/2X,5F12.4)
69              THETAA=THETAA/C
70      C       CALCULATE CD USING COSINE RULE
71              CD=SQRT(AB**2+AD**2-2.0*AB*AD*COS(THETAA))
72      C       CALCULATE PHID AND THETAD
73              AA = (CD**2 + AD**2 - AB**2)/(2.0*CD*AD)
74              IF(ABS(AA).LE.1.0) GO TO 11
75              IF(AA.LT.0.0) AA=-1.0
76              IF(AA.GT.1.0) AA= 1.0
77           11 PHID = C*ACOS(AA)
78              THETAA = THETAA*C
79              IF(THETAA.LT.180.0)GO TO 111
80              PHID = -PHID
81          111 IF(ABS(PHID) .LT. 0.01)PHID=0.0
82              THETAD = 180.0 - PHID
83              THETAA = THETAA/C
84              WRITE(6,901)
85          901 FORMAT( /8X,'PHI(3)'4X'THETA(3)')
86              WRITE(6,101)PHID,THETAD
87              THETAD=THETAD/C
88              SNB=90.0
89              SNC=90.0-THETAD*C+THETAA*C
90              SNBC=THETAD*C-THETAA*C
91              WRITE(6,902)
92          902 FORMAT(/12X,'CD'9X'SNB'9X'SNC'8X'SNBC'10X'AA')
93              WRITE(6,888)CD,SNB,SNC,SNBC,AA
94          888 FORMAT(/2X,4F12.4,F12.9)
95              SMB=90.0/C
96              SMC=SNC/C
97              SMBC=SNBC/C
98              VB=AB*OMEGAA
99              VC=VB*SIN(SMC)
100             VBC=VB*SIN(SMBC)
101             WRITE(6,903)
102         903 FORMAT(/10X,'LINEAR VELOCITIES')
103             WRITE(6,904)
104         904 FORMAT(/10X,'V(B)'8X'V(C)'7X'V(BC)')
105             WRITE(6,101)VB,VC,VBC
106             THETAX=THETAA+90.0/C
107             THETAY=THETAD+90.0/C
108             X=COS(THETAX)
109             Y=SIN(THETAX)
110             CVB=VB*CMPLX(X,Y)
111             ABSVB=CABS(CVB)
112             PHZVB=C*ATAN2(AIMAG(CVB),REAL(CVB))
113             WRITE(6,905)
114         905 FORMAT(/10X,'REAL'8X'IMAG'9X'ABS'7X'PHASE')
115             WRITE(6,301)CVB,ABSVB,PHZVB
116             X=COS(THETAY)
117             Y=SIN(THETAY)
118             CVC=VC*CMPLX(X,Y)
119             ABSVC=CABS(CVC)
120     C       TEST VEL OF C
121             IF (ABSVC.GT.0.0001)GO TO 20
122             PHZVC = 0.0
123             GO TO 25
124          20 PHZVC=C*ATAN2(AIMAG(CVC),REAL(CVC))
125          25 CONTINUE
```

```
126              WRITE(6,302)CVC,ABSVC,PHZVC
127              X=COS(THETAD)
128              Y=SIN(THETAD)
129              CVBC=VBC*CMPLX(X,Y)
130              ABSVBC=CABS(CVBC)
131       C      TEST ABS VEL OF B WRT C
132              IF(ABSVBC.GT.0.0001)GO TO 30
133              PHZVBC=0.0
134              GO TO 35
135         30 PHZVBC=C*ATAN2(AIMAG(CVBC),REAL(CVBC))
136         35 CONTINUE
137              WRITE(6,303)CVBC,ABSVBC,PHZVBC
138       C      CALCULATE ATC AND ATBC
139              A1=COS(THETAY)
140              A2=SIN(THETAY)
141              B1=COS(THETAD)
142              B2=SIN(THETAD)
143              C1=-VB**2/AB*COS(THETAA)+AB*ALPHAA*COS(THETAX)-2*VBC*VC/CD*COS(THE
144             1TAY)+VC**2/CD*COS(THETAD)
145              C2=-VB**2/AB*SIN(THETAA)+AB*ALPHAA*SIN(THETAX)-2*VBC*VC/CD*SIN(THE
146             1TAY)+VC**2/CD*SIN(THETAD)
147              ATC=(C1*B2-C2*B1)/(A1*B2-A2*B1)
148              ATBC=(A1*C2-A2*C1)/(A1*B2-A2*B1)
149       C      CALCULATE ACCELERATION COMPONENTS
150              X=COS(THETAA)
151              Y=SIN(THETAA)
152              ACCNB=-VB**2/AB*CMPLX(X,Y)
153              ABSNB=CABS(ACCNB)
154              PHZNB=C*ATAN2(AIMAG(ACCNB),REAL(ACCNB))
155              X=COS(THETAX)
156              Y=SIN(THETAX)
157              ACCTB=AB*ALPHAA *CMPLX(X,Y)
158              ABSTB=CABS(ACCTB)
159       C      TEST ABS TAN ACC OF B
160              IF(ABSTB.GT.0.0001)GO TO 40
161              PHZTB = 0.0
162              GO TO 45
163         40 PHZTB=C*ATAN2(AIMAG(ACCTB),REAL(ACCTB))
164         45 CONTINUE
165              WRITE(6,906)
166        906 FORMAT(/10X,'NORMAL AND CORIOLIS ACCELERATIONS' 26X,'TANGENTIAL AC
167             2CELERATIONS')
168              WRITE(6,915)
169        915 FORMAT(/10X,'REAL'8X'IMAG'9X'ABS'7X'PHASE'19X,'REAL'8X'IMAG'9X'ABS
170             2'7X'PHASE')
171              WRITE(6,301)ACCNB,ABSNB,PHZNB,ACCTB,ABSTB,PHZTB
172              X=COS(THETAD)
173              Y=SIN(THETAD)
174              ACCNC=-VC**2/CD* CMPLX(X,Y)
175              ABSNC=CABS(ACCNC)
176       C      TEST ABS NOR ACC OF C
177              IF(ABSNC.GT.0.0001)GO TO 440
178              PHZNC = 0.0
179              GO TO 445
180        440 PHZNC=C*ATAN2(AIMAG(ACCNC),REAL(ACCNC))
181        445 CONTINUE
182              X=COS(THETAY)
183              Y=SIN(THETAY)
184              ACCTC=ATC*CMPLX(X,Y)
185              ABSTC=CABS(ACCTC)
186       C      TEST ABS TAN ACC OF C
187              IF(ABSTC.GT.0.0001)GO TO 50
188              PHZTC = 0.0
189              GO TO 55
190         50 PHZTC=C*ATAN2(AIMAG(ACCTC),REAL(ACCTC))
191         55 CONTINUE
192              WRITE(6,302)ACCNC,ABSNC,PHZNC,ACCTC,ABSTC,PHZTC
193              ACCNBC = 0.0
194              ABSNBC = 0.0
195              PHZNBC = 0.0
```

```
196            X=COS(THETAD)
197            Y=SIN(THETAD)
198            ACCTBC=ATBC*CMPLX(X,Y)
199            ABSTBC=CABS(ACCTBC)
200     C      TEST TAN  ACC B WRT C
201            IF(ABSTBC.GT.0.0001)GO TO 70
202            PHZTBC=0.0
203            GO TO 75
204         70 PHZTBC=C*ATAN2(AIMAG(ACCTBC),REAL(ACCTBC))
205         75 CONTINUE
206            WRITE(6,303)ACCNBC,ABSNBC,PHZNBC,ACCTBC,ABSTBC,PHZTBC
207            X=COS(THETAY)
208            Y=SIN(THETAY)
209            ACCCOR=(2*VBC*VC/CD)*CMPLX(X,Y)
210            ABSCOR=CABS(ACCCOR)
211     C      TEST ABS CORIOLIS ACC
212            IF(ABSCOR.GT.0.0001)GO TO 60
213            PHZCOR=0.0
214            GO TO 65
215         60 PHZCOR=C*ATAN2(AIMAG(ACCCOR),REAL(ACCCOR))
216         65 CONTINUE
217            WRITE(6,304)ACCCOR,ABSCOR,PHZCOR
218            ACCB=ACCNB+ACCTB
219            ABSAB=CABS(ACCB)
220            PHZAB=C*ATAN2(AIMAG(ACCB),REAL(ACCB))
221            WRITE(6,907)
222        907 FORMAT(/10X,'ABSOLUTE ACCELERATIONS')
223            WRITE(6,905)
224            WRITE(6,301)ACCB,ABSAB,PHZAB
225            ACCC=ACCNC+ACCTC
226            ABSAC=CABS(ACCC)
227     C      TEST ABS ACC OF C
228            IF(ABSAC.GT.0.0001)GO TO 770
229            PHZAC = 0.0
230            GO TO 775
231        770 PHZAC=C*ATAN2(AIMAG(ACCC),REAL(ACCC))
232        775 CONTINUE
233            WRITE(6,302)ACCC,ABSAC,PHZAC
234            ACCBC=ACCCOR+ACCTBC+ACCNBC
235            ABSABC=CABS(ACCBC)
236     C      TEST ABS ACC B WRT C
237            IF(ABSABC.GT.0.0001)GO TO 80
238            PHZABC=0.0
239            GO TO 85
240         80 PHZABC=C*ATAN2(AIMAG(ACCBC),REAL(ACCBC))
241         85 CONTINUE
242            WRITE(6,303)ACCBC,ABSABC,PHZABC
243            THETAA=THETAA*C
244            THETAD=THETAD*C
245            THETAX=THETAX*C
246            THETAY=THETAY*C
247            WRITE(6,908)
248        908 FORMAT( /6X,'THETA(X)'4X'THETA(Y)')
249            WRITE(6,201)THETAX,THETAY
250        201 FORMAT (/2X,4F12.4,12X,4F12.4)
251            WRITE(6,909)
252        909 FORMAT(/12X,'A1'10X'A2'10X'B1'10X'B2'10X'C1'10X'C2'9X'ATC'8X'ATBC'
253           1)
254            WRITE(6,101)A1,A2,B1,B2,C1,C2,ATC,ATBC
255        101 FORMAT(/2X,8F12.4)
256        301 FORMAT(/2X,4F12.4,8X'(B) ',4F12.4)
257        302 FORMAT(/2X,4F12.4,8X'(C) ',4F12.4)
258        303 FORMAT(/2X,4F12.4,8X'(BC)',4F12.4)
259        304 FORMAT(/2X,4F12.4,8X'(COR)',4F12.4)
260            GO TO 1
261        999 STOP
262            END
```

```
PROBLEM DATA
     AB          AD      THETA(2)    OMEGA(2)    ALPHA(2)
   .1666       .3333     30.0000     62.8300      .0000
 PHI(3)      THETA(3)
 23.7828     156.2172
     CD          SNB        SNC        SNBC         AA
   .2066      90.0000    -36.2172    126.2172   .915080868
     LINEAR VELOCITIES
    V(B)        V(C)       V(BC)
  10.4675     -6.1847     8.4450
    REAL        IMAG        ABS        PHASE
  -5.2337      9.0651    10.4675    120.0000      (B)
   2.4941      5.6595     6.1847     66.2172      (C)
  -7.7278      3.4056     8.4450    156.2172     (BC)
```

NORMAL AND CORIOLIS ACCELERATIONS					TANGENTIAL ACCELERATIONS

REAL	IMAG	ABS	PHASE		REAL	IMAG	ABS	PHASE
-569.5603	-328.8358	657.6716	-150.0000	(B)	.0000	.0000	.0000	.0000
169.4519	-74.6764	185.1770	-23.7828	(C)	-417.9108	-948.3019	1036.3040	-113.7828
.0000	.0000	.0000	.0000	(BC)	-525.0377	231.3808	573.7610	156.2172
203.9362	462.7617	505.7058	66.2172	(COR)				

```
     ABSOLUTE ACCELERATIONS
    REAL        IMAG        ABS        PHASE
 -569.5603   -328.8358   657.6716   -150.0000     (B)
 -248.4589  -1022.9783  1052.7186   -103.6515     (C)
 -321.1015    694.1425   764.8137    114.8247    (BC)
 THETA(X)    THETA(Y)
 120.0000    246.2172
      A1          A2          B1          B2          C1          C2         ATC         ATBC
   -.4033      -.9151      -.9151       .4033    -942.9485   -716.9210   1036.3040    573.7610
```

PROBLEM DATA

AB	AD	THETA(2)	OMEGA(2)	ALPHA(2)
.2500	.1666	60.0000	30.0000	.0000

PHI(3)	THETA(3)
79.1236	100.8764

CD	SNB	SNC	SNBC	AA
.2205	90.0000	49.1236	40.8764	.188690612

LINEAR VELOCITIES

V(B)	V(C)	V(BC)
7.5000	5.6709	4.9082

REAL	IMAG	ABS	PHASE	
-6.4952	3.7500	7.5000	150.0000	(B)
-5.5691	-1.0700	5.6709	-169.1236	(C)
-.9261	4.8201	4.9082	100.8764	(BC)

NORMAL AND CORIOLIS ACCELERATIONS

TANGENTIAL ACCELERATIONS

REAL	IMAG	ABS	PHASE		REAL	IMAG	ABS	PHASE
-112.5000	-194.8557	225.0000	-120.0000	(B)	.0000	.0000	.0000	.0000
27.5242	-143.2493	145.8696	-79.1236	(C)	103.3647	19.8607	105.2554	10.8764
.0000	.0000	.0000	.0000	(BC)	4.5773	-23.8224	24.2581	-79.1236
-247.9662	-47.6447	252.5020	-169.1236	(COR)				

ABSOLUTE ACCELERATIONS

REAL	IMAG	ABS	PHASE	
-112.5000	-194.8557	225.0000	-120.0000	(B)
130.8889	-123.3886	179.8795	-43.3105	(C)
-243.3889	-71.4671	253.6646	-163.6360	(BC)

THETA(X)	THETA(Y)
150.0000	190.8764

A1	A2	B1	B2	C1	C2	ATC	ATBC
-.9820	-.1887	-.1887	.9820	107.9420	-3.9617	-105.2554	-24.2581

SLIDING COUPLER: SIMPLIFIED VECTOR METHOD

```
LINKAGE+ANALYSIS(1).SLG-CPLR/ANALYSIS(3)
    1       C       SLIDING-COUPLER MECHANISM - SIMPLIFIED VECTOR METHOD
    2       C       AB,BC,AC = LINK LENGTHS
    3       C       C        = DEGREES/RADIAN(CONSTANT)
    4       C       PHIC     = PHI(C)
    5       C       THETAA   = THETA(A), POSITION ANGLE OF LINK AB
    6       C       ALPHAA   = ALPHA(A),ANGULAR ACCELERATION OF CRANK
    7       C       OMEGAA   = OMEGA(A),ANGULAR VELOCITY OF CRANK
    8       C       VB       = VELOCITY OF B
    9       C       VC       = VELOCITY OF C
   10       C       VCB      = VELOCITY OF C RELATIVE TO B
   11       C       CVB      = VEL OF B (COMPLEX)
   12       C       CVC      = VEL OF C (COMPLEX)
   13       C       CVCB     = VEL OF C RELATIVE TO BE (COMPLEX)
   14       C       ABSVB    = ABSOLUTE VELOCITY OF B
   15       C       ABXVC    = ABSOLUTE VELOCITY OF C
   16      .C       ABSVCB   = ABSOLUTE VELOCITY OF C RELATIVE TO B
   17      ·C       ACCNB    = NORMAL ACC OF B (COMPLEX)
   18       C       ACCTB    = TANGENTIAL ACC OF B (COMPLEX)
   19       C       ACCNC    = NORMAL ACC OF C (COMPLEX)
   20       C       ACCTC    = TANGENTIAL ACC OF C (COMPLEX)
   21       C       ACCNCB   = NORMAL ACC OF C RELATIVE TO B(COMPLEX)
   22       C       ACCTCB   = TANGENTIAL ACC OF C RELATIVE TO B(COMPLEX)
   23       C       ATC      = TANGENTIAL ACC OF C
   24       C       ATCB     = TAGENTIAL ACC OF C RELATIVE TO B
   25       C       ACCB     = ACC OF B (COMPLEX)
   26       C       ACCC     = ACC OF C (COMPLEX)
   27       C       ACCCB    = ACC OF C RELATIVE TO B (COMPLEX)
   28       C       ABSNB    = ABSOLUTE NORMAL ACC OF B
   29       C       ABSTB    = ABSOLUTE TANGENTIAL ACC OF B
   30       C       ABSNC    = ABSOLUTE NORMAL ACC OF C
   31       C       ABSTB    = ABSOLUTE TANGENTIAL ACC OF C
   32       C       ABSNCB   = ABSOLUTE NORMAL ACC OF C RELATIVE TO B
   33       C       ABSTCB   = ABSOLUTE TANGENTIAL ACC OF C RELATIVE TO B
   34       C       ABSAB    = ABSOLUTE ACCELERATION OF B
   35       C       ABSAC    = ABSOLUTE ACCELERATION OF C
   36       C       ABSACB   = ABSOLUTE ACC OF C RELATIVE TO B
   37       C       PHZVB    = PHASE ANGLE OF VELOCITY OF B
   38       C       PHZVC    = PHASE ANGLE OF VELOCITY OF C
   39       C       PHZVCB   = PHASE ANGLE OF VELOCITY OF C RELATIVE TO B
   40       C       PHZNB    = PHASE ANGLE OF NORMAL ACC OF B
   41       C       PHZTB    = PHASE ANGLE OF TANGENTIAL ACC OF B
   42       C       PHZNC    = PHASE ANGLE OF NORMAL ACC OF C
   43       C       PHZTC    = PHASE ANGLE OF TANGENTIAL ACC OF C
   44       C       PHZNCB   = PHASE ANGLE OF NORMAL ACC OF C RELATIVE TO B
   45       C       PHZTCB   = PHASE ANGLE OF TANGENTIAL ACC OF C RELATIVE TO B
   46       C       PHZAB    = PHASE ANGLE OF ABSOLUTE ACC OF B
   47       C       PHZAC    = PHASE ANGLE OF ABSOLUTE ACC OF C
   48       C       PHZACB   = PHASE ANGLE OF ABSOLUTE ACC OF C RELATIVE TO B
   49       C
   50               C = 57.29578
   51               COMPLEX CVB,CVC,CVCB
   52               COMPLEX ACCNB,ACCTB,ACCB,ACCNC,ACCTC,ACCCOR,ACCC
   53               COMPLEX ACCNCB,ACCTCB,ACCCB
   54             1 READ(5,100)AB,AC,THETAA,OMEGAA,ALPHAA
   55           100 FORMAT(7F10.4)
   56               IF(AB.EQ.0.0) GO TO 999
   57               WRITE(6,104)
   58           104 FORMAT(1H1,9X,'PROBLEM DATA')
```

```
59              WRITE(6,105)
60        105 FORMAT(//12X,'AB'10X'AC'4X'THETA(A)'4X'OMEGA(A)'4X'ALPHA(A)')
61              WRITE(6,106)AB,AC,THETAA,OMEGAA,ALPHAA
62        106 FORMAT(/2X,8F12.4)
63              THETAA=THETAA/C
64              BC=SQRT(AB**2+AC**2-2*AB*AC*COS(THETAA))
65              AA = (BC**2 + AC**2 - AB**2)/(2.0*BC*AC)
66              IF(ABS(AA).LE.1.0) GO TO 11
67              IF(AA.LT.0.0) AA=-1.0
68              IF(AA.GT.1.0) AA= 1.0
69         11 PHI  = C*ACOS(AA)
70              PHIC = -PHI
71              THETAA=THETAA*C
72              IF(THETAA.LT.180.0)GO TO 111
73              PHIC = PHI
74        111 IF(ABS(PHIC) .LT. 0.01)PHIC=0.0
75              WRITE(6,901)
76        901 FORMAT(/11X,'PHI'6X'PHI(C)')
77              WRITE(6,101)PHI,PHIC
78              SNB=90.0
79              SNC=(THETAA-PHIC)
80              SNCB=(90.0-THETAA+PHIC)
81              WRITE(6,902)
82        902 FORMAT(/12X,'BC'9X'SNB'9X'SNC'8X'SNCB'10X'AA')
83              WRITE(6,888)BC,SNB,SNC,SNCB,AA
84        888 FORMAT(/2X,4F12.4,F12.9)
85              VB=AB*OMEGAA
86              VC = -VB*SIN(SNC/C)/SIN(SNB/C)
87              VCB = -VB*SIN(SNCB/C)/SIN(SNB/C)
88              X=COS((THETAA+90.0)/C)
89              Y=SIN((THETAA+90.0)/C)
90              CVB=VB*CMPLX(X,Y)
91              ABSVB = CABS(CVB)
92              PHZVB=C*ATAN2(AIMAG(CVB),REAL(CVB))
93              X=COS(PHIC/C)
94              Y=SIN(PHIC/C)
95              CVC=VC*CMPLX(X,Y)
96              ABSVC = CABS(CVC)
97    C         TEST ABSVC
98              IF(ABSVC.GT. 0.0001)GO TO 80
99              PHZVC=0.0
100             GO TO 85
101         80 PHZVC=C*ATAN2(AIMAG(CVC),REAL(CVC))
102         85 CONTINUE
103             X=COS((PHIC+90.0)/C)
104             Y=SIN((PHIC+90.0)/C)
105             CVCB=VCB*CMPLX(X,Y)
106             ABSVCB = CABS(CVCB)
107   C         TEST ABSVCB
108             IF(ABSVCB.GT. 0.0001)GO TO 90
109             PHZVCB=0.0
110             GO TO 95
111         90 PHZVCB=C*ATAN2(AIMAG(CVCB),REAL(CVCB))
112         95 CONTINUE
113             WRITE(6,903)
114        903 FORMAT(/10X,'LINEAR VELOCITIES')
115             WRITE(6,904)
116        904 FORMAT(/10X'V(B)'8X'V(C)'7X'V(CB)')
117             WRITE(6,107)VB,VC,VCB
```

```
118                 WRITE(6,905)
119            905 FORMAT(/10X,'REAL'8X'IMAG'9X'ABS'7X'PHASE')
120                 WRITE(6,301)CVB,ABSVB,PHZVB
121                 WRITE(6,302)CVC,ABSVC,PHZVC
122                 WRITE(6,303)CVCB,ABSVCB,PHZVCB
123                 A1=COS(PHIC/C)
124                 A2=SIN(PHIC/C)
125                 B1=-COS((PHIC+90.0)/C)
126                 B2=-SIN((PHIC+90.0)/C)
127                 C1=-OMEGAA**2*AB*COS((THETAA/C)+ALPHAA*AB*COS((THETAA+90.0)/C)
128                1-VCB**2/BC*COS(PHIC/C)-2.0*VC*VCB/BC*COS((PHIC+90.0)/C)
129                 C2=-OMEGAA**2*AB*SIN(THETAA/C)+ALPHAA*AB*SIN((THETAA+90.0)/C)
130                1-VCB**2/BC*SIN(PHIC/C)-2.0*VC*VCB/BC*SIN((PHIC+90.0)/C)
131                 ATC=(C1*B2-C2*B1)/(A1*B2-A2*B1)
132                 ATCB=(A1*C2-A2*C1)/(A1*B2-A2*B1)
133                 X=COS(THETAA/C)
134                 Y=SIN(THETAA/C)
135                 ACCNB=-OMEGAA**2*AB*CMPLX(X,Y)
136                 ABSNB=CABS(ACCNB)
137                 PHZNB = C*ATAN2(AIMAG(ACCNB),REAL(ACCNB))
138                 X=COS((THETAA+90.0)/C)
139                 Y=SIN((THETAA+90.0)/C)
140                 ACCTB=ALPHAA*AB*CMPLX(X,Y)
141                 ABSTB=CABS(ACCTB)
142      C          TEST ABSTB
143                 IF(ABSTB.GT.0.0001)GO TO 10
144                 PHZTB = 0.0
145                 GO TO 15
146             10 PHZTB = C*ATAN2(AIMAG(ACCTB),REAL(ACCTB))
147             15 CONTINUE
148                 ACCNC = 0.0
149                 PHZNC=0.0
150                 ABSNC=0.0
151                 X=COS(PHIC/C)
152                 Y=SIN(PHIC/C)
153                 ACCTC=ATC*CMPLX(X,Y)
154                 ABSTC=CABS(ACCTC)
155      C          TEST ABSTC
156                 IF(ABSTC.GT. 0.0001)GO TO 30
157                 PHZTC=0.0
158                 GO TO 35
159             30 PHZTC = C*ATAN2(AIMAG(ACCTC),REAL(ACCTC))
160             35 CONTINUE
161                 X=COS((PHIC+90.0)/C)
162                 Y=SIN((PHIC+90.0)/C)
163                 ACCCOR=2.0*VC*VCB/BC*CMPLX(X,Y)
164                 ABSCOR=CABS(ACCCOR)
165      C          TEST ABS CORIOLIS ACC
166                 IF(ABSCOR.GT.0.0001)GO TO 110
167                 PHZCOR=0.0
168                 GO TO 115
169            110 PHZCOR=C*ATAN2(AIMAG(ACCCOR),REAL(ACCCOR))
170            115 CONTINUE
171                 X=COS(PHIC/C)
172                 Y=SIN(PHIC/C)
173                 ACCNCB=-VCB**2/BC*CMPLX(X,Y)
174                 ABSNCB=CABS(ACCNCB)
175      C          TEST ABSNCB
176                 IF(ABSNCB.GT.0.0001)GO TO 40
```

```
177            PHZNCB=0.0
178            GO TO 45
179         40 PHZNCB = C*ATAN2(AIMAG(ACCNCB),REAL(ACCNCB))
180         45 CONTINUE
181            X=COS((PHIC+90.0)/C)
182            Y=SIN((PHIC+90.0)/C)
183            ACCTCB=ATCB*CMPLX(X,Y)
184            ABSTCB=CABS(ACCTCB)
185     C      TEST ABSTCB
186            IF(ABSTCB.GT.0.0001)GO TO 50
187            PHZTCB=0.0
188            GO TO 55
189         50 PHZTCB = C*ATAN2(AIMAG(ACCTCB),REAL(ACCTCB))
190         55 CONTINUE
191            WRITE(6,906)
192        906 FORMAT(/10X,'NORMAL AND CORIOLIS ACCELERATIONS'26X'TANGENTIAL
193           1ACCELERATIONS')
194            WRITE(6,915)
195        915 FORMAT(/10X,'REAL'8X'IMAG'9X'ABS'7X'PHASE'18X'REAL'8X'IMAG'
196           19X'ABS'7X'PHASE')
197            WRITE(6,301)ACCNB,ABSNB,PHZNB,ACCTB,ABSTB,PHZTB
198            WRITE(6,302)ACCNC,ABSNC,PHZNC,ACCTC,ABSTC,PHZTC
199            WRITE(6,303)ACCNCB,ABSNCB,PHZNCB,ACCTCB,ABSTCB,PHZTCB
200            WRITE(6,304)ACCCOR,ABSCOR,PHZCOR
201            ACCB = ACCNB + ACCTB
202            ABSAB = CABS(ACCB)
203            PHZAB = C*ATAN2(AIMAG(ACCB),REAL(ACCB))
204            ACCC= ACCNC + ACCTC + ACCCOR
205            ABSAC = CABS(ACCC)
206     C      TEST ABSAC
207            IF(ABSAC.GT.0.0001)GO TO 60
208            PHZAC=0.0
209            GO TO 65
210         60 PHZAC = C*ATAN2(AIMAG(ACCC),REAL(ACCC))
211         65 CONTINUE
212            ACCCB = ACCNCB + ACCTCB
213            ABSACB = CABS(ACCCB)
214     C      TEST ABSACB
215            IF(ABSACB.GT.0.0001)GO TO 70
216            PHZACB=0.0
217            GO TO 75
218         70 PHZACB = C*ATAN2(AIMAG(ACCCB),REAL(ACCCB))
219         75 CONTINUE
220            WRITE(6,907)
221        907 FORMAT(/10X'ABSOLUTE ACCELERATIONS')
222            WRITE(6,905)
223            WRITE(6,301)ACCB,ABSAB,PHZAB
224            WRITE(6,302)ACCC,ABSAC,PHZAC
225            WRITE(6,303)ACCCB,ABSACB,PHZACB
226            WRITE(6,909)
227        909 FORMAT(//12X,'A1'10X'A2',10X'B1'10X'B2'10X'C1'10X'C2'9X'ATC'
228           18X'ATCB')
229            WRITE(6,107)A1,A2,B1,B2,C1,C2,ATC,ATCB
230        101 FORMAT(/2X,8F12.4)
231        107 FORMAT(/2X,8F12.4)
232        301 FORMAT(/2X,4F12.4,8X'B ',4F12.4)
233        302 FORMAT(/2X,4F12.4,8X'C ',4F12.4)
234        303 FORMAT(/2X,4F12.4,8X'CB',4F12.4)
235        304 FORMAT(/2X,4F12.4,8X'COR',4F12.4)
236            GO TO 1
237        999 STOP
238            END
```

PROBLEM DATA

AB	AC	THETA(A)	OMEGA(A)	ALPHA(A)
.6667	.8333	120.0000	18.0000	.0000

PHI	PHI(C)
26.3310	-26.3310

BC	SNB	SNC	SNCB	AA
1.3017	90.0000	146.3310	-56.3310	.896246806

LINEAR VELOCITIES

V(B)	V(C)	V(CB)
12.0006	-6.6531	9.9875

REAL	IMAG	ABS	PHASE	
-10.3928	-6.0003	12.0006	-150.0000	B
-5.9628	2.9510	6.6531	153.6690	C
4.4300	8.9513	9.9875	63.6690	CB

NORMAL AND CORIOLIS ACCELERATIONS   TANGENTIAL   ACCELERATIONS

REAL	IMAG	ABS	PHASE		REAL	IMAG	ABS	PHASE
108.0054	-187.0708	216.0108	-60.0000	B	.0000	.0000	.0000	.0000
.0000	.0000	.0000	.0000	C	92.4432	-45.7505	103.1448	-26.3310
-68.6803	33.9901	76.6310	153.6690	CB	7.8340	15.8293	17.6618	63.6690
-45.2841	-91.5009	102.0934	-116.3310	COR				

ABSOLUTE ACCELERATIONS

REAL	IMAG	ABS	PHASE	
108.0054	-187.0708	216.0108	-60.0000	B
47.1591	-137.2514	145.1273	-71.0375	C
-60.8463	49.8195	78.6400	140.6902	CB

A1	A2	B1	B2	C1	C2	ATC	ATCB
.8962	-.4436	-.4436	-.8962	84.6092	-61.5798	103.1448	17.6618

PROBLEM DATA

AB	AC	THETA(A)	OMEGA(A)	ALPHA(A)
3.0000	2.0000	210.0000	30.0000	.0000

PHI	PHI(C)
18.0675	18.0675

BC	SNB	SNC	SNCB	AA
4.8366	90.0000	191.9325	-101.9325	.950691596

LINEAR VELOCITIES

V(B)	V(C)	V(CB)
90.0000	18.6083	88.0553

REAL	IMAG	ABS	PHASE	
45.0000	-77.9423	90.0000	-60.0000	B
17.6907	5.7711	18.6083	18.0675	C
-27.3093	83.7134	88.0553	108.0675	CB

NORMAL AND CORIOLIS ACCELERATIONS   TANGENTIAL   ACCELERATIONS

REAL	IMAG	ABS	PHASE		REAL	IMAG	ABS	PHASE
2338.2687	1349.9998	2700.0000	30.0000	B	.0000	.0000	.0000	.0000
.0000	.0000	.0000	.0000	C	987.3007	322.0806	1038.5079	18.0675
-1524.1017	-497.1977	1603.1505	-161.9325	CB	-37.0066	113.4394	119.3230	108.0675
-210.1404	644.1609	677.5709	108.0675	COR				

ABSOLUTE ACCELERATIONS

REAL	IMAG	ABS	PHASE	
2338.2687	1349.9998	2700.0000	30.0000	B
777.1604	966.2415	1240.0004	51.1898	C
-1561.1083	-383.7583	1607.5850	-166.1892	CB

A1	A2	B1	B2	C1	C2	ATC	ATCB
.9507	.3101	.3101	-.9507	1024.3073	208.6412	1038.5079	119.3230

SLIDER–CRANK: MODIFIED VECTOR METHOD

```
LINKAGE*ANALYSIS(1).SL-CRANK/ANALYSIS-2(2)
    1     C        SLIDER-CRANK MECHANISM ANALYSIS-MODIFIED VECTOR METHOD
    2     C        C      = DEGREES/RADIAN(CONSTANT)
    3     C        AB     = CRANK
    4     C        BC     = CONNECTING ROD
    5     C        PHI    = PHI(C),ANGLE BETWEEN CONNECTING ROD AND DEAD CENTER
    6     C        THETAA = POSITION ANGLE OF LINK AB
    7     C        THETAC = POSITION ANGLE OF LINK BC
    8     C        OMEGAA = ANGULAR VELOCITY OF LINK AB
    9     C        ALPHAA = ANGULAR ACC OF LINK AB
   10     C        VB     = VELOCITY OF B
   11     C        VC     = VELOCITY OF C
   12     C        VBC    = VELOCITY OF B RELATIVE TO C
   13     C        CVB    = VEL OF B (COMPLEX)
   14     C        CVC    = VEL OF C (COMPLEX)
   15     C        CVBC   = VEL OF B RELATIVE TO C (COMPLEX)
   16     C        ABSVB  = ABSOLUTE VELOCITY OF B
   17     C        ABSVC  = ABSOLUTE VELOCITY OF C
   18     C        ABSVBC = ABSOLUTE VELOCITY OF B RELATIVE TO C
   19     C        ACCNB  = NORMAL ACC OF B (COMPLEX)
   20     C        ACCTB  = TANGENTIAL ACC OF B (COMPLEX)
   21     C        ACCNC  = NORMAL ACC OF C (COMPLEX)
   22     C        ACCTC  = TANGENTIAL ACC OF C (COMPLEX)
   23     C        ACCB   = ACC OF B (COMPLEX)
   24     C        ACCC   = ACC OF C (COMPLEX)
   25     C        ACCBC  = ACC OF B RELATIVE TO C (COMPLEX)
   26     C        ABSNB  = ABSOLUTE NORMAL ACC OF B
   27     C        ABSTB  = ABSOLUTE TANGENTIAL ACC OF B
   28     C        ABSNC  = ABSOLUTE NORMAL ACC OF C
   29     C        ABSTC  = ABSOLUTE TANGENTIAL ACC OF C
   30     C        ABSAB  = ABSOLUTE ACC OF B
   31     C        ABSAC  = ABSOLUTE ACC OF C
   32     C        ABSABC = ABSOLUTE ACC OF B RELATIVE TO C
   33     C        PHZVB  = PHASE ANGLE OF VELOCITY OF B
   34     C        PHZVC  = PHASE ANGLE OF VELOCITY OF C
   35     C        PHZVBC = PHASE ANGLE OF VELOCITY OF B RELATIVE T C
   36     C        PHZNB  = PHASE ANGLE OF NORMAL ACC OF B
   37     C        PHZTB  = PHASE ANGLE OF TANGENTIAL ACC OF B
   38     C        PHZNC  = PHASE ANGLE OF NORMAL ACC OF C
   39     C        PHZTC  = PHASE ANGLE OF TANGENTIAL ACC OF C
   40     C        PHZAB  = PHASE ANGLE OF ABSOLUTE ACC OF B
   41     C        PHZAC  = PHASE ANGLE OF ABSOLUTE ACC OF C
   42     C        PHZABC = PHASE ANGLE OF ABSOLUTE ACC OF B RELATIVE TO C
   43              REAL NUMR1,NUMR2,NUMR3
   44              COMPLEX CVB,CVC,CVBC,ACCC,ACCNC,ACCTC,CXY
   45              COMPLEX ACCNB,ACCTB,ACCB,ACC1BC,ACC2BC,ACCBC
   46              C=57.29578
   47            1 READ(5,99)AB,BC,THETAA,OMEGAA,ALPHAA
   48           99 FORMAT(5F12.4)
   49              IF(AB.EQ.0.0) GO TO 999
   50              WRITE(6,899)
   51          899 FORMAT(1H1,9X,'PROBLEM DATA')
   52              WRITE(6,900)
   53          900 FORMAT(//12X,'AB'10X'BC'4X'THETA(A)'4X'OMEGA(A)'4X'ALPHA(A)')
   54              WRITE(6,100)AB,BC,THETAA,OMEGAA,ALPHAA
   55          100 FORMAT(/2X,5F12.4)
   56              THETAA=THETAA/C
   57              PHI=ASIN(AB/BC*SIN(THETAA))
   58              PHI=PHI*C
   59              WRITE(6,901)
   60          901 FORMAT(//8X,'PHI(C)')
   61              WRITE(6,101)PHI
   62              THETAA=THETAA*C
   63              SNB=90.0-PHI
   64              SNC=THETAA+PHI
   65              SNBC=90.0-THETAA
   66              WRITE(6,902)
   67          902 FORMAT(/11X,'SNB'9X'SNC'8X'SNBC')
   68              WRITE(6,101)SNB,SNC,SNBC
   69              VB=AB*OMEGAA
   70              VC=-VB*SIN(SNC/C)/SIN(SNB/C)
   71              VBC=-VB*SIN(SNBC/C)/SIN(SNB/C)
   72              THETAA=THETAA/C
   73              PHI=PHI/C
```

```
74               WRITE(6,903)
75         903 FORMAT(/10X,'LINEAR VELOCITIES')
76               WRITE(6,904)
77         904 FORMAT(/10X,'V(B)'8X'V(C)'7X'V(BC)')
78               WRITE(6,101)VB,VC,VBC
79               X=SIN(THETAA)
80               Y=-COS(THETAA)
81               CXY=CMPLX(X,Y)
82               CVB=-AB*OMEGAA*(CXY)
83               ABSVB=CABS(CVB)
84               PHZVB=C*ATAN2(AIMAG(CVB),REAL(CVB))
85               WRITE(6,905)
86         905 FORMAT(/10X,'REAL'8X'IMAG'9X'ABS'7X'PHASE')
87               WRITE(6,301)CVB,ABSVB,PHZVB,
88               CVC=VC
89               ABSVC=CABS(CVC)
90      C       TEST FOR ABS VEL OF C
91               IF(ABSVC.GT.0.0001)GO TO 20
92               PHZVC=0.0
93               GO TO 25
94          20 PHZVC=C*ATAN2(AIMAG(CVC),REAL(CVC))
95          25 CONTINUE
96               WRITE(6,302)CVC,ABSVC,PHZVC
97               X=SIN(THETAA)-SIN(SNC/C)/SIN(SNB/C)
98               Y=-COS(THETAA)
99               CXY=CMPLX(X,Y)
100              CVBC=-AB*OMEGAA*(CXY)
101              ABSVBC=CABS(CVBC)
102     C       TEST VEL OF B REL TO C
103              IF(ABSVBC.GT.0.0001) GO TO 30
104              PHZVBC=0.0
105              GO TO 35
106         30 PHZVBC=C*ATAN2(AIMAG(CVBC),REAL(CVBC))
107         35 CONTINUE
108              WRITE(6,303)CVBC,ABSVBC,PHZVBC
109              WRITE(6,906)
110        906 FORMAT(/10X,'NORMAL ACCELERATIONS'40X,'TANGENTIAL ACCELERATIONS')
111              WRITE(6,915)
112        915 FORMAT(/10X,'REAL'8X'IMAG'9X'ABS'7X'PHASE'20X,'REAL'8X'IMAG'
113              19X'ABS'7X'PHASE')
114              X=COS(THETAA)
115              Y=SIN(THETAA)
116              CXY=CMPLX(X,Y)
117              ACCNB=-AB*OMEGAA**2*(CXY)
118              ABSNB=CABS(ACCNB)
119              PHZNB=C*ATAN2(AIMAG(ACCNB),REAL(ACCNB))
120              X=SIN(THETAA)
121              Y=-COS(THETAA)
122              CXY=CMPLX(X,Y)
123              ACCTB=-AB*ALPHAA*(CXY)
124              ABSTB=CABS(ACCTB)
125     C       TEST ABS TAN ACC OF B
126              IF(ABSTB.GT.0.0001) GO TO 40
127              PHZTB=0.0
128              GO TO 45
129         40 PHZTB=C*ATAN2(AIMAG(ACCTB),REAL(ACCTB))
130         45 CONTINUE
131              WRITE(6,301)ACCNB,ABSNB,PHZNB,ACCTB,ABSTB,PHZTB
132              NUMR1 = AB*COS(THETAA)*SIN(THETAA)
133              NUMR2=AB*(COS(THETAA)**2-SIN(THETAA)**2)
134              NUMR3=AB**3* COS(THETAA)**2*SIN(THETAA)**2
135              DNUMR=SQRT(BC**2-AB**2*(SIN(THETAA))**2)
136              ACCC=-AB*OMEGAA**2*(COS(THETAA)+NUMR2/DNUMR+NUMR3/DNUMR**3)-AB*
137              1ALPHAA*(SIN(THETAA)+NUMR1/DNUMR)
138              ACCNC=0.0
139              ABSNC=0.0
140              PHZNC=0.0
141              ACCTC=ACCC
142              ABSTC=CABS(ACCTC)
143     C       TEST FOR TAN ACC OF C
144              IF(ABSTC.GT.0.0001)GO TO 90
145              PHZTC= 0.0
146              GO TO 95
147         90 PHZTC=C*ATAN2(AIMAG(ACCTC),REAL(ACCTC))
```

```
148           95  CONTINUE
149               WRITE(6,302)ACCNC,ABSNC,PHZNC,ACCTC,ABSTC,PHZTC
150               X=NUMR2/DNUMR+NUMR3/DNUMR**3
151               Y=-SIN(THETAA)
152               CXY=CMPLX(X,Y)
153               ACC1BC=AB*OMEGAA**2*(CXY)
154               ABS1BC=CABS(ACC1BC)
155    C          TEST FOR TERM 1 OF ABS ACC OF B REL TO C
156               IF(ABS1BC.GT.O.0001) GO TO 50
157               PHZ1BC=O.0
158               GO TO 55
159           50  PHZ1BC=C*ATAN2(AIMAG(ACC1BC),REAL(ACC1BC))
160           55  CONTINUE
161               X=NUMR1/DNUMR
162               Y=COS(THETAA)
163               CXY=CMPLX(X,Y)
164               ACC2BC=AB*ALPHAA*(CXY)
165               ABS2BC=CABS(ACC2BC)
166    C          TEST TERM 2 OF ACC OF B REL TO C
167               IF(ABS2BC.GT.O.0001) GO TO 60
168               PHZ2BC=O.0
169               GO TO 65
170           60  PHZ2BC=C*ATAN2(AIMAG(ACC2BC),REAL(ACC2BC))
171           65  CONTINUE
172               WRITE(6,907)
173          907  FORMAT(/1OX,'ABSOLUTE ACCELERATIONS')
174               WRITE(6,905)
175               ACCB=ACCNB+ACCTB
176               ABSAB=CABS(ACCB)
177               PHZAB=C*ATAN2(AIMAG(ACCB),REAL(ACCB))
178               WRITE(6,301)ACCB,ABSAB,PHZAB
179               ACCC=ACCNC+ACCTC
180               ABSAC=CABS(ACCC)
181    C          TEST FOR ABS ACC OF C
182               IF(ABSAC.GT.O.0001)GO TO 10
183               PHZAC=O.0
184               GO TO 15
185           10  PHZAC=C*ATAN2(AIMAG(ACCC),REAL(ACCC))
186           15  CONTINUE
187               WRITE(6,302)ACCC,ABSAC,PHZAC
188               ACCBC=ACC1BC+ACC2BC
189               ABSABC=CABS(ACCBC)
190    C          TEST ABS ACC OF B REL TO C
191               IF(ABSABC.GT.O.0001) GO TO 70
192               PHZABC=O.0
193               GO TO 75
194           70  PHZABC=C*ATAN2(AIMAG(ACCBC),REAL(ACCBC))
195           75  CONTINUE
196               WRITE(6,303)ACCBC,ABSABC,PHZABC
197               THETAA=THETAA*C
198               PHI=PHI*C
199          201  FORMAT(/2X,4F12.4,12X,4F12.4)
200          101  FORMAT(/2X,8F12.4)
201          301  FORMAT(/2X,4F12.4,8X'(B) ',4F12.4)
202          302  FORMAT(/2X,4F12.4,8X'(C) ',4F12.4)
203          303  FORMAT(/2X,4F12.4,8X'(BC)',4F12.4)
204               GO TO 1
205          999  STOP
206               END
```

PROBLEM DATA

AB	BC	THETA(A)	OMEGA(A)	ALPHA(A)
1.5000	3.0000	120.0000	1.0000	.0000

PHI(C)

25.6589

SNB	SNC	SNBC
64.3411	145.6589	-30.0000

LINEAR VELOCITIES

V(B)	V(C)	V(BC)
1.5000	-.9387	.8321

REAL	IMAG	ABS	PHASE	
-1.2990	-.7500	1.5000	-150.0000	(B)
-.9387	.0000	.9387	180.0000	(C)
-.3603	-.7500	.8321	-115.6589	(BC)

NORMAL ACCELERATIONS

REAL	IMAG	ABS	PHASE	
.7500	-1.2990	1.5000	-60.0000	(B)
.0000	.0000	.0000	.0000	(C)

TANGENTIAL ACCELERATIONS

REAL	IMAG	ABS	PHASE
.0000	.0000	.0000	.0000
1.1180	.0000	1.1180	.0000

ABSOLUTE ACCELERATIONS

REAL	IMAG	ABS	PHASE	
.7500	-1.2990	1.5000	-60.0000	(B)
1.1180	.0000	1.1180	.0000	(C)
-.3680	-1.2990	1.3502	-105.8176	(BC)

PROBLEM DATA

AB	BC	THETA(A)	OMEGA(A)	ALPHA(A)
1.5000	3.0000	30.0000	-1.0000	.0000

PHI(C)

14.4775

SNB	SNC	SNBC
75.5225	44.4775	60.0000

LINEAR VELOCITIES

V(B)	V(C)	V(BC)
-1.5000	1.0854	1.3416

REAL	IMAG	ABS	PHASE	
.7500	-1.2990	1.5000	-60.0000	(B)
1.0854	.0000	1.0854	.0000	(C)
-.3354	-1.2990	1.3416	-104.4775	(BC)

NORMAL ACCELERATIONS   TANGENTIAL ACCELERATIONS

REAL	IMAG	ABS	PHASE		REAL	IMAG	ABS	PHASE
-1.2990	-.7500	1.5000	-150.0000	(B)	.0000	.0000	.0000	.0000
.0000	.0000	.0000	.0000	(C)	-1.7251	.0000	1.7251	-180.0000

ABSOLUTE ACCELERATIONS

REAL	IMAG	ABS	PHASE	
-1.2990	-.7500	1.5000	-150.0000	(B)
-1.7251	.0000	1.7251	180.0000	(C)
.4260	-.7500	.8626	-60.4018	(BC)

FOUR-BAR: SIMPLIFIED VECTOR METHOD

```
01◆LBL "4-BAR L"          51 RCL 12
02 "  4-BAR LINKAGE"      52 X↑2
03 XEQ 09                 53 -
04 "  ____  _____"      54 RCL 11
05 XEQ 04                 55 /
06 FIX 4                  56 2
07 SF 04                  57 /
08 "AB?"                  58 RCL 18
09 PROMPT                 59 /
10 STO 10                 60 ACOS
11 "BC?"                  61 STO 19
12 PROMPT                 62 RCL 18
13 STO 11                 63 X↑2
14 "CD?"                  64 RCL 12
15 PROMPT                 65 X↑2
16 STO 12                 66 +
17 "AD?"                  67 RCL 11
18 PROMPT                 68 X↑2
19 STO 13                 69 -
20 "THETA A?"             70 RCL 18
21 PROMPT                 71 /
22 STO 14                 72 2
23 "OMEGA A?"             73 /
24 PROMPT                 74 RCL 12
25 STO 15                 75 /
26 "ALPHA A?"             76 ACOS
27 PROMPT                 77 STO 23
28 STO 16                 78 RCL 14
29 RCL 14                 79 SIN
30 COS                    80 STO 20
31 STO 17                 81 RCL 10
32 RCL 10                 82 *
33 *                      83 RCL 18
34 RCL 13                 84 /
35 *                      85 ASIN
36 2                      86 STO 21
37 *                      87 CHS
38 CHS                    88 RCL 23
39 RCL 10                 89 -
40 X↑2                    90 180
41 +                      91 +
42 RCL 13                 92 STO 24
43 X↑2                    93 RCL 19
44 +                      94 RCL 21
45 SQRT                   95 -
46 STO 18                 96 STO 25
47 X↑2                    97 RCL 15
48 RCL 11                 98 RCL 10
49 X↑2                    99 *
50 +                     100 STO 26
```

```
101 RCL 25                          151 "--- --- ---"
102 RCL 14                          152 XEQ 04
103 -                               153 RCL 24
104 STO 69                          154 90
105 SIN                             155 +
106 STO 27                          156 STO 34
107 RCL 25                          157 XEQ 05
108 RCL 24                          158 RCL 30
109 -                               159 XEQ 07
110 STO 70                          160 XEQ 08
111 SIN                             161 ":C="
112 ST/ 27                          162 ASTO 33
113 STO 28                          163 XEQ 03
114 RCL 24                          164 RCL 14
115 RCL 14                          165 90
116 -                               166 +
117 STO 68                          167 STO 32
118 SIN                             168 XEQ 05 ⮦
119 RCL 28                          169 RCL 26
120 /                               170 XEQ 07
121 STO 28                          171 XEQ 08
122 RCL 26                          172 ":B="
123 *                               173 ASTO 33
124 STO 29                          174 XEQ 03
125 RCL 27                          175 RCL 25
126 RCL 26                          176 90
127 *                               177 +
128 STO 30                          178 STO 35
129 "IN/SEC"                        179 XEQ 05
130 ASTO 31                         180 RCL 29
131 "LINEAR VELOCITY"               181 XEQ 07
132 XEQ 09                          182 XEQ 08
133 "------ --------"               183 ":C/B="
134 XEQ 04                          184 ASTO 33
135 "V:C="                          185 XEQ 03
136 ARCL 30                         186 RCL 34
137 ARCL 31                         187 COS
138 XEQ 04                          188 STO 36
139 "V:B="                          189 LASTX
140 ARCL 26                         190 SIN
141 ARCL 31                         191 STO 37
142 XEQ 04                          192 RCL 35
143 "V:C/B="                        193 COS
144 ARCL 29                         194 STO 38
145 ARCL 31                         195 CHS
146 XEQ 04                          196 STO 39
147 "IPS∠"                          197 LASTX
148 ASTO 71                         198 SIN
149 "VEC LIN VEL"                   199 STO 40
150 XEQ 09                          200 CHS
```

```
201 STO 41          251 +
202 RCL 24          252 SIN
203 COS             253 STO 52
204 STO 42          254 RCL 46
205 STO 43          255 *
206 RCL 30          256 ST+ 51
207 X↑2             257 RCL 25
208 RCL 12          258 SIN
209 /               259 STO 63
210 STO 44          260 RCL 48
211 ST* 43          261 *
212 RCL 15          262 ST- 51
213 X↑2             263 RCL 43
214 RCL 10          264 RCL 41
215 *               265 *
216 STO 45          266 STO 54
217 RCL 17          267 RCL 51
218 *               268 RCL 39
219 ST- 43          269 *
220 RCL 16          270 ST- 54
221 RCL 10          271 RCL 36
222 *               272 RCL 41
223 STO 46          273 *
224 RCL 32          274 STO 55
225 COS             275 RCL 37
226 STO 47          276 RCL 39
227 *               277 *
228 ST+ 43          278 ST- 55
229 RCL 29          279 RCL 36
230 X↑2             280 RCL 51
231 RCL 11          281 *
232 /               282 STO 56
233 STO 48          283 RCL 37
234 RCL 25          284 RCL 43
235 COS             285 *
236 STO 49          286 ST- 56
237 *               287 RCL 54
238 ST- 43          288 RCL 55
239 RCL 24          289 /
240 SIN             290 STO 57
241 STO 50          291 RCL 56
242 RCL 44          292 RCL 55
243 *               293 /
244 STO 51          294 STO 58
245 RCL 20          295 "NORM ACC"
246 RCL 45          296 XEQ 09
247 *               297 "---- ---"
248 ST- 51          298 XEQ 04
249 RCL 14          299 "IPS↑2⊾"
250 90              300 ASTO 63
```

```
301 RCL 24              351 XEQ 07
302 XEQ 05              352 XEQ 08
303 RCL 44              353 ":C/B="
304 CHS                 354 ASTO 33
305 XEQ 07              355 XEQ 16
306 XEQ 08              356 "ABS ACC"
307 ":C="               357 XEQ 09
308 ASTO 33             358 "--- ---"
309 XEQ 16              359 XEQ 04
310 RCL 14              360 RCL 24
311 XEQ 05              361 XEQ 05
312 RCL 45              362 RCL 59
313 CHS                 363 STO 64
314 XEQ 07              364 RCL 60
315 XEQ 08              365 STO 65
316 ":B/C="             366 RCL 27
317 ASTO 33             367 RCL 26
318 XEQ 16              368 *
319 RCL 25              369 X↑2
320 XEQ 05              370 RCL 12
321 RCL 48              371 /
322 CHS                 372 CHS
323 XEQ 07              373 ST* 64
324 XEQ 08              374 ST* 65
325 ":C/B="             375 RCL 34
326 ASTO 33             376 XEQ 05
327 XEQ 16              377 RCL 59
328 "TAN ACC"           378 STO 66
329 XEQ 09              379 RCL 60
330 "--- ---"           380 STO 67
331 XEQ 04              381 RCL 57
332 RCL 34              382 ST* 66
333 XEQ 05              383 ST* 67
334 RCL 57              384 XEQ 06
335 XEQ 07              385 1
336 XEQ 08              386 XEQ 07
337 ":C="              387 XEQ 08
338 ASTO 33             388 ":C="
339 XEQ 16              389 ASTO 33
340 RCL 32              390 XEQ 16
341 XEQ 05              391 RCL 14
342 RCL 46              392 XEQ 05
343 XEQ 07              393 RCL 59
344 XEQ 08              394 STO 64
345 ":B="              395 RCL 60
346 ASTO 33             396 STO 65
347 XEQ 16              397 RCL 26
348 RCL 35              398 RCL 15
349 XEQ 05              399 *
350 RCL 58              400 CHS
```

```
401 ST* 64          451 XEQ 09
402 ST* 65          452 "——————— ———"
403 RCL 32          453 XEQ 04
404 XEQ 05          454 "AB="
405 RCL 59          455 ARCL 10
406 STO 66          456 XEQ 04
407 RCL 60          457 "BC="
408 STO 67          458 ARCL 11
409 RCL 46          459 XEQ 04
410 ST* 66          460 "CD="
411 ST* 67          461 ARCL 12
412 XEQ 06          462 XEQ 04
413 1               463 "AD="
414 XEQ 07          464 ARCL 13
415 XEQ 08          465 XEQ 04
416 ":B="           466 "THETA A="
417 ASTO 33         467 ARCL 14
418 XEQ 16          468 XEQ 04
419 RCL 25          469 "OMEGA A="
420 XEQ 05          470 ARCL 15
421 RCL 59          471 XEQ 04
422 STO 64          472 "ALPHA A="
423 RCL 60          473 ARCL 16
424 STO 65          474 XEQ 04
425 RCL 28          475 "BD="
426 RCL 26          476 ARCL 18
427 *               477 XEQ 04
428 X↑2             478 "PHI D="
429 RCL 11          479 ARCL 21
430 /               480 XEQ 04
431 CHS             481 "GAMMA B="
432 ST* 64          482 ARCL 19
433 ST* 65          483 XEQ 04
434 RCL 35          484 "GAMMA D="
435 XEQ 05          485 ARCL 23
436 RCL 59          486 XEQ 04
437 STO 66          487 "THETA B="
438 RCL 60          488 ARCL 25
439 STO 67          489 XEQ 04
440 RCL 58          490 "THETA C="
441 ST* 66          491 ARCL 24
442 ST* 67          492 XEQ 04
443 XEQ 06          493 "SMB="
444 1               494 ARCL 70
445 XEQ 07          495 XEQ 04
446 XEQ 08          496 "SMC="
447 ":C/B="         497 ARCL 69
448 ASTO 33         498 XEQ 04
449 XEQ 16          499 "SMCB="
450 "PROBLEM DATA"  500 ARCL 68
```

```
501 XEQ 04          551 RTN
502 "A 1="          552◆LBL 06
503 ARCL 36         553 RCL 65
504 XEQ 04          554 RCL 64
505 "A 2="          555 RCL 67
506 ARCL 37         556 RCL 66
507 XEQ 04          557 XROM "C+"
508 "B 1="          558 STO 59
509 ARCL 39         559 X<>Y
510 XEQ 04          560 STO 60
511 "B 2="          561 RTN
512 ARCL 41         562◆LBL 07
513 XEQ 04          563 ST* 59
514 "C 1="          564 ST* 60
515 ARCL 43         565 RCL 60
516 XEQ 04          566 RCL 59
517 "C 2="          567 XROM "MAGZ"
518 ARCL 51         568 STO 61
519 XEQ 04          569 RCL 59
520 "ATC="          570 X=0?
521 ARCL 57         571 GTO 12
522 XEQ 04          572 RCL 60
523 "ATCB="         573 RCL 59
524 ARCL 58         574 /
525 XEQ 04          575 ATAN
526 XEQ 13          576 STO 62
527◆LBL 05          577 RTN
528 57.296          578◆LBL 12
529 /               579 0
530 0               580 ATAN
531 XROM "e↑Z"      581 STO 62
532 STO 59          582 RTN
533 X<>Y            583◆LBL 04
534 STO 60          584 AVIEW
535 RTN             585 ADV
536◆LBL 08          586 CLA
537 RCL 59          587 RTN
538 X<0?            588◆LBL 09
539 GTO 14          589 AVIEW
540 RTN             590 RTN
541◆LBL 14          591◆LBL 03
542 RCL 60          592 FIX 2
543 X<0?            593 26
544 GTO 15          594 ACCHR
545 180             595 CLA
546 ST+ 62          596 ARCL 33
547 RTN             597 ARCL 61
548◆LBL 15          598 ARCL 71
549 180             599 ARCL 62
550 ST- 62          600 ACA
```

```
601 PRBUF
602 ADV
603 FIX 4
604 RTN
605◆LBL 16
606 FIX 2
607 22
608 ACCHR
609 CLA
610 ARCL 33
611 ARCL 61
612 ARCL 63
613 ARCL 62
614 ACA
615 PRBUF
616 ADV
617 FIX 4
618 RTN
619◆LBL 13
620 AOFF
621 STOP
622 .END.
```

PROBLEM DATA
_____ ___

AB=1.5000

BC=3.0000

CD=3.0000

AD=4.0000

THETA A=30.0000

OMEGA A=2.0000

ALPHA A=1.0000

BD=2.8032

PHI D=15.5188

GAMMA B=62.1478

GAMMA D=62.1478

THETA B=46.6289

THETA C=102.3334

SMB=-55.7045

SMC=16.6289

SMCB=72.3334

A 1=-0.9769

A 2=-0.2136

B 1=0.7269

B 2=-0.6867

C 1=-8.7635

C 2=-4.2502

ATC=11.0244

ATCB=2.7601

4-BAR LINKAGE
____ _____

LINEAR VELOCITY
_____ _____

V:C=-1.0392IN/SEC

V:B=3.0000IN/SEC

V:C/B=-3.4601IN/SEC

VEC LIN VEL
___ ___ ___

$\bar{v}$:C=1.04IPS∠12.33

$\bar{v}$:B=3.00IPS∠120.00

$\bar{v}$:C/B=3.46IPS∠-43.37

NORM ACC
____ ___

$\bar{a}$:C=0.36IPS↑2∠-77.67

$\bar{a}$:B/C=6.00IPS↑2∠-150.00

$\bar{a}$:C/B=3.99IPS↑2∠-133.37

TAN ACC
___ ___

$\bar{a}$:C=11.02IPS↑2∠-167.67

$\bar{a}$:B=1.50IPS↑2∠120.00

$\bar{a}$:C/B=2.76IPS↑2∠136.63

ABS ACC
___ ___

$\bar{a}$:C=11.03IPS↑2∠-165.80

$\bar{a}$:B=6.18IPS↑2∠-164.04

$\bar{a}$:C/B=4.85IPS↑2∠-168.04

SLIDER-CRANK: SIMPLIFIED VECTOR METHOD

```
01♦LBL "SL-CR V"              51 RCL 17
02 "  SLIDER CR ANL"         52 -
03 XEQ 09                    53 SIN
04 " ------- -- ---"         54 STO 20
05 XEQ 04                    55 RCL 19
06 FIX 4                     56 RCL 20
07 SF 04                     57 /
08 "AB?"                     58 STO 21
09 PROMPT                    59 RCL 18
10 STO 10                    60 *
11 "BC?"                     61 STO 22
12 PROMPT                    62 90
13 STO 11                    63 RCL 12
14 "THETA 2?"                64 -
15 PROMPT                    65 SIN
16 STO 12                    66 STO 23
17 "OMEGA 2?"                67 RCL 20
18 PROMPT                    68 /
19 STO 13                    69 STO 24
20 "ALPHA 2?"                70 RCL 18
21 PROMPT                    71 *
22 STO 09                    72 STO 25
23 "ECC?"                    73 "LINEAR VELOCITY"
24 PROMPT                    74 XEQ 09
25 STO 14                    75 "------ --------"
26 RCL 12                    76 XEQ 04
27 SIN                       77 "IN/SEC"
28 STO 15                    78 ASTO 26
29 RCL 10                    79 "V:B="
30 *                         80 ARCL 18
31 RCL 14                    81 ARCL 26
32 -                         82 XEQ 04
33 RCL 11                    83 "V:C="
34 /                         84 ARCL 22
35 ASIN                      85 ARCL 26
36 STO 16                    86 XEQ 04
37 180                       87 "V:B/C="
38 RCL 16                    88 ARCL 25
39 -                         89 ARCL 26
40 STO 17                    90 XEQ 04
41 RCL 10                    91 RCL 12
42 RCL 13                    92 90
43 *                         93 +
44 STO 18                    94 STO 27
45 RCL 17                    95 RCL 17
46 RCL 12                    96 90
47 -                         97 +
48 SIN                       98 STO 28
49 STO 19                    99 "VEC LIN VEL"
50 90                        100 XEQ 09
```

101 `"— — —"`	151 RCL 10
102 XEQ 04	152 *
103 RCL 27	153 STO 37
104 XEQ 05	154 RCL 27
105 RCL 18	155 COS
106 XEQ 07	156 STO 38
107 XEQ 08	157 RCL 37
108 `"IPS∠"`	158 *
109 ASTO 29	159 ST+ 36
110 `":B="`	160 RCL 17
111 ASTO 30	161 COS
112 XEQ 03	162 STO 39
113 0	163 RCL 25
114 XEQ 05	164 X↑2
115 RCL 22	165 RCL 11
116 XEQ 07	166 /
117 XEQ 08	167 STO 43
118 `":C="`	168 RCL 39
119 ASTO 30	169 *
120 XEQ 03	170 ST+ 36
121 RCL 28	171 RCL 15
122 XEQ 05	172 RCL 40
123 RCL 25	173 *
124 XEQ 07	174 CHS
125 XEQ 08	175 STO 41
126 `":B/C="`	176 RCL 27
127 ASTO 30	177 SIN
128 XEQ 03	178 STO 42
129 1	179 RCL 37
130 STO 31	180 *
131 0	181 ST+ 41
132 STO 32	182 RCL 17
133 RCL 28	183 SIN
134 COS	184 STO 44
135 STO 33	185 RCL 43
136 LASTX	186 *
137 SIN	187 ST+ 41
138 STO 34	188 RCL 36
139 RCL 12	189 RCL 34
140 COS	190 *
141 STO 35	191 STO 45
142 RCL 18	192 RCL 41
143 RCL 13	193 RCL 33
144 *	194 *
145 STO 40	195 ST- 45
146 RCL 35	196 RCL 31
147 *	197 RCL 34
148 CHS	198 *
149 STO 36	199 STO 46
150 RCL 09	200 RCL 32

```
201 RCL 33
202 *
203 ST- 46
204 RCL 31
205 RCL 41
206 *
207 STO 47
208 RCL 32
209 RCL 36
210 *
211 ST- 47
212 RCL 45
213 RCL 46
214 /
215 STO 48
216 RCL 47
217 RCL 46
218 /
219 STO 49
220 "NORM ACC"
221 XEQ 09
222 "--- ---"
223 XEQ 04
224 RCL 12
225 XEQ 05
226 RCL 40
227 CHS
228 XEQ 07
229 XEQ 08
230 "IPS↑2∠"
231 ASTO 50
232 ":B="
233 ASTO 30
234 XEQ 16
235 0
236 STO 59
237 STO 60
238 1
239 XEQ 07
240 XEQ 08
241 ":C="
242 ASTO 30
243 XEQ 16
244 RCL 17
245 XEQ 05
246 RCL 43
247 CHS
248 XEQ 07
249 XEQ 08
250 ":B/C="
```

```
251 ASTO 30
252 XEQ 16
253 "TAN ACC"
254 XEQ 09
255 "--- ---"
256 XEQ 04
257 RCL 27
258 XEQ 05
259 RCL 37
260 XEQ 07
261 XEQ 08
262 ":B="
263 ASTO 30
264 XEQ 16
265 0
266 XEQ 05
267 RCL 48
268 XEQ 07
269 XEQ 08
270 ":C="
271 ASTO 30
272 XEQ 16
273 RCL 28
274 XEQ 05
275 RCL 49
276 XEQ 07
277 XEQ 08
278 ":B/C="
279 ASTO 30
280 XEQ 16
281 "ABS ACC"
282 XEQ 09
283 "--- ---"
284 XEQ 04
285 RCL 12
286 XEQ 05
287 RCL 59
288 STO 51
289 RCL 60
290 STO 52
291 RCL 40
292 CHS
293 ST* 51
294 ST* 52
295 RCL 27
296 XEQ 05
297 RCL 59
298 STO 53
299 RCL 60
300 STO 54
```

```
301 RCL 37              351 XEQ 04
302 ST* 53              352 "AB="
303 ST* 54              353 ARCL 10
304 XEQ 06              354 XEQ 04
305 1                   355 "BC="
306 XEQ 07              356 ARCL 11
307 XEQ 08              357 XEQ 04
308 ":B="               358 "THETA 2="
309 ASTO 30            359 ARCL 12
310 XEQ 16              360 XEQ 04
311 0                   361 "OMEGA 2="
312 XEQ 05              362 ARCL 13
313 RCL 48              363 XEQ 04
314 XEQ 07              364 "ALPHA 2="
315 XEQ 08              365 ARCL 09
316 ":C="               366 XEQ 04
317 ASTO 30            367 "ECC="
318 XEQ 16              368 ARCL 14
319 RCL 17              369 XEQ 04
320 XEQ 05              370 "PHI 3="
321 RCL 59              371 ARCL 16
322 STO 51              372 XEQ 04
323 RCL 60              373 "THETA 3="
324 STO 52              374 ARCL 17
325 RCL 25              375 XEQ 04
326 X↑2                 376 90
327 RCL 11              377 RCL 17
328 /                   378 -
329 CHS                 379 STO 55
330 ST* 51              380 RCL 17
331 ST* 52              381 RCL 12
332 RCL 28              382 -
333 XEQ 05              383 STO 56
334 RCL 59              384 90
335 STO 53              385 RCL 12
336 RCL 60              386 -
337 STO 54              387 STO 57
338 RCL 49              388 "SNB="
339 ST* 53              389 ARCL 55
340 ST* 54              390 XEQ 04
341 XEQ 06              391 "SNC="
342 1                   392 ARCL 56
343 XEQ 07              393 XEQ 04
344 XEQ 08              394 "SNBC="
345 ":B/C="             395 ARCL 57
346 ASTO 30            396 XEQ 04
347 XEQ 16              397 "A 1="
348 "PROBLEM DATA"      398 ARCL 31
349 XEQ 09              399 XEQ 04
350 "------ ----"       400 "A 2="
```

```
401 ARCL 32              451 RCL 53
402 XEQ 04               452 XROM "C+"
403 "B 1="               453 STO 59
404 ARCL 33              454 X<>Y
405 XEQ 04               455 STO 60
406 "B 2="               456 RTN
407 ARCL 34              457*LBL 07
408 XEQ 04               458 ST* 59
409 "C 1="               459 ST* 60
410 ARCL 36              460 RCL 60
411 XEQ 04               461 RCL 59
412 "C 2="               462 XROM "MAGZ"
413 ARCL 41              463 STO 61
414 XEQ 04               464 RCL 59
415 "ATC="               465 X=0?
416 ARCL 48              466 GTO 12
417 XEQ 04               467 RCL 60
418 "ATBC="              468 RCL 59
419 ARCL 49              469 /
420 XEQ 04               470 ATAN
421 XEQ 13               471 STO 62
422*LBL 05               472 RTN
423 57.296               473*LBL 12
424 /                    474 0
425 0                    475 ATAN
426 XROM "e↑Z"           476 STO 62
427 STO 59               477 RTN
428 X<>Y                 478*LBL 04
429 STO 60               479 AVIEW
430 RTN                  480 ADV
431*LBL 08               481 CLA
432 RCL 59               482 RTN
433 X<0?                 483*LBL 09
434 GTO 14               484 AVIEW
435 RTN                  485 RTN
436*LBL 14               486*LBL 03
437 RCL 60               487 FIX 2
438 X<0?                 488 26
439 GTO 15               489 ACCHR
440 180                  490 CLA
441 ST+ 62               491 ARCL 30
442 RTN                  492 ARCL 61
443*LBL 15               493 ARCL 29
444 180                  494 ARCL 62
445 ST- 62               495 ACA
446 RTN                  496 PRBUF
447*LBL 06               497 ADV
448 RCL 52               498 FIX 4
449 RCL 51               499 RTN
450 RCL 54               500*LBL 16
```

```
501 FIX 2
502 22
503 ACCHR
504 CLA
505 ARCL 30
506 ARCL 61
507 ARCL 50
508 ARCL 62
509 ACA
510 PRBUF
511 ADV
512 FIX 4
513 RTN
514◆LBL 13
515 AOFF
516 STOP
517 .END.
```

PROBLEM DATA	SLIDER CR ANL
------ ----	------ -- ---

PROBLEM DATA

AB=1.5000

BC=3.0000

THETA 2=150.0000

OMEGA 2=1.0000

ALPHA 2=0.0000

ECC=0.0000

PHI 3=14.4775

THETA 3=165.5225

SNB=-75.5225

SNC=15.5225

SNBC=-60.0000

A 1=1.0000

A 2=0.0000

B 1=-0.2500

B 2=-0.9682

C 1=0.7181

C 2=-0.6000

ATC=0.8730

ATBC=0.6197

SLIDER CR ANL

LINEAR VELOCITY

V:B=1.5000IN/SEC

V:C=-0.4146IN/SEC

V:B/C=1.3416IN/SEC

VEC LIN VEL

$\bar{u}$:B=1.50IPS∠-120.00

$\bar{u}$:C=0.41IPS∠180.00

$\bar{u}$:B/C=1.34IPS∠-104.48

NORM ACC

$\bar{a}$:B=1.50IPS↑2∠-30.00

$\bar{a}$:C=0.00IPS↑2∠0.00

$\bar{a}$:B/C=0.60IPS↑2∠-14.48

TAN ACC

$\bar{a}$:B=0.00IPS↑2∠0.00

$\bar{a}$:C=0.87IPS↑2∠0.00

$\bar{a}$:B/C=0.62IPS↑2∠-104.48

ABS ACC

$\bar{a}$:B=1.50IPS↑2∠-30.00

$\bar{a}$:C=0.87IPS↑2∠0.00

$\bar{a}$:B/C=0.86IPS↑2∠-60.40

PROBLEM DATA
------ ---

AB=1.5000

BC=3.0000

THETA 2=150.0000

OMEGA 2=1.0000

ALPHA 2=0.0000

ECC=0.5000

PHI 3=4.7802

THETA 3=175.2198

SMB=-85.2198

SMC=25.2198

SMBC=-60.0000

A 1=1.0000

A 2=0.0000

B 1=-0.0833

B 2=-0.9965

C 1=0.7346

C 2=-0.7028

ATC=0.7933

ATBC=0.7053

SLIDER CR ANL
------ -- ---

LINEAR VELOCITY
------ --------

V:B=1.5000IN/SEC

V:C=-0.6414IN/SEC

V:B/C=1.3036IN/SEC

VEC LIN VEL
--- --- ---

v̄:B=1.50IPS∠-120.00

v̄:C=0.64IPS∠180.00

v̄:B/C=1.30IPS∠-94.78

NORM ACC
---- ---

ā:B=1.50IPS↑2∠-30.00

ā:C=0.00IPS↑2∠0.00

ā:B/C=0.57IPS↑2∠-4.78

TAN ACC
--- ---

ā:B=0.00IPS↑2∠0.00

ā:C=0.79IPS↑2∠0.00

ā:B/C=0.71IPS↑2∠-94.78

ABS ACC
--- ---

ā:B=1.50IPS↑2∠-30.00

ā:C=0.79IPS↑2∠0.00

ā:B/C=0.90IPS↑2∠-56.01

PROBLEM DATA
------- ----

AB=1.5000

BC=3.0000

THETA 2=150.0000

OMEGA 2=1.0000

ALPHA 2=0.0000

ECC=-0.5000

PHI 3=24.6243

THETA 3=155.3757

SNB=-65.3757

SNC=5.3757

SNBC=-60.0000

A 1=1.0000

A 2=0.0000

B 1=-0.4167

B 2=-0.9091

C 1=0.6803

C 2=-0.4664

ATC=0.8940

ATBC=0.5130

SLIDER CR ANL
------ -- ---

LINEAR VELOCITY
------ --------

V:B=1.5000IN/SEC

V:C=-0.1546IN/SEC

V:B/C=1.4290IN/SEC

VEC LIN VEL
--- --- ---

$\bar{v}$:B=1.50IPS∠-120.00

$\bar{v}$:C=0.15IPS∠180.00

$\bar{v}$:B/C=1.43IPS∠-114.63

NORM ACC
---- ---

$\bar{a}$:B=1.50IPS↑2∠-30.00

$\bar{a}$:C=0.00IPS↑2∠0.00

$\bar{a}$:B/C=0.68IPS↑2∠-24.62

TAN ACC
--- ---

$\bar{a}$:B=0.00IPS↑2∠0.00

$\bar{a}$:C=0.89IPS↑2∠0.00

$\bar{a}$:B/C=0.51IPS↑2∠-114.63

ABS ACC
--- ---

$\bar{a}$:B=1.50IPS↑2∠-30.00

$\bar{a}$:C=0.89IPS↑2∠0.00

$\bar{a}$:B/C=0.85IPS↑2∠-61.63

QUICK-RETURN: SIMPLIFIED VECTOR METHOD

```
01♦LBL "Q-R MEC"          51 -
02 " Q-R MECHANISM"       52 2
03 XEQ 09                 53 /
04 " --- --------"        54 RCL 15
05 XEQ 04                 55 /
06 FIX 4                  56 RCL 11
07 SF 04                  57 /
08 "AB?"                  58 ACOS
09 PROMPT                 59 STO 65
10 12                     60 180
11 /                      61 RCL 12
12 STO 10                 62 X<=Y?
13 "AD?"                  63 GTO 10
14 PROMPT                 64 -1
15 12                     65 ST* 65
16 /                      66♦LBL 10
17 STO 11                 67 180
18 "THETA 2?"             68 RCL 65
19 PROMPT                 69 -
20 STO 12                 70 STO 17
21 "OMEGA 2?"             71 RCL 12
22 PROMPT                 72 SIN
23 STO 13                 73 STO 16
24 "ALPHA 2?"             74♦LBL 02
25 PROMPT                 75 RCL 10
26 STO 14                 76 RCL 13
27 RCL 12                 77 *
28 COS                    78 STO 09
29 STO 08                 79 RCL 17
30 RCL 11                 80 RCL 12
31 *                      81 -
32 RCL 10                 82 STO 18
33 *                      83 COS
34 2                      84 STO 19
35 *                      85 RCL 09
36 CHS                    86 *
37 RCL 11                 87 STO 20
38 X↑2                    88 RCL 18
39 +                      89 SIN
40 RCL 10                 90 STO 21
41 X↑2                    91 RCL 09
42 +                      92 *
43 SQRT                   93 STO 22
44 STO 15                 94 "LIN VELOCITY"
45 X↑2                    95 XEQ 09
46 RCL 11                 96 "--- --------"
47 X↑2                    97 XEQ 04
48 +                      98 FIX 2
49 RCL 10                 99 "FPS"
50 X↑2                    100 ASTO 23
```

```
101 "V:B="              151 COS
102 ARCL 09             152 STO 28
103 ARCL 23             153 LASTX
104 XEQ 04              154 SIN
105 "V:C="              155 STO 29
106 ARCL 20             156 RCL 17
107 ARCL 23             157 SIN
108 XEQ 04              158 STO 30
109 "V:B/C="            159 LASTX
110 ARCL 22             160 COS
111 ARCL 23             161 STO 31
112 XEQ 04              162 RCL 20
113 FIX 4               163 X↑2
114 "VEC LIN VEL"       164 *
115 XEQ 09              165 RCL 15
116 "--- --- ---"       166 /
117 XEQ 04              167 STO 32
118 RCL 12              168 RCL 28
119 90                  169 RCL 20
120 +                   170 *
121 STO 24              171 RCL 15
122 XEQ 05              172 /
123 RCL 09              173 RCL 22
124 XEQ 07              174 *
125 XEQ 08              175 2
126 ":B="              176 *
127 ASTO 25             177 ST- 32
128 "FPS∠"             178 RCL 24
129 ASTO 26             179 COS
130 XEQ 03              180 STO 33
131 RCL 17              181 RCL 10
132 90                  182 *
133 +                   183 RCL 14
134 STO 27              184 *
135 XEQ 05              185 ST+ 32
136 RCL 20              186 RCL 08
137 XEQ 07              187 RCL 10
138 XEQ 08              188 *
139 ":C="              189 RCL 13
140 ASTO 25             190 X↑2
141 XEQ 03              191 *
142 RCL 17              192 ST- 32
143 XEQ 05              193 RCL 30
144 RCL 22              194 RCL 20
145 XEQ 07              195 X↑2
146 XEQ 08              196 *
147 ":B/C="            197 RCL 15
148 ASTO 25             198 /
149 XEQ 03              199 STO 34
150 RCL 27              200 RCL 29
```

201 RCL 20
202 *
203 RCL 15
204 /
205 RCL 22
206 *
207 2
208 *
209 ST- 34
210 RCL 24
211 SIN
212 STO 35
213 RCL 10
214 *
215 RCL 14
216 *
217 ST+ 34
218 RCL 16
219 RCL 10
220 *
221 RCL 13
222 X↑2
223 *
224 ST- 34
225 RCL 32
226 RCL 30
227 *
228 STO 36
229 RCL 34
230 RCL 31
231 *
232 ST- 36
233 RCL 28
234 RCL 30
235 *
236 STO 37
237 RCL 29
238 RCL 31
239 *
240 ST- 37
241 RCL 36
242 RCL 37
243 /
244 STO 38
245 RCL 28
246 RCL 34
247 *
248 STO 39
249 RCL 29
250 RCL 32

251 *
252 ST- 39
253 RCL 39
254 RCL 37
255 /
256 STO 40
257 "NORM ACC"
258 XEQ 09
259 "---- ---"
260 XEQ 04
261 RCL 12
262 XEQ 05
263 RCL 09
264 RCL 13
265 *
266 STO 41
267 CHS
268 XEQ 07
269 XEQ 08
270 ":B="
271 ASTO 42
272 "FPS↑2↙"
273 ASTO 43
274 XEQ 16
275 RCL 17
276 XEQ 05
277 RCL 20
278 X↑2
279 RCL 15
280 /
281 STO 44
282 CHS
283 XEQ 07
284 XEQ 08
285 ":C="
286 ASTO 42
287 XEQ 16
288 RCL 27
289 XEQ 05
290 0
291 XEQ 07
292 XEQ 08
293 ":B/C="
294 ASTO 42
295 XEQ 16
296 "TANG ACC"
297 XEQ 09
298 "---- ---"
299 XEQ 04
300 RCL 24

```
301 XEQ 05        351 STO 48
302 RCL 10        352 RCL 60
303 RCL 14        353 STO 49
304 *             354 RCL 45
305 STO 45        355 ST* 48
306 XEQ 07        356 ST* 49
307 XEQ 08        357 RCL 12
308 ":B="         358 XEQ 05
309 ASTO 42       359 RCL 59
310 XEQ 16        360 STO 50
311 RCL 27        361 RCL 60
312 XEQ 05        362 STO 51
313 RCL 38        363 RCL 41
314 XEQ 07        364 CHS
315 XEQ 08        365 ST* 50
316 ":C="         366 ST* 51
317 ASTO 42       367 XEQ 06
318 XEQ 16        368 1
319 RCL 17        369 XEQ 07
320 XEQ 05        370 XEQ 08
321 RCL 40        371 ":B="
322 XEQ 07        372 ASTO 42
323 XEQ 08        373 XEQ 16
324 ":B/C="       374 RCL 27
325 ASTO 42       375 XEQ 05
326 XEQ 16        376 RCL 59
327 "COR ACC"     377 STO 48
328 XEQ 09        378 RCL 60
329 "--- ---"     379 STO 49
330 XEQ 04        380 RCL 38
331 RCL 27        381 ST* 48
332 XEQ 05        382 ST* 49
333 RCL 20        383 RCL 17
334 RCL 15        384 XEQ 05
335 /             385 RCL 59
336 RCL 22        386 STO 50
337 *             387 RCL 60
338 2             388 STO 51
339 *             389 RCL 19
340 STO 46        390 RCL 10
341 XEQ 07        391 *
342 XEQ 08        392 RCL 13
343 XEQ 16        393 *
344 "ABS ACC"     394 X↑2
345 XEQ 09        395 RCL 15
346 "--- ---"     396 /
347 XEQ 04        397 CHS
348 RCL 24        398 ST* 50
349 XEQ 05        399 ST* 51
350 RCL 59        400 XEQ 06
```

```
401 1
402 XEQ 07
403 XEQ 08
404 ":C="
405 ASTO 42
406 XEQ 16
407 RCL 17
408 XEQ 05
409 RCL 59
410 STO 48
411 RCL 60
412 STO 49
413 RCL 48
414 ST* 48
415 ST* 49
416 RCL 27
417 XEQ 05
418 RCL 59
419 STO 50
420 RCL 60
421 STO 51
422 RCL 19
423 RCL 21
424 *
425 RCL 41
426 *
427 RCL 10
428 *
429 2
430 *
431 RCL 15
432 /
433 ST* 50
434 ST* 51
435 XEQ 06
436 1
437 XEQ 07
438 XEQ 08
439 ":B/C="
440 ASTO 42
441 XEQ 16
442 "PROBLEM DATA"
443 XEQ 09
444 "------- ----"
445 XEQ 04
446 "AB="
447 ARCL 10
448 XEQ 04
449 "AD="
450 ARCL 11
```

```
451 XEQ 04
452 "THETA 2="
453 ARCL 12
454 XEQ 04
455 "OMEGA 2="
456 ARCL 13
457 XEQ 04
458 "ALPHA 2="
459 ARCL 14
460 XEQ 04
461 "PHI 3="
462 ARCL 65
463 XEQ 04
464 "THETA 3="
465 ARCL 17
466 XEQ 04
467 "CD="
468 ARCL 15
469 XEQ 04
470 "A 1="
471 ARCL 28
472 XEQ 04
473 "A 2="
474 ARCL 29
475 XEQ 04
476 "B 1="
477 ARCL 31
478 XEQ 04
479 "B 2="
480 ARCL 30
481 XEQ 04
482 "C 1="
483 ARCL 32
484 XEQ 04
485 "C 2="
486 ARCL 34
487 XEQ 04
488 "ATC="
489 ARCL 38
490 XEQ 04
491 "ATBC="
492 ARCL 40
493 XEQ 04
494 XEQ 13
495+LBL 05
496 57.296
497 /
498 0
499 XROM "e↑Z"
500 STO 59
```

```
501 X<>Y
502 STO 60
503 RTN
504*LBL 08
505 RCL 59
506 X<0?
507 GTO 14
508 RTN
509*LBL 14
510 RCL 60
511 X<0?
512 GTO 15
513 180
514 ST+ 62
515 RTN
516*LBL 15
517 180
518 ST- 62
519 RTN
520*LBL 06
521 RCL 49
522 RCL 48
523 RCL 51
524 RCL 50
525 XROM "C+"
526 STO 59
527 X<>Y
528 STO 60
529 RTN
530*LBL 07
531 ST* 59
532 ST* 60
533 RCL 60
534 RCL 59
535 XROM "MAGZ"
536 STO 61
537 RCL 59
538 X=0?
539 GTO 12
540 RCL 60
541 RCL 59
542 /
543 ATAN
544 STO 62
545 RTN
546*LBL 12
547 0
548 ATAN
549 STO 62
550 RTN
```

```
551*LBL 04
552 AVIEW
553 ADV
554 CLA
555 RTN
556*LBL 09
557 AVIEW
558 RTN
559*LBL 03
560 FIX 1
561 26
562 ACCHR
563 CLA
564 ARCL 25
565 ARCL 61
566 ARCL 26
567 ARCL 62
568 ACA
569 PRBUF
570 ADV
571 FIX 4
572 RTN
573*LBL 16
574 FIX 1
575 22
576 ACCHR
577 CLA
578 ARCL 42
579 ARCL 61
580 ARCL 43
581 ARCL 62
582 ACA
583 PRBUF
584 ADV
585 FIX 4
586 RTN
587*LBL 13
588 AOFF
589 .END.
```

PROBLEM DATA
------- ----

AB=0.1667

AD=0.3333

THETA 2=30.0000

OMEGA 2=62.8300

ALPHA 2=0.0000

PHI 3=23.7940

THETA 3=156.2060

CD=0.2066

A 1=-0.4034

A 2=-0.9150

B 1=-0.9150

B 2=0.4034

C 1=-943.4515

C 2=-717.2896

ATC=1,036.9562

ATBC=573.8702

Q-R MECHANISM
--- ---------

LIN VELOCITY
--- --------

V:B=10.47FPS

V:C=-6.19FPS

V:B/C=8.45FPS

VEC LIN VEL
--- --- ---

$\bar{v}$:B=10.5FPS∠120.0

$\bar{v}$:C=6.2FPS∠66.2

$\bar{v}$:B/C=8.4FPS∠156.2

NORM ACC
---- ---

$\bar{a}$:B=657.9FPS↑2∠-150.0

$\bar{a}$:C=185.2FPS↑2∠-23.8

$\bar{a}$:B/C=0.0FPS↑2∠0.0

TANG ACC
---- ---

$\bar{a}$:B=0.0FPS↑2∠0.0

$\bar{a}$:C=1,037.0FPS↑2∠-113.8

$\bar{a}$:B/C=573.9FPS↑2∠156.2

COR ACC
--- ---

$\bar{a}$:B/C=506.1FPS↑2∠66.2

ABS ACC
--- ---

$\bar{a}$:B=657.9FPS↑2∠-150.0

$\bar{a}$:C=1,053.4FPS↑2∠-103.7

$\bar{a}$:B/C=765.1FPS↑2∠114.8

PROBLEM DATA
------- ----

AB=0.2500

AD=0.1667

THETA 2=60.0000

OMEGA 2=30.0000

ALPHA 2=0.0000

PHI 3=79.1066

THETA 3=100.8934

CD=0.2205

A 1=-0.9820

A 2=-0.1890

B 1=-0.1890

B 2=0.9820

C 1=107.9082

C 2=-3.9766

ATC=-105.2122

ATBC=-24.2977

Q-R MECHANISM
--- ---------

LIN VELOCITY
--- --------

V:B=7.50FPS

V:C=5.67FPS

V:B/C=4.91FPS

VEC LIN VEL
--- --- ---

$\bar{v}$:B=7.5FPS∠150.0

$\bar{v}$:C=5.7FPS∠-169.1

$\bar{v}$:B/C=4.9FPS∠100.9

NORM ACC
---- ---

$\bar{a}$:B=225.0FPS↑2∠-120.0

$\bar{a}$:C=145.8FPS↑2∠-79.1

$\bar{a}$:B/C=0.0FPS↑2∠0.0

TANG ACC
---- ---

$\bar{a}$:B=0.0FPS↑2∠0.0

$\bar{a}$:C=105.2FPS↑2∠10.9

$\bar{a}$:B/C=24.3FPS↑2∠-79.1

COR ACC
--- ---

$\bar{a}$:B/C=252.5FPS↑2∠-169.1

ABS ACC
--- ---

$\bar{a}$:B=225.0FPS↑2∠-120.0

$\bar{a}$:C=179.8FPS↑2∠-43.3

$\bar{a}$:B/C=253.7FPS↑2∠-163.6

SLIDING COUPLER: SIMPLIFIED VECTOR METHOD

```
01◆LBL "SL-CPLR"          51 X↑2
02 "SLIDING COUPLER"      52 RCL 11
03 XEQ 09                 53 X↑2
04 "------- -------"      54 +
05 XEQ 04                 55 RCL 10
06 SF 04                  56 X↑2
07 FIX 4                  57 -
08 "AB?"                  58 2
09 PROMPT                 59 /
10 12                     60 RCL 17
11 /                      61 /
12 STO 10                 62 RCL 11
13 "AC?"                  63 /
14 PROMPT                 64 ACOS
15 12                     65 CHS
16 /                      66 STO 18
17 STO 11                 67 180
18 "THETA A?"             68 ENTER↑
19 PROMPT                 69 RCL 12
20 STO 12                 70 X<=Y?
21 "OMEGA A?"             71 GTO 02
22 PROMPT                 72 -1
23 STO 13                 73 ST* 18
24 "ALPHA A?"             74 GTO 02
25 PROMPT                 75◆LBL 01
26 STO 14                 76 180
27 RCL 12                 77 ENTER↑
28 SIN                    78 RCL 12
29 STO 15                 79 X<=Y?
30 LASTX                  80 GTO 18
31 COS                    81 0
32 STO 16                 82 ACOS
33 RCL 11                 83 CHS
34 *                      84 STO 18
35 RCL 10                 85◆LBL 18
36 *                      86 0
37 2                      87 ACOS
38 *                      88 STO 18
39 CHS                    89◆LBL 02
40 RCL 11                 90 RCL 13
41 X↑2                    91 RCL 10
42 +                      92 *
43 RCL 10                 93 STO 19
44 X↑2                    94 RCL 12
45 +                      95 RCL 18
46 SQRT                   96 -
47 STO 17                 97 STO 20
48 X=0?                   98 SIN
49 GTO 01                 99 STO 21
50 RCL 17                 100 RCL 19
```

```
101 *                          151 ":C="
102 CHS                        152 ASTO 28
103 STO 22                     153 XEQ 03
104 RCL 20                     154 RCL 18
105 COS                        155 90
106 STO 23                     156 +
107 RCL 19                     157 STO 29
108 *                          158 XEQ 05
109 CHS                        159 RCL 24
110 STO 24                     160 XEQ 07
111 "LIN VELOCITY"             161 XEQ 08
112 XEQ 09                     162 ":C/B="
113 "--- --------"             163 ASTO 28
114 XEQ 04                     164 XEQ 03
115 "FPS"                      165 RCL 18
116 ASTO 25                    166 COS
117 "V:B="                     167 STO 30
118 ARCL 19                    168 LASTX
119 ARCL 25                    169 SIN
120 XEQ 04                     170 STO 31
121 "V:C="                     171 RCL 29
122 ARCL 22                    172 COS
123 ARCL 25                    173 CHS
124 XEQ 04                     174 STO 32
125 "V:C/B="                   175 LASTX
126 ARCL 24                    176 SIN
127 ARCL 25                    177 CHS
128 XEQ 04                     178 STO 33
129 "VEC LIN VEL"              179 RCL 13
130 XEQ 09                     180 X†2
131 "--- --- ---"              181 RCL 10
132 XEQ 04                     182 *
133 RCL 12                     183 STO 34
134 90                         184 RCL 24
135 +                          185 X†2
136 STO 26                     186 RCL 17
137 XEQ 05                     187 /
138 RCL 19                     188 STO 35
139 XEQ 07                     189 RCL 14
140 XEQ 08                     190 RCL 10
141 "FPS∠"                     191 *
142 ASTO 27                    192 STO 36
143 ":B="                      193 RCL 24
144 ASTO 28                    194 RCL 17
145 XEQ 03                     195 /
146 RCL 18                     196 STO 37
147 XEQ 05                     197 RCL 22
148 RCL 22                     198 *
149 XEQ 07                     199 2
150 XEQ 08                     200 *
```

201 STO 38
202 RCL 16
203 RCL 34
204 *
205 CHS
206 STO 39
207 RCL 26
208 COS
209 STO 40
210 RCL 36
211 *
212 ST+ 39
213 RCL 30
214 RCL 35
215 *
216 ST- 39
217 RCL 32
218 CHS
219 RCL 38
220 *
221 ST- 39
222 RCL 15
223 RCL 34
224 *
225 CHS
226 STO 41
227 RCL 26
228 SIN
229 STO 42
230 RCL 36
231 *
232 ST+ 41
233 RCL 31
234 RCL 35
235 *
236 ST- 41
237 RCL 33
238 CHS
239 RCL 38
240 *
241 ST- 41
242 RCL 39
243 RCL 33
244 *
245 STO 43
246 RCL 41
247 RCL 32
248 *
249 ST- 43
250 RCL 30

251 RCL 33
252 *
253 STO 44
254 RCL 31
255 RCL 32
256 *
257 ST- 44
258 RCL 30
259 RCL 41
260 *
261 STO 45
262 RCL 31
263 RCL 39
264 *
265 ST- 45
266 RCL 43
267 RCL 44
268 /
269 STO 46
270 RCL 45
271 RCL 44
272 /
273 STO 47
274 "NORM ACC"
275 XEQ 09
276 "---- ---"
277 XEQ 04
278 RCL 12
279 XEQ 05
280 RCL 34
281 CHS
282 XEQ 07
283 XEQ 08
284 "FPS↑2∠"
285 ASTO 48
286 ":B="
287 ASTO 49
288 XEQ 16
289 RCL 29
290 XEQ 05
291 0
292 XEQ 07
293 XEQ 08
294 ":C="
295 ASTO 49
296 XEQ 16
297 RCL 18
298 XEQ 05
299 RCL 35
300 CHS

```
301 XEQ 07          351 XEQ 05
302 XEQ 08          352 RCL 59
303 ":C/B="         353 STO 50
304 ASTO 49         354 RCL 60
305 XEQ 16          355 STO 51
306 "TANG ACC"      356 RCL 34
307 XEQ 09          357 CHS
308 "---- ---"      358 ST* 50
309 XEQ 04          359 ST* 51
310 RCL 26          360 RCL 26
311 XEQ 05          361 XEQ 05
312 RCL 36          362 RCL 59
313 XEQ 07          363 STO 52
314 XEQ 08          364 RCL 60
315 ":B="           365 STO 53
316 ASTO 49         366 RCL 36
317 XEQ 16          367 ST* 52
318 RCL 18          368 ST* 53
319 XEQ 05          369 XEQ 06
320 RCL 46          370 1
321 XEQ 07          371 XEQ 07
322 XEQ 08          372 XEQ 08
323 ":C="           373 ":B="
324 ASTO 49         374 ASTO 49
325 XEQ 16          375 XEQ 16
326 RCL 29          376 RCL 29
327 XEQ 05          377 XEQ 05
328 RCL 47          378 RCL 59
329 XEQ 07          379 STO 50
330 XEQ 08          380 RCL 60
331 ":C/B="         381 STO 51
332 ASTO 49         382 RCL 21
333 XEQ 16          383 RCL 23
334 "CORR ACC"      384 *
335 XEQ 09          385 RCL 34
336 "---- ---"      386 *
337 XEQ 04          387 RCL 10
338 RCL 29          388 *
339 XEQ 05          389 2
340 RCL 38          390 *
341 XEQ 07          391 RCL 17
342 XEQ 08          392 /
343 ":C="           393 ST* 50
344 ASTO 49         394 ST* 51
345 XEQ 16          395 RCL 18
346 "ABS ACC"       396 XEQ 05
347 XEQ 09          397 RCL 59
348 "--- ---"       398 STO 52
349 XEQ 04          399 RCL 60
350 RCL 12          400 STO 53
```

```
401 RCL 46
402 ST* 52
403 ST* 53
404 XEQ 06
405 1
406 XEQ 07
407 XEQ 08
408 ":C="
409 ASTO 49
410 XEQ 16
411 RCL 18
412 XEQ 05
413 RCL 59
414 STO 50
415 RCL 60
416 STO 51
417 RCL 23
418 X↑2
419 RCL 34
420 *
421 RCL 10
422 *
423 CHS
424 RCL 17
425 /
426 ST* 50
427 ST* 51
428 RCL 29
429 XEQ 05
430 RCL 59
431 STO 52
432 RCL 60
433 STO 53
434 RCL 47
435 ST* 52
436 ST* 53
437 XEQ 06
438 1
439 XEQ 07
440 XEQ 08
441 ":C/B="
442 ASTO 49
443 XEQ 16
444 "PROBLEM DATA"
445 XEQ 09
446 "------- ----"
447 XEQ 04
448 "AB="
449 ARCL 10
450 XEQ 04
```

```
451 "AC="
452 ARCL 11
453 XEQ 04
454 "THETA A="
455 ARCL 12
456 XEQ 04
457 "OMEGA A="
458 ARCL 13
459 XEQ 04
460 "ALPHA A="
461 ARCL 14
462 XEQ 04
463 "PHI C="
464 ARCL 18
465 XEQ 04
466 "A 1="
467 ARCL 30
468 XEQ 04
469 "A 2="
470 ARCL 31
471 XEQ 04
472 "B 1="
473 ARCL 32
474 XEQ 04
475 "B 2="
476 ARCL 33
477 XEQ 04
478 "C 1="
479 ARCL 39
480 XEQ 04
481 "C 2="
482 ARCL 41
483 XEQ 04
484 "ATC="
485 ARCL 46
486 XEQ 04
487 "ATBC="
488 ARCL 47
489 XEQ 04
490 XEQ 13
491◆LBL 05
492 57.296
493 /
494 0
495 XROM "e↑Z"
496 STO 59
497 X<>Y
498 STO 60
499 RTN
500◆LBL 08
```

```
501 RCL 59              551 RTN
502 X<0?                552*LBL 09
503 GTO 14              553 AVIEW
504 RTN                 554 RTN
505*LBL 14              555*LBL 03
506 RCL 60              556 FIX 2
507 X<0?                557 26
508 GTO 15              558 ACCHR
509 180                 559 CLA
510 ST+ 62              560 ARCL 28
511 RTN                 561 ARCL 61
512*LBL 15              562 ARCL 27
513 180                 563 ARCL 62
514 ST- 62              564 ACA
515 RTN                 565 PRBUF
516*LBL 06              566 ADV
517 RCL 51              567 FIX 4
518 RCL 50              568 RTN
519 RCL 53              569*LBL 16
520 RCL 52              570 FIX 1
521 XROM "C+"           571 22
522 STO 59              572 ACCHR
523 X<>Y                573 CLA
524 STO 60              574 ARCL 49
525 RTN                 575 ARCL 61
526*LBL 07              576 ARCL 48
527 ST* 59              577 ARCL 62
528 ST* 60              578 ACA
529 RCL 60              579 PRBUF
530 RCL 59              580 ADV
531 XROM "MAGZ"         581 FIX 4
532 STO 61              582 RTN
533 RCL 59              583*LBL 13
534 X=0?                584 AOFF
535 GTO 12              585 .END.
536 RCL 60
537 RCL 59
538 /
539 ATAN
540 STO 62
541 RTN
542*LBL 12
543 0
544 ATAN
545 STO 62
546 RTN
547*LBL 04
548 AVIEW
549 ADV
550 CLA
```

SLIDING COUPLER
------- -------

LIN VELOCITY
--- --------

V:B=12.0000FPS

V:C=-6.6530FPS

V:C/B=9.9869FPS

VEC LIN VEL
--- --- ---

ū:B=12.00FPS∠-150.00

ū:C=6.65FPS∠153.67

ū:C/B=9.99FPS∠63.67

NORM ACC
---- ---

ā:B=216.0FPS↑2∠-60.0

ā:C=0.0FPS↑2∠0.0

ā:C/B=76.6FPS↑2∠153.7

TANG ACC
---- ---

ā:B=0.0FPS↑2∠0.0

ā:C=103.1FPS↑2∠-26.3

ā:C/B=17.7FPS↑2∠63.7

CORR ACC
---- ---

ā:C=102.1FPS↑2∠-116.3

ABS ACC
--- ---

ā:B=216.0FPS↑2∠-60.0

ā:C=145.1FPS↑2∠-71.0

ā:C/B=78.6FPS↑2∠140.7

PROBLEM DATA
------- ----

AB=0.6667

AC=0.8333

THETA A=120.0000

OMEGA A=18.0000

ALPHA A=0.0000

PHI C=-26.3295

A 1=0.8963

A 2=-0.4435

B 1=-0.4435

B 2=-0.8963

C 1=84.6063

C 2=-61.5830

ATC=103.1432

ATBC=17.6686

SLIDING COUPLER
------- -------

LIN VELOCITY
--- --------

V:B=7.5000FPS

V:C=1.5507FPS

V:C/B=7.3379FPS

VEC LIN VEL
--- --- ---

ū:B=7.50FPS∠-60.00

ū:C=1.55FPS∠18.07

ū:C/B=7.34FPS∠108.07

NORM ACC
---- ---

ā:B=225.0FPS↑2∠30.0

ā:C=0.0FPS↑2∠0.0

ā:C/B=133.6FPS↑2∠-161.9

TANG ACC
---- ---

ā:B=0.0FPS↑2∠0.0

ā:C=86.5FPS↑2∠18.1

ā:C/B=9.9FPS↑2∠108.1

CORR ACC
---- ---

ā:C=56.5FPS↑2∠108.1

ABS ACC
--- ---

ā:B=225.0FPS↑2∠30.0

ā:C=103.3FPS↑2∠51.2

ā:C/B=134.0FPS↑2∠-166.2

PROBLEM DATA
------- ----

AB=0.2500

AC=0.1667

THETA A=210.0000

OMEGA A=30.0000

ALPHA A=0.0000

PHI C=18.0675

A 1=0.9507

A 2=0.3101

B 1=0.3101

B 2=-0.9507

C 1=85.3589

C 2=17.3868

ATC=86.5423

ATBC=9.9436

SLIDER-CRANK: MODIFIED VECTOR METHOD

```
01•LBL "SL-CR C"           51 RCL 18
02 "  SL-CR ANAL CL"       52 *
03 XEQ 09                  53 RCL 17
04 "  ____ ___ __"         54 *
05 XEQ 04                  55 CHS
06 FIX 4                   56 STO 20
07 SF 04                   57 90
08 "AB?"                   58 RCL 13
09 PROMPT                  59 -
10 STO 10                  60 SIN
11 "BC?"                   61 STO 21
12 PROMPT                  62 90
13 STO 11                  63 RCL 15
14 "OMEGA?"                64 -
15 PROMPT                  65 SIN
16 STO 12                  66 STO 22
17 "THETA A?"              67 RCL 21
18 PROMPT                  68 RCL 22
19 STO 13                  69 /
20 SIN                     70 RCL 17
21 STO 14                  71 *
22 "ALPHA A?"              72 CHS
23 PROMPT                  73 STO 23
24 STO 09                  74 "IN/SEC"
25 RCL 14                  75 ASTO 24
26 RCL 10                  76 "LINEAR VELOCITY"
27 *                       77 XEQ 09
28 RCL 11                  78 "_____ _____"
29 /                       79 XEQ 04
30 ASIN                    80 "V:B="
31 STO 15                  81 ARCL 17
32 CHS                     82 ARCL 24
33 180                     83 XEQ 04
34 +                       84 "V:C="
35 STO 16                  85 ARCL 20
36 RCL 10                  86 ARCL 24
37 RCL 12                  87 XEQ 04
38 *                       88 "V:B/C="
39 STO 17                  89 ARCL 23
40 RCL 15                  90 ARCL 24
41 RCL 13                  91 XEQ 04
42 +                       92 "VEC LIN VEL"
43 SIN                     93 XEQ 09
44 STO 18                  94 "___ ___ ___"
45 90                      95 XEQ 04
46 RCL 15                  96 RCL 13
47 -                       97 XEQ 05
48 SIN                     98 RCL 60
49 STO 19                  99 STO 28
50 1/X                     100 RCL 59
```

```
101 STO 27          151 RCL 12
102 RCL 17          152 X↑2
103 CHS             153 *
104 XEQ 07          154 STO 31
105 XEQ 08          155 CHS
106 "IPS∠"          156 ST* 27
107 ASTO 25         157 ST* 28
108 ":B="           158 RCL 13
109 ASTO 26         159 XEQ 05
110 XEQ 03          160 RCL 59
111 0               161 STO 29
112 XEQ 16          162 RCL 60
113 RCL 59          163 STO 30
114 STO 29          164 RCL 10
115 RCL 60          165 RCL 09
116 STO 30          166 *
117 RCL 20          167 STO 32
118 XEQ 07          168 CHS
119 XEQ 08          169 ST* 30
120 ":C="           170 ST* 29
121 ASTO 26         171 XEQ 06
122 XEQ 03          172 1
123 RCL 17          173 XEQ 07
124 CHS             174 XEQ 08
125 ST* 27          175 "IPS↑2∠"
126 ST* 28          176 ASTO 33
127 RCL 20          177 ":B="
128 ST* 29          178 ASTO 34
129 ST* 30          179 XEQ 17
130 XEQ 06          180 RCL 49
131 1               181 X>0?
132 XEQ 07          182 GTO 20
133 XEQ 08          183 0
134 ":B/C="         184 STO 61
135 ASTO 26         185 STO 62
136 XEQ 03          186 ":C="
137 -5              187 ASTO 34
138 STO 49          188 XEQ 17
139 "NORM ACC"      189 "TAN ACC"
140 XEQ 09          190 XEQ 09
141 "--- ---"       191 "--- ---"
142 XEQ 04          192 XEQ 04
143*LBL 18          193 0
144 RCL 13          194 STO 61
145 XEQ 16          195 STO 62
146 RCL 59          196 ":B="
147 STO 27          197 ASTO 34
148 RCL 60          198 XEQ 17
149 STO 28          199*LBL 19
150 RCL 10          200 RCL 13
```

201 COS	251 X↑2
202 STO 35	252 *
203 LASTX	253 STO 44
204 SIN	254 RCL 43
205 STO 36	255 /
206 X↑2	256 STO 45
207 CHS	257 ST+ 41
208 RCL 35	258 RCL 41
209 X↑2	259 RCL 31
210 +	260 CHS
211 STO 37	261 *
212 RCL 10	262 STO 27
213 *	263 0
214 STO 38	264 STO 28
215 RCL 36	265 RCL 10
216 X↑2	266 RCL 35
217 RCL 10	267 *
218 X↑2	268 RCL 36
219 *	269 *
220 CHS	270 STO 46
221 RCL 11	271 RCL 47
222 X↑2	272 /
223 +	273 STO 48
224 STO 39	274 RCL 36
225 SQRT	275 +
226 STO 47	276 RCL 32
227 1/X	277 CHS
228 RCL 38	278 *
229 *	279 STO 29
230 STO 40	280 0
231 RCL 35	281 STO 30
232 +	282 XEQ 06
233 STO 41	283 1
234 3	284 XEQ 07
235 ENTER↑	285 XEQ 08
236 2	286 ":C="
237 /	287 ASTO 34
238 STO 42	288 XEQ 17
239 RCL 39	289 RCL 49
240 RCL 42	290 X>0?
241 Y↑X	291 GTO 21
242 STO 43	292 "ABS ACC"
243 RCL 10	293 XEQ 09
244 ENTER↑	294 "— —-"
245 3	295 XEQ 04
246 Y↑X	296 10
247 RCL 35	297 STO 49
248 X↑2	298 XEQ 18
249 *	299♦LBL 20
250 RCL 36	300 XEQ 19

```
301*LBL 21              351 ARCL 09
302 RCL 38              352 XEQ 04
303 RCL 47              353 "PHI C="
304 /                   354 ARCL 15
305 STO 50              355 XEQ 04
306 RCL 44              356 90
307 RCL 43              357 RCL 15
308 /                   358 -
309 ST+ 50              359 STO 51
310 RCL 31              360 RCL 13
311 ST* 50              361 RCL 15
312 RCL 50              362 +
313 STO 27              363 STO 52
314 RCL 31              364 90
315 RCL 36              365 RCL 13
316 *                   366 -
317 CHS                 367 STO 53
318 STO 28              368 "SNB="
319 RCL 48              369 ARCL 51
320 RCL 32              370 XEQ 04
321 *                   371 "SNC="
322 STO 29              372 ARCL 52
323 RCL 35              373 XEQ 04
324 RCL 32              374 "SNBC="
325 *                   375 ARCL 53
326 STO 30              376 XEQ 04
327 XEQ 01              377 GTO 13
328 1                   378*LBL 05
329 XEQ 07              379 57.296
330 XEQ 08              380 /
331 ":B/C="             381 0
332 ASTO 34             382 XROM "etZ"
333 XEQ 17              383 CHS
334 "PROBLEM DATA"      384 STO 60
335 XEQ 09              385 X<>Y
336 "----- ---"         386 STO 59
337 XEQ 04              387 RTN
338 "AB="               388*LBL 08
339 ARCL 10             389 RCL 59
340 XEQ 04              390 X<0?
341 "BC="               391 GTO 14
342 ARCL 11             392 RTN
343 XEQ 04              393*LBL 14
344 "THETA A="          394 RCL 60
345 ARCL 13             395 X<0?
346 XEQ 04              396 GTO 15
347 "OMEGA A="          397 180
348 ARCL 12             398 ST+ 62
349 XEQ 04              399 RTN
350 "ALPHA A=" '        400*LBL 15
```

```
401 100                          451 RTN
402 ST- 62                       452*LBL 01
403 RTN                          453 RCL 28
404*LBL 06                       454 RCL 27
405 RCL 28                       455 RCL 30
406 RCL 27                       456 RCL 29
407 RCL 30                       457 XROM "C+"
408 RCL 29                       458 STO 59
409 XROM "C-"                    459 X<>Y
410 STO 59                       460 STO 60
411 X<>Y                         461 RTN
412 STO 60                       462*LBL 03
413 RTN                          463 FIX 2
414*LBL 07                       464 26
415 ST* 59                       465 ACCHR
416 ST* 60                       466 CLA
417 RCL 60                       467 ARCL 26
418 RCL 59                       468 ARCL 61
419 XROM "MAGZ"                  469 ARCL 25
420 STO 61                       470 ARCL 62
421 RCL 59                       471 ACA
422 X=0?                         472 PRBUF
423 GTO 12                       473 ADV
424 RCL 60                       474 FIX 4
425 RCL 59                       475 RTN
426 /                           476*LBL 17
427 ATAN                         477 FIX 2
428 STO 62                       478 22
429 RTN                          479 ACCHR
430*LBL 12                       480 CLA
431 0                           481 ARCL 34
432 ATAN                         482 ARCL 61
433 STO 62                       483 ARCL 33
434 RTN                          484 ARCL 62
435*LBL 04                       485 ACA
436 AVIEW                        486 PRBUF
437 ADV                          487 ADV
438 CLA                          488 FIX 4
439 RTN                          489 RTN
440*LBL 09                       490*LBL 13
441 AVIEW                        491 AOFF
442 RTN                          492 STOP
443*LBL 16                       493 END
444 57.296
445 /
446 0
447 XROM "etZ"
448 STO 59
449 X<>Y
450 STO 60
```

SL-CR ANAL CL
———— ——— ——

LINEAR VELOCITY
—————— ————————

V:B=1.5000IN/SEC

V:C=-0.9387IN/SEC

V:B/C=0.8321IN/SEC

VEC LIN VEL
——— ——— ———

ū:B=1.50IPS∠-150.00

ū:C=0.94IPS∠180.00

ū:B/C=0.83IPS∠-115.66

NORM ACC
———— ———

ē:B=1.50IPS↑2∠-60.00

ē:C=0.00IPS↑2∠0.00

TAN ACC
——— ———

ā:B=0.00IPS↑2∠0.00

ā:C=1.12IPS↑2∠0.00

ABS ACC
——— ———

ā:B=1.50IPS↑2∠-60.00

ā:C=1.12IPS↑2∠0.00

ā:B/C=1.35IPS↑2∠-105.82

PROBLEM DATA
———————— ————

AB=1.5000

BC=3.0000

THETA A=120.0000

OMEGA A=1.0000

ALPHA A=0.0000

PHI C=25.6589

SNB=64.3411

SNC=145.6589

SNBC=-30.0000

PROBLEM DATA
------- ----

AB=1.5000

BC=3.0000

THETA A=30.0000

OMEGA A=-1.0000

ALPHA A=0.0000

PHI C=14.4775

SMB=75.5225

SMC=44.4775

SMBC=60.0000

SL-CR ANAL CL
----- ---- --

LINEAR VELOCITY
------ --------

V:B=-1.5000IN/SEC

V:C=1.0854IN/SEC

V:B/C=1.3416IN/SEC

VEC LIN VEL
--- --- ---

ū:B=1.50IPS∠-60.00

ū:C=1.09IPS∠0.00

ū:B/C=1.34IPS∠-104.48

NORM ACC
---- ---

ā:B=1.50IPS↑2∠-150.00

ā:C=0.00IPS↑2∠0.00

TAN ACC
--- ---

ā:B=0.00IPS↑2∠0.00

ā:C=1.73IPS↑2∠180.00

ABS ACC
--- ---

ā:B=1.50IPS↑2∠-150.00

ā:C=1.73IPS↑2∠180.00

ā:B/C=0.86IPS↑2∠-60.40

Appendix B

B.1   NOMENCLATURE

a	linear acceleration
$\bar{a}$	average linear acceleration
a_B	linear acceleration of point B
a, b, c, etc.	termini of velocity vectors $\bar{V}_A$, $\bar{V}_B$, $\bar{V}_C$, etc. on velocity polygon
a', b', c', etc.	termini of acceleration vectors $\bar{A}_A$, $\bar{A}_B$, $\bar{A}_C$, etc. on acceleration polygon
A	linear acceleration (magnitude)
$\bar{A}$	linear acceleration vector (magnitude and direction)
A_B	linear acceleration of point B
$A_{B/C}$	linear acceleration of point B relative to point C
A^{CD}	effective component of acceleration along CD
A^{Cor}	Coriolis acceleration
A^N	normal acceleration
A^r	rotational component of acceleration
A^t	translational component of acceleration
A^T	tangential acceleration
A, B, C, etc.	pivot points on a linkage
CCW	counterclockwise direction
CW	clockwise direction

e	eccentricity of a slider–crank mechanism
i	$\sqrt{-1}$
I	instant center
k_a	acceleration scale (actual acceleration represented by unit length of acceleration vector or acceleration axis of motion curve)
k_s	space scale (actual length of machine member or displacement represented by unit length of vector or displacement axis of motion curve
k_t	time scale (actual time represented) by unit length on time axis of motion curve
k_v	velocity scale (actual velocity represented by unit length of velocity vector or velocity axis of motion curve)
n	number of links of a mechanism
N	number of instant centers; number of revolutions per minute
o	pole or origin of velocity polygon
o'	pole or origin of acceleration polygon
P	point of contact between two sliding bodies
P(C)	contact point P on C
P(F)	contact point P on F
P(C)/P(F)	point P on C relative to point P on F
R, r	radius
s	linear displacement
t	time
v	linear velocity
$\bar{v}$	average linear velocity, unit vector
v_1	initial linear velocity
v_2	final linear velocity
v_B	linear velocity of point B
V	linear velocity (magnitude)
$\bar{V}$	linear velocity vector (magnitude and direction)

V_B	linear velocity of point B
$V_{B/C}$	linear velocity of point B relative to point C
V^{CD}	effective component of velocity along CD
V^r	rotational component of velocity
V^t	translational component of velocity
V_A^i	initial velocity of point A
V_A^f	final velocity of point A initial velocity of point B
V_{B_i}	initial velocity of point B
V_{B_f}	final velocity of point B
α (alpha)	angular acceleration, or other angle
α_2	angular acceleration of link 2
α_{AB}	angular acceleration of link AB
β (beta)	angle
γ (gamma)	angle
ϕ (phi)	angle
θ (theta)	angular displacement
ω (omega)	angular velocity
ω_1	initial angular velocity
ω_2	angular velocity of link 2, final angular velocity
ω_{AB}	angular velocity of link AB
1, 2, 3, etc.	links 1, 2, 3, etc.
23	instant center of links 2 and 3
$\perp$	perpendicular to

B.2   TRIGONOMETRY REVIEW

Functions of a Right Triangle (Figure B.1)

$$\sin \alpha = \frac{\text{opposite}}{\text{hypotenuse}} = \frac{a}{c}$$

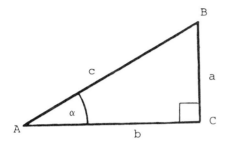

Figure B.1   Right triangle.

$$\cos \alpha = \frac{\text{adjacent}}{\text{hypotenuse}} = \frac{b}{c}$$

$$\tan \alpha = \frac{\text{opposite}}{\text{adjacent}} = \frac{a}{b} = \frac{\sin \alpha}{\cos \alpha}$$

$$\operatorname{cosec} \alpha = \frac{1}{\sin \alpha} = \frac{c}{a}$$

$$\sec \alpha = \frac{1}{\cos \alpha} = \frac{c}{b}$$

$$\cot \alpha = \frac{1}{\tan \alpha} = \frac{b}{a} = \frac{\cos \alpha}{\sin \alpha}$$

Functions of an Angle in the Interval $0° < \theta < 360°$

If a vector AB is rotated through the four quadrants as shown in Figure B.2, any function of the angle θ is numerically equal to the same function of the acute angle α between the terminal side of the vector and the x axis. That is,

Function of $\theta = \pm$ same function α

where the positive or negative sign depends on the quadrant in which the angle α falls. Signs are determined as shown in Figure B.2.

These signs of the functions of θ in the four quadrants may be summarized as follows:

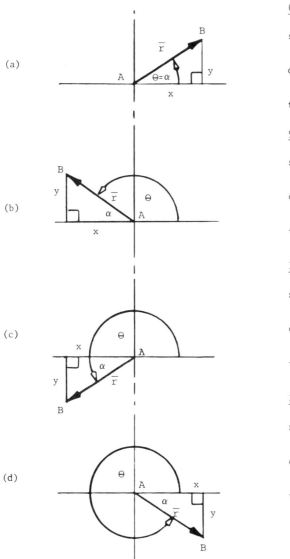

(a)

(b)

(c)

(d)

$$0° \angle \Theta \angle 90°$$

$$\sin \Theta = \frac{+y}{r} = \frac{y}{r}$$

$$\cos \Theta = \frac{+x}{r} = \frac{x}{r}$$

$$\tan \Theta = \frac{+y}{x} = \frac{y}{x}$$

$$90° \angle \Theta \angle 180°$$

$$\sin \Theta = \frac{+y}{r} = \frac{y}{r}$$

$$\cos \Theta = \frac{-x}{r} = -\frac{x}{r}$$

$$\tan \Theta = \frac{+y}{-x} = -\frac{y}{x}$$

$$180° \angle \Theta \angle 270°$$

$$\sin \Theta = \frac{-y}{r} = -\frac{y}{r}$$

$$\cos \Theta = \frac{-x}{r} = -\frac{x}{r}$$

$$\tan \Theta = \frac{-y}{-x} = \frac{y}{x}$$

$$270° \angle \Theta \angle 360°$$

$$\sin \Theta = \frac{-y}{r} = -\frac{y}{r}$$

$$\cos \Theta = \frac{+x}{r} = \frac{x}{r}$$

$$\tan \Theta = \frac{-y}{x} = -\frac{y}{x}$$

Figure B.2   Vector in rotation.

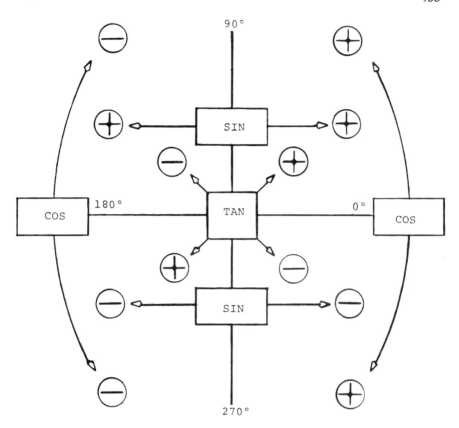

Figure B.3 Sign diagram.

$0° < \theta < 90°$: $\alpha = \theta$ sin θ is positive
                                                cos θ is positive
                                                tan θ is positive

$90° < \theta < 180°$: $\alpha = 180° - \theta$ sin θ is positive
                                                cos θ is negative
                                                tan θ is negative

$180° < \theta < 270°$: $\alpha = \theta - 180°$ sin θ is negative
                                                cos θ is negative
                                                tan θ is positive

$270° < \theta < 360°$: $\alpha = 360° - \theta$ sin θ is negative
                                                cos θ is positive
                                                tan θ is negative

Signs of the functions of θ are most conveniently remembered using Figure B.3.

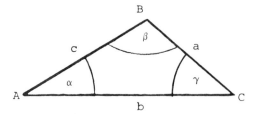

Figure B.4   Oblique triangle.

<u>Laws for Oblique Triangles</u>

Laws of Cosines

In any triangle, the square of any side is equal to the sum of the squares of the other sides minus twice their product times the cosine of their included angle. For example, in triangle ABC in Figure B.4,

$$a^2 = b^2 + c^2 - 2bc \cos \alpha$$

$$b^2 = a^2 + c^2 - 2ac \cos \beta$$

$$c^2 = a^2 + b^2 - 2ab \cos \gamma$$

where

$$\alpha + \beta + \gamma = 180°$$

Law of Sines

In any triangle, any two sides are proportional to the sides of the opposite angles. In triangle ABC in Figure B.4,

$$\frac{a}{\sin \alpha} = \frac{b}{\sin \beta} = \frac{c}{\sin \gamma}$$

Laws of Tangents

In any triangle, the difference of the opposite angles divided by their sum equals the tangent of one-half the difference of the opposite angles divided by the tangent of one-half their sum. In triangle ABC in Figure B.4,

$$\frac{a - b}{a + b} = \frac{\tan (1/2)(\alpha - \beta)}{\tan (1/2)(\alpha + \beta)}$$

$$\frac{a - c}{a + c} = \frac{\tan (1/2)(\alpha - \gamma)}{\tan (1/2)(\alpha + \gamma)}$$

$$\frac{b - c}{b + c} = \frac{\tan (1/2)(\beta - \gamma)}{\tan (1/2)(\beta + \gamma)}$$

Other Useful Relationships

$$\sin^2 \alpha + \cos^2 \alpha = 1$$

$$\sin (\alpha + \beta) = \sin \alpha \cos \beta + \cos \alpha \sin \beta$$

$$\sin (\alpha - \beta) = \sin \alpha \cos \beta - \cos \alpha \sin \beta$$

$$\cos (\alpha + \beta) = \cos \alpha \cos \beta - \sin \alpha \sin \beta$$

$$\cos (\alpha - \beta) = \cos \alpha \cos \beta + \sin \alpha \sin \beta$$

$$\tan (\alpha + \beta) = \frac{\tan \alpha + \tan \beta}{1 - \tan \alpha \tan \beta}$$

$$\tan (\alpha - \beta) = \frac{\tan \alpha - \tan \beta}{1 + \tan \alpha \tan \beta}$$

$$\sin 2\alpha = 2 \sin \alpha \cos \alpha$$

$$\cos 2\alpha = \cos^2 \alpha - \sin^2 \alpha$$

$$\tan 2\alpha = \frac{2 \tan \alpha}{1 - \tan^2 \alpha}$$

$$\sin \frac{\alpha}{2} = \sqrt{\frac{1 - \cos \alpha}{2}}$$

$$\cos \frac{\alpha}{2} = \sqrt{\frac{1 + \cos \alpha}{2}}$$

$$\tan \frac{\alpha}{2} = \sqrt{\frac{\sin \alpha}{1 + \cos \alpha}}$$

$$\sin \alpha + \sin \beta = 2 \sin \frac{\alpha + \beta}{2} \cos \frac{\alpha - \beta}{2}$$

$$\sin \alpha - \sin \beta = 2 \cos \frac{\alpha + \beta}{2} \sin \frac{\alpha - \beta}{2}$$

$$\cos \alpha + \cos \beta = 2 \cos \frac{\alpha + \beta}{2} \cos \frac{\alpha - \beta}{2}$$

$$\cos \alpha - \cos \beta = -2 \sin \frac{\alpha + \beta}{2} \sin \frac{\alpha - \beta}{2}$$

$$\sin(90° - \theta) = +\cos\theta$$

$$\sin(90° + \theta) = +\cos\theta$$

$$\sin(180° - \theta) = +\sin\theta$$

$$\sin(180° + \theta) = -\sin\theta$$

$$\sin(270° - \theta) = -\cos\theta$$

$$\sin(270° + \theta) = -\cos\theta$$

$$\sin(360° - \theta) = -\sin\theta$$

$$\sin(360° + \theta) = +\sin\theta$$

$$\sin(-\theta) = -\sin\theta$$

$$\cos(90° - \theta) = +\sin\theta$$

$$\cos(90° + \theta) = -\sin\theta$$

$$\cos(180° - \theta) = -\cos\theta$$

$$\cos(180° + \theta) = -\cos\theta$$

$$\cos(270° - \theta) = -\sin\theta$$

$$\cos(270° + \theta) = +\sin\theta$$

$$\cos(360° - \theta) = +\cos\theta$$

$$\cos(360° + \theta) = +\cos\theta$$

$$\cos(-\theta) = +\cos\theta$$

B.3   TABLE OF TRIGONOMETRIC FUNCTIONS

deg	rad	sin	cos	tan	deg	rad	sin	cos	tan
0	.000	.000	1.000	.000					
1	.017	.017	1.000	.017	46	.803	.719	.695	1.036
2	.035	.035	.999	.035	47	.820	.731	.682	1.072
3	.052	.052	.999	.052	48	.838	.743	.669	1.111
4	.070	.070	.998	.070	49	.855	.755	.656	1.150
5	.087	.087	.996	.087	50	.873	.766	.643	1.192
6	.105	.105	.995	.105	51	.890	.777	.629	1.235
7	.122	.122	.993	.123	52	.908	.788	.616	1.280
8	.140	.139	.990	.141	53	.925	.799	.602	1.327
9	.157	.156	.988	.158	54	.942	.809	.588	1.376
10	.175	.174	.985	.176	55	.960	.819	.574	1.428
11	.192	.191	.982	.194	56	.977	.829	.559	1.483
12	.209	.208	.978	.213	57	.995	.839	.545	1.540
13	.227	.225	.974	.231	58	1.012	.848	.530	1.600
14	.244	.242	.970	.249	59	1.030	.857	.515	1.664
15	.262	.259	.966	.268	60	1.047	.866	.500	1.732
16	.279	.276	.961	.287	61	1.065	.875	.485	1.804
17	.297	.292	.956	.306	62	1.082	.883	.470	1.881
18	.314	.309	.951	.325	63	1.100	.891	.454	1.963
19	.332	.326	.946	.344	64	1.117	.899	.438	2.050
20	.349	.342	.940	.364	65	1.134	.906	.423	2.145
21	.367	.358	.934	.384	66	1.152	.914	.407	2.246
22	.384	.375	.927	.404	67	1.169	.921	.391	2.356
23	.401	.391	.921	.424	68	1.187	.927	.375	2.475
24	.419	.407	.914	.445	69	1.204	.934	.358	2.605
25	.436	.423	.906	.466	70	1.222	.940	.342	2.747
26	.454	.438	.899	.488	71	1.239	.946	.326	2.904
27	.471	.454	.891	.510	72	1.257	.951	.309	3.078
28	.489	.470	.883	.532	73	1.274	.956	.292	3.271
29	.506	.485	.875	.554	74	1.292	.961	.276	3.487
30	.524	.500	.866	.577	75	1.309	.966	.259	3.732
31	.541	.515	.857	.601	76	1.326	.970	.242	4.011
32	.559	.530	.848	.625	77	1.344	.974	.225	4.331
33	.576	.545	.839	.649	78	1.361	.978	.208	4.705
34	.593	.559	.829	.675	79	1.379	.982	.191	5.145
35	.611	.574	.819	.700	80	1.396	.985	.174	5.671
36	.628	.588	.809	.727	81	1.414	.988	.156	6.314
37	.646	.602	.799	.754	82	1.431	.990	.139	7.115
38	.663	.616	.788	.781	83	1.449	.993	.122	8.144
39	.681	.629	.777	.810	84	1.466	.995	.105	9.514
40	.698	.643	.766	.839	85	1.484	.996	.087	11.430
41	.716	.656	.755	.869	86	1.501	.998	.070	14.301
42	.733	.669	.743	.900	87	1.518	.999	.052	19.081
43	.751	.682	.731	.933	88	1.536	.999	.035	28.636
44	.768	.695	.719	.966	89	1.553	1.000	.017	57.290
45	.785	.707	.707	1.000	90	1.571	1.000	.000	—

Selected References

Albert, C. D., and F. S. Rogers, Kinematics of Machinery, Wiley, New York, 1931.

Annand, W. J. D., Mechanics of Machines, Chemical Publishing, New York, 1966.

Barton, L. O., "Finding Slider-Crank Acceleration Graphically," Machine Design, Vol. 50, No. 28, December 7, 1978.

Barton, L. O., "Simplified Slider-Crank Equations," Machine Design, Vol. 51, No. 8, April 1979.

Barton, L. O., "The Acceleration Polygon—A Generalized Procedure," Engineering Design Graphics Journal, Vol. 43, No. 2, Spring 1979.

Barton, L. O., "Painless Analysis of Four-Bar Linkages," Machine Design, Vol. 51, No. 17, July 26, 1979.

Barton, L. O., "Simplifying Velocity Analysis for Mechanisms," Machine Design, Vol. 53, No. 13, June 11, 1981.

Barton, L. O., "Simplified Analysis of Quick Return Mechanisms," Machine Design, Vol. 52, No. 18, August 7, 1980.

Barton, L. O., "A Diagrammatic Representation of the Basic Motion Equations," Engineering Design Graphics Journal, Vol. 44, No. 3, Fall 1980.

Barton, L. O., "A New Way to Analyze Slider-Cranks," Machine Design, Vol. 17, July 22, 1982.

Barton, L. O., "Simplifying the Analysis of Sliding Coupler Mechanisms," Machine Design, Vol. 55, No. 19, August 25, 1983.

Beggs, J. S., Mechanism, McGraw-Hill, New York, 1955.

Bickford, J. H., Mechanisms for Product Design, Industrial Press, New York, 1972.

Billings, J. H., Applied Kinematics, 2nd ed., Van Nostrand Reinhold, New York, 1943.

Chironis, N. P., Mechanisms, Linkages and Mechanical Controls, McGraw-Hill, 1965.

Chironis, N. P., Machine Devices and Instrumentation, McGraw-Hill, New York, 1966.

Dent, J. A., and A. C. Harper, Kinematics and Kinetics of Machinery, Wiley, New York, 1921.

Durley, R. J., Kinematics of Machines, Wiley, New York, 1911.

Esposito, A., Kinematics for Technology, Charles E. Merrill, Columbus, Ohio, 1973.

Guillet, G. L., Kinematics of Machines, Wiley, New York, 1934.

Hain, K., Applied Kinematics, ed. by D. P. Adams and T. P. Goodman, McGraw-Hill, New York, 1967.

Hall Jr., A. S., Kinematics and Linkage Design, Prentice-Hall, Englewood Cliffs, N.J., 1961.

Ham, C. W., E. J. Crane, and W. L. Rogers, Mechanics of Machinery, 4th ed., McGraw-Hill, New York, 1958.

Heck, R. C. H., Elementary Kinematics, D. Van Nostrand, New York, 1910.

Hinkle, R. T., Kinematics of Machines, 2nd ed., Prentice-Hall, Englewood Cliffs, N.J., 1960.

Hirschhorn, J., Kinematics and Dynamics of Plane Mechanisms, McGraw-Hill, New York, 1962.

Holowenko, A. R., Dynamics of Machinery, Wiley, New York, 1955.

Horton, H. L., Mathematics at Work, Industrial Press, New York, 1949.

Hunt, K. H., Mechanisms and Motion, Wiley, New York, 1959.

James, W. H., and M. C. Mackenzie, Principles of Mechanism, Wiley, New York, 1918.

Keown, R. McA., and V. M. Faires, Mechanism, McGraw-Hill, New York, 1939.

Kepler, H. B., Basic Graphical Kinematics, 2nd ed., McGraw-Hill, New York, 1973.

Klein, A. W., _Kinematics of Machinery_, McGraw-Hill, New York, 1917.

Kolstee, H. M., _Motion and Power_, Prentice-Hall, Englewood Cliffs, N.J., 1982.

Lent, D., _Analysis and Design of Mechanisms_, 2nd ed., Prentice-Hall, Englewood Cliffs, N.J., 1970.

Mabie, H. H., and F. W. Ocvirk, _Mechanisms and Dynamics of Machinery_, Wiley, New York, 1975.

Martin, G. H., _Kinematics and Dynamics of Machines_, 2nd ed., McGraw-Hill, New York, 1982.

Maxwell, R. L., _Kinematics and Dynamics of Machinery_, Prentice-Hall, Englewood Cliffs, N.J., 1960.

Nielsen, K. L., _Modern Trigonometry_, Barnes and Noble, New York, 1966.

Patton, W. J., _Kinematics_, Reston, Reston, Va., 1979.

Paul, B., _Kinematics and Dynamics of Planar Machinery_, Prentice-Hall, Englewood Cliffs, N.J., 1979.

Pearce, C. E., _Principles of Mechanism_, Wiley, New York, 1934.

Prageman, I. H., _Mechanism_, International Textbook Press, Scranton, Pa., 1943.

Ramous, A. J., _Applied Kinematics_, Prentice-Hall, Englewood Cliffs, N.J., 1972.

Reuleaux, F., _Kinematics of Machinery_ (trans. and ed. by A. B. W. Kennedy), reprinted by Dover, New York, 1963.

Rosenauer, N., and A. H. Willis, _Kinematics of Mechanisms_, Associated General Publications, Sydney, Australia, 1953.

Sahag, L. M., _Kinematics of Machines_, Ronald Press, New York, 1952.

Schwamb, P., A. L. Merrill, et al., _Elements of Mechanism_, 6th ed., Wiley, New York, 1947.

Shigley, J. E., _Kinematic Analysis of Mechanisms_, 2nd ed., McGraw-Hill, New York, 1969.

Shigley, J. E., and J. J. Vicker, Jr., _Theory of Machines and Mechanisms_, McGraw-Hill, New York, 1980.

Smith, W. G., _Engineering Kinematics_, McGraw-Hill, New York, 1923.

Soni, A. H., _Mechanism Synthesis and Analysis_, McGraw-Hill, New York, 1974.

Tao, D. C., _Fundamentals of Applied Kinematics_, Addison-Wesley, Reading, Mass., 1967.

Tuttle, S. B., Mechanisms for Engineering Design, Wiley, New York, 1967.

Walker, J. D., Applied Mechanics, English Universities Press, London, 1959.

Wilson, C. E., Mechanism Design-Oriented Kinematics, American Technical Society, Chicago, 1969.

Woods, A. T., and A. W. Stahl, Kinematics, revised and rewritten by Philip K. Slaymaker, D. Van Nostrand, New York, 1926.

Index